Advances in Soil Science

Advances in Soil Science

B.A. Stewart, Editor

Advances in Soil Science

Volume 11
Soil Degradation

Edited by R. Lal and B.A. Stewart

With Contributions by
I.P. Abrol, P. Alvo, F. De Coninck, H. Eswaran, N.R. Fausey,
R.K. Gupta, R. Lal, T.J. Logan, D.A. MacLeod, E. McKyes,
C.E. Mullins, K.H. Northcote, G.S.V. Raghavan, G.K. Sims,
B.A. Stewart, J.M. Tisdall, T. Varghese, and I.M. Young

With 74 Illustrations

Springer-Verlag
New York Berlin Heidelberg
London Paris Tokyo Hong Kong

R. Lal
Department of Agronomy, Ohio State University
Columbus, Ohio 43210-2001, U.S.A.

B.A. Stewart
USDA Conservation & Production Research Laboratory
Bushland, Texas 79012, U.S.A.

ISSN: 0176-9340

Printed on acid-free paper.

Softecover reprint of the hardcover 1st edition 1990

Typeset by Asco Trade Typesetting Ltd., Hong Kong.

9 8 7 6 5 4 3 2 1

ISBN-13: 978-1-4612-7966-2 e-ISBN-13: 978-1-4612-3322-0
DOI: 10.1007/978-1-4612-3322-0

Preface

The purpose of *Advances in Soil Science* is to provide a forum for leading scientists to analyze and summarize the available scientific information on a subject, assessing its importance and identifying additional research needs. A wide array of subjects has been addressed by authors from many countries in the initial ten volumes of the series. The quick acceptance of the series by both authors and readers has been very gratifying and confirms our perception that a need did exist for a medium to fill the gap between the scientific journals and the comprehensive reference books.

This volume is the first of the series devoted entirely to a single topic—soil degradation. Future volumes will include both single-topic volumes as well as volumes containing reviews of different topics of soil science, as in the case of the first ten volumes.

There are increasing concern and attention about managing natural resources, particularly soil and water. Soil degradation is clearly one of the most pressing problems facing mankind. Although the spotlight regarding soil degradation in recent years has focused on Africa, concern about the degradation of soil and water resources is worldwide. The widespread concern about global environmental change is also being linked to severe problems of soil degradation. Therefore, we are indeed pleased that the first volume of the series devoted to a single topic addresses such an important issue.

The current volume is also the first of the series involving a guest editor. Dr. R. Lal was invited to serve as the lead editor for organizing the volume, selecting the authors, and coordinating their efforts. Dr. Lal is eminently qualified because of his wide background of experience. Dr. Lal has worked extensively in India, Australia, North America, and Africa and has traveled widely in many countries. He is currently a Professor of Soil Science at Ohio State University and is clearly one of the world's most knowledgeable scientists concerned with soil degradation. It has been a stimulating and rewarding experience to work with Dr. Lal in editing this volume.

I also want to thank the authors for their excellent contributions and cooperation and the Springer-Verlag staff for their kind assistance and counsel. Finally, and most importantly, I thank the readers for their acceptance and use of *Advances in Soil Science*.

B.A. Stewart
Series Editor

Contents

Contributors

I.P. ABROL, ICAR, Krishi Bhawan, New Delhi 110001, India.

P. ALVO, Department of Agricultural Engineering, Macdonald College, McGill University, 21111 Lakeshore Road, Ste. Anne de Bellevue, Quebec H9X 1CO, Canada.

F. DE CONINCK, Geological Institute, University of Gent, 281 Krijgslaan, 9000 Gent, Belgium.

H. ESWARAN, Soil Management Support Services, P.O. Box 2890, Washington, D.C. 20013, U.S.A.

N.R. FAUSEY, USDA, ARS, The Ohio State University, 2021 Coffey Road, Columbus, Ohio 43210, U.S.A.

R.K. GUPTA, Central Soil Salinity Research Institute, Karnal 132001, Haryana, India.

R. LAL, Department of Agronomy, The Ohio State University, 2021 Coffey Road, Columbus, Ohio 43210, U.S.A.

T.J. LOGAN, Department of Agronomy, The Ohio State University, 2021 Coffey Road, Columbus, Ohio 43210, U.S.A.

D.A. MACLEOD, Agronomy and Soil Science Department, University of New England, Armidale, NSW 2351, Australia.

E. MCKYES, Department of Agricultural Engineering, Macdonald College, McGill University, 21111 Lakeshore Road, Ste. Anne de Bellevue, Quebec H9X 1CO, Canada.

C.E. MULLINS, Plant and Soil Science Department, University of Aberdeen, Aberdeen AB9 2UE, Scotland.

K.H. NORTHCOTE, Formerly with Division of Soils, CSIRO, Glen Osmond, SA 5064, Australia (presently retired).

G.S.V. RAGHAVAN, Department of Agricultural Engineering, Macdonald College, McGill University, 21111 Lakeshore Road, Ste. Anne de Bellevue, Quebec H9X 1CO, Canada.

G.K. SIMS, Department of Agronomy, The Ohio State University, 2021 Coffey Road, Columbus, Ohio 43210, U.S.A.

B.A. STEWART, USDA Conservation and Production Research Laboratory, P.O. Drawer 10, Bushland, Texas 79012, U.S.A.

J.M. Tisdall, Institute for Irrigation and Salinity Research, Tatura, Victoria 3616, Australia.
T. Varghese, Soil Science Department, Kerala Agricultural University, Trivandrum, Kerala, India.
I.M. Young, Scottish Crop Research Institute, Dundee DD2 5DA, United Kingdom.

Soil Degradation: A Global Threat

R. Lal and B.A. Stewart

Introduction

Soil degradation is one of the greatest challenges facing mankind. Although the problem is as old as the settled agriculture, its extent and impact on human welfare and global environment are more now than ever before. It is a major concern for at least two reasons. First, soil degradation undermines the productive capacity of an ecosystem. Second, it affects global climate through alterations in water and energy balances and disruptions in cycles of carbon, nitrogen, sulfur, and other elements. Through its impact on agricultural productivity and environment, soil degradation leads to political and social instability, enhanced rate of deforestation, intensive use of marginal and fragile lands, accelerated runoff and soil erosion, pollution of natural waters, and emission of greenhouse gases into the atmosphere. In fact, soil degradation affects the mere fabric of mankind.

Soil degradation is defined as "the decline in soil quality caused through its misuse by humans." It is a broad and vague term, however, and refers to a decline in the soil's productivity through adverse changes in nutrient status and soil organic matter, structural attributes, and concentrations of electrolytes and toxic chemicals. In other words, it refers to a diminution of the soil's current and/or potential capability to produce quantitative or qualitative goods or services as a result of one or more degradative processes (UNEP, 1982).

Geographic Extent of Soil Degradation

Soil degradation has plagued mankind for thousands of years. Most ancient civilizations flourished on fertile soils. However, as the soil's productivity declined, so did cultures and civilizations that depended on it. Archaeological evidence has shown that soil degradation was responsible for the extinction of the Harappan civilization in western India, Mesopotamia in western Asia, and the Mayan culture in Central America (Olson, 1981). It is estimated that over the millennia as much as two billion hectares of land that were once biologically productive have been rendered unproductive through soil degradation (UNEP, 1986). Furthermore, the current rate of soil degradation is estimated at five to seven million hectares per year, and the annual rate may climb to ten million hectares by the turn of the century (FAO/UNEP, 1983). If these statistics are nearly correct, there is a need for immediate action by the international community to do something about it. Planners must develop policies to restore the disturbed ecosystems and ensure that conservation-effective measures are taken to preserve and enhance the productivity of existing lands.

One of the problems of these statistics is the ambiguity of the term *soil degradation*. Ambiguity can be avoided if we can precisely define critical limits of those index soil properties beyond which crops would not grow. These limits vary among soils, land use and farming systems, climatic conditions, and agro-ecological environments. For example, we do not precisely know the critical level of humified organic matter content for major soils of the world beyond which soil structure collapses. It is the lack of precise knowledge about the critical limits of principal properties that makes one skeptical about the validity of statistical data on the global extent of soil degradation. Nonetheless, the problem is extremely severe and worthy of our immediate attention.

Processes, Factors, and Causes of Soil Degradation

Soil degradation is an outcome of depletive human activities and their interaction with natural environments. Processes of soil degradation are the mechanisms responsible for the decline in soil quality. There are three principal types of degradation: physical, chemical, and biological. Each of these types has different processes of soil degradation (Figure 1).

Physical Degradation

Physical degradation refers to the deterioration of the physical properties of soil. This includes the following:

Compaction and Hardsetting. Densification of soil is caused by the elimination or reduction of structural pores. Increase in soil bulk density is caused by natural and man-induced factors. Hardsetting is a problem in soils of low-activity clays and soils that contain low organic matter content. Soils prone to compaction and hardsetting are susceptible to accelerated runoff and erosion.

Soil Erosion and Sedimentation. Worldwide erosion of topsoil by wind and water exceeds formation at an alarming rate. Desertification, the spread of desert-like conditions, is a direct consequence of wind erosion. Eroded

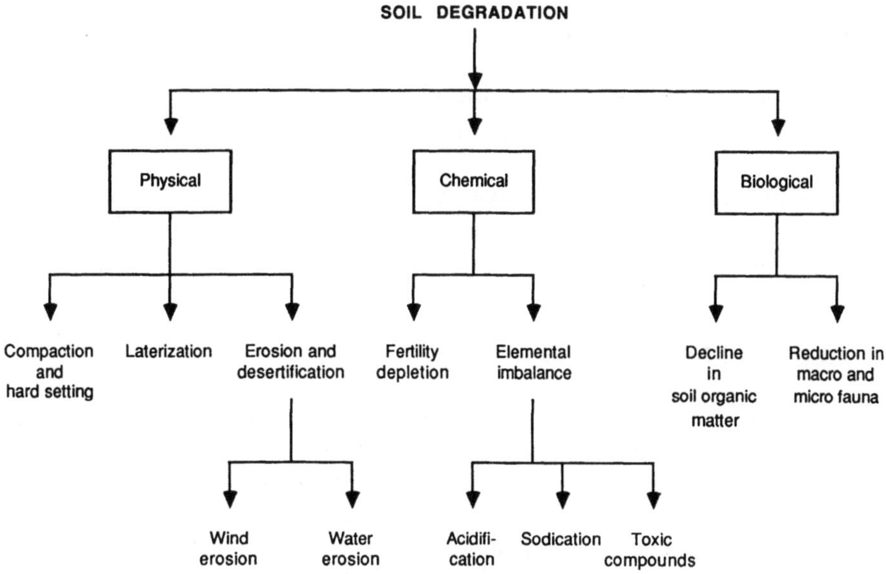

Figure 1. Types and processes of soil degradation.

soil, usually containing two to five times more organic matter and colloidal fraction than the original soil, causes severe on-site and off-site effects.

Laterization. Laterite is a hard sheet of iron and aluminum-rich duricrust. *Laterization*, therefore, refers to the desiccation and hardening of plinthitic material on exposure and desiccation.

Biological Degradation

Reduction in soil organic matter content, decline in biomass carbon, and decrease in activity and diversity of soil fauna are ramifications of biological degradation. Because of prevailing high soil and air temperatures, biological degradation of soil is more severe in the tropics than in the temperate zone. Biological degradation can also be caused by indiscriminate and excessive use of chemicals and soil pollutants.

Chemical Degradation

Nutrient depletion is a major cause of chemical degradation. In addition, excessive leaching of cations in soils with low-activity clays causes a decline in soil pH and a reduction in base saturation. Chemical degradation is also caused by the buildup of some toxic chemicals and an elemental imbalance that is injurious to plant growth.

The causes of soil degradation are anthropogenic perturbations related to socioeconomic pressures and population growth. These causes trigger human activities or factors responsible for soil degradation. Some important factors are deforestation, cultivation of marginal lands, intensive farming, excessive and indiscriminate use of chemicals, excessive grazing with high stocking rate, population transmigration and infrastructure development in ecologically sensitive areas, etc. These causes and factors are especially rampant in many developing countries of Asia, Africa, and tropical America.

Impact of Soil Degradation

An obvious economic impact of soil degradation is on agricultural productivity. Soil supplies essential nutrients and water, and degradation of capacity and intensity factors of water and nutrient availability affects plant growth. However, the environmental consequences of soil degradation have not been given the emphasis they deserve. Although scientists have recognized the dangers of global warming caused by the burning of fossil fuel, the emission of CO_2 and other greenhouse gases through soil degradation has hitherto been ignored. Radiatively active greenhouse gases emitted by soil-related processes are CO_2, CH_4, CO, N_2O, and NO.

Soil organic matter is a major active reservoir in the global carbon cycle. However, there are few data on its size and turnover related to intensification of diverse farming systems. An immediate response to deforestation and intensive cropping and grazing, especially in the tropics, is the very rapid mineralization of soil organic matter. Structural deterioration and accelerated erosion that follow result in the transport of carbon and nutrients out of an ecosystem. When these nutrients reach rivers, lakes, and streams, they cause pollution and eutrophication. Sediment related carbon is easily ejected into the atmosphere as CO_2. Evaluating the precise magnitude of sediment transport and its impact on the global environment remain to be major challenges to soil scientists, hydrologists, and geographers.

Future Considerations

This volume is an effort to collate the up-to-date information and comprehensive reviews on the principal processes of soil degradation. The volume, comprising eight review papers, is provisionally divided into three sections. The first five chapters address the processes of physical degradation, namely, soil compaction, hardsetting, plinthite and laterization, soil erosion, and soil wetness and anaerobioses. There are two chapters on chemical degradation related to nutrient imbalance and salt buildup. The last chapter is a comprehensive review of biological degradation.

This volume is an important step toward creating awareness of the magnitude and severity of the problem. In addition, however, there is also a need to indicate the urgency of developing methodologies for restoring the productivity of lands that have been rendered unproductive by past mismanagement. The production base can be expanded vastly even if only 50% of the supposedly two billion hectares of new degraded lands can be brought under cultivation. Restoration of degraded lands will also decrease the pressure on bringing new lands under cultivation. A logical followup of this volume may be another volume to address this important aspect of "Restoration of Degraded Soils."

References

FAO/UNEP. 1983. Guidelines for the control of soil degradation. FAO, Rome, Italy.

Olson, G.W. 1981. Archaeology: lessons on future soil use. *J. Soil Water Conserv.* 36:261–264.

UNEP. 1982. World soil policy. United Nations Envir. Program, Nairobi, Kenya.

UNEP. 1986. Farming systems principles for improved forest production and the control of soil degradation in the arid, semi-arid and humid tropics. Expert meeting, sponsored by the UNEP, 20–30 June 1983, ICRISAT, Hyderabad, India.

Soil Compaction in Agriculture: A View Toward Managing the Problem

G.S.V. Raghavan, P. Alvo, and E. McKyes*

I. Introduction

A. Background

Nature induces man to live preferentially in certain climatic zones that also happen to be those best suited for agriculture. In Canada, 60% of the best agricultural land is situated within 80 km of major urban centers (Dumanski, 1980). Urban and industrial expansion place further demands on the total land resource, such that agricultural land is affected directly through

*Department of Agricultural Engineering, Macdonald College, McGill University, Ste. Anne de Bellevue, Quebec H9X 1CO, Canada.

conversion to other uses, or indirectly by serving as a sink for sometimes hazardous by-products of industry (Coote, 1980). The changing face of human endeavor has created urban opportunities at the expense of the farm labor force. The diminishing land and labor resource, coupled with the increasing demand for food and higher input costs, have put pressure on science and technology to increase the productivity of existing agricultural soil and the efficiency of agricultural production.

On the one hand, the need for higher yields has been met by improvements such as subsurface drainage, irrigation, weed and pest control, and breeding of resistant crop cultivars. On the other, increasing mechanization and the trend toward larger and more efficient farm vehicles have led to a gradual densification of soils and a corresponding reduction of productivity in many regions. The susceptibility of soils to compaction has also risen due to extensive use of inorganic fertilizers and overtillage, practices that have led to a general reduction of organic matter in soils.

B. Economic Impact

The total economic impact of soil compaction is difficult to assess due to the vast number of interrelated factors involved. Corn yield reductions of up to 50% were attributed to compaction in clay soils in Quebec (Raghavan et al., 1978a). Results from a large number of experiments on cereals and oilseeds in Sweden showed that reductions of 25% were not uncommon (Eriksson et al., 1974); 35% higher fuel costs for tilling compacted soils were reported by Voorhees (1980). Lindstrom et al. (1981) attributed higher runoff and erosion rates to topsoil compaction. Chancellor (1976) noted higher operational costs of irrigation due to poor infiltration and, presumably, higher evaporative losses on compacted soils. Less efficient use of fertilizer, future costs of restoring soil structure, and a number of other factors have been mentioned in the literature.

On-farm losses due to yield reductions were estimated at $1.18 billion in the United States alone (Gill, 1971). This was based on an assumed 10% loss on acreage where winter freezing is limited to the top 25 cm of soil and a 1% loss where freezing reaches the subsoil. There has since been evidence that losses in northern regions were underestimated. Various studies have contradicted the once-held view that natural freezing/thawing and wetting/drying cycles sufficiently alleviate compaction caused by machinery traffic in a seeding-to-seeding cycle (Voorhees et al., 1985). The persistence of subsoil compaction in particular has been noted by several authors (Blake et al., 1976; Hakansson, 1982; Gameda et al., 1987a). In Canada, on-farm effects of soil degradation in general were estimated to range from $698 to $915 million in 1984 (Girt, 1986). Mehuys (1984) attributed 85% of the economic impact of soil degradation in Quebec to compaction. He estimated average compaction-related yield reductions to be 15% of potential yields, representing a province-wide total loss in farm revenue of $100 million.

C. Objectives

Although soil compaction has been associated with yield reductions since the 1930s and perhaps earlier, it is now clear that soil compaction has sufficient economic impact to justify the increasing volume of literature devoted to it over the last 20 years. In some respects, soil compaction is understood well enough that a framework for managing the physical condition of soils at the farm level can be formulated. The objectives of this paper are therefore to provide a conceptual outline of machine-soil-plant relationships based on recent research and to offer some views on practical aspects of managing soils with respect to compaction. Some aspects of tillage research will necessarily enter the discussion.

Before continuing, we would like to point the reader to a few of the more recent reviews on soil compaction as well as to some relevant related works: on compaction, Barnes et al. (1971), Eriksson et al. (1974), Chancellor (1976), Bowen (1981), Soane et al. (1981a, 1981b, 1982), Taylor and Gill (1984), Soane (1985), Raghavan (1985), and Hakansson et al. (1988); on subsoil compaction, Gameda et al. (1985); on tillage, Van Doren et al. (1982); on modeling machine-soil-crop interaction, Hadas et al. (1988); on socioeconomics, Lovejoy and Napier, (1986).

II. The Machine-Soil-Crop System

A. Conceptualization

In modern mechanized agriculture, it is important to understand changes in soil condition caused by machinery traffic as well as crop response to changes in the root-zone environment. A categorical view is presented in Figure 1. The order of the three main elements—machine, soil, and crop—reflects an action-reaction-output approach. The first element is subdivided into compaction and tillage machinery. The second is divided into physical, biological, and chemical soil characteristics. The third reflects the water, air, and nutrient needs of crops. The connecting lines are meant to represent the possible interactions between subelements. Most of the compaction-related literature is concerned with such interactions, in particular with those including soil physical status.

The complexity does not stop there. Finer divisions of each subelement are also related and give rise to new sets of interactions. Time-related effects, which include weather and natural alleviation of compaction (including the effects of root action on soil structure), sequences of soil manipulations, developmental aspects of crop growth, and time and economic constraints on production, are also interdependent. There are also a number of possible cultural and tillage systems to consider: monocultural, rotational, intercropping, zero-till, reduced tillage, etc. The literature on machine-soil-crop relationships is diverse and full of as yet unresolved

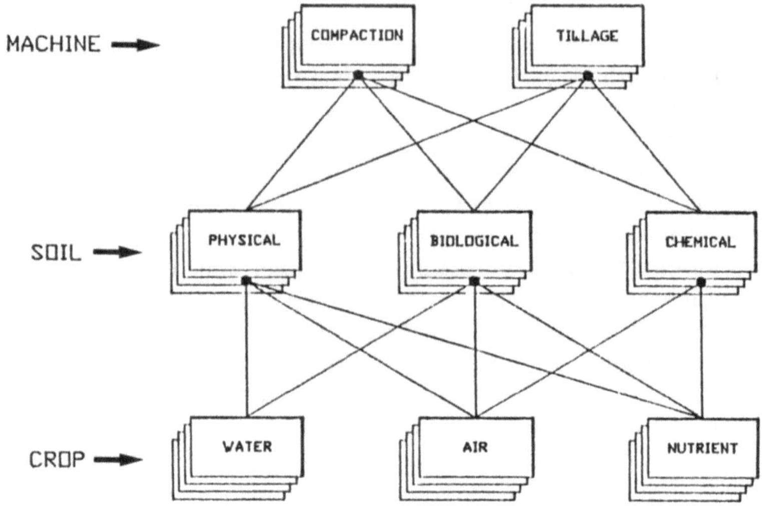

Figure 1. Conceptual representation of machine-soil-crop system with subelements and interactions.

theoretical and practical problems; however, the possibility of developing strategies for sustainable agriculture in the near future is clear.

To put things in perspective, the compaction process is briefly described. Crop response to soil physical status and physical response to machinery traffic are then discussed. It is acknowledged that the latter relationships are used as indicators of changes in soil environment in general. Crop response is viewed as an integration of changes in soil condition and processes as a whole, i.e., including the effects of changes in soil environment on water balance, microbial ecology, nutrient release, and so on. Changes in soil physical structure are indicative of the integrated effect of sequences of mechanical manipulations and natural forces on soil environment over a period of time.

B. The Process of Soil Compaction

Soil compaction may be defined as the compression of a mass of soil into a smaller volume. Changes in bulk properties are accompanied by changes in structural properties, thermal and hydraulic conductivity, and gaseous transfer characteristics. These in turn affect chemical and biological balances. In a word, the soil environment is changed in such a way as to affect all soil processes to a greater or lesser extent, depending on the degree of compaction.

The degree of compaction is usually expressed in terms of dry bulk density, porosity, or penetration resistance; however, the characterization of

Figure 2. Clay profile after 15 tractor passes with contact pressure of 62 kPa. Divisions on meter stick at left indicate 0.30-m intervals.

compaction is a subject of controversy. The parameters that best describe compaction by wheel traffic are not directly coupled to crop response; neither are changes due to decompactive processes (tillage and natural alleviation) well described in terms of compaction-related parameters. Soane (1985) and Soane and Boone (1986) suggest that new structural indexes be developed to provide stronger links between traffic-soil and soil-plant interactions. Efforts in this direction have been made by Boone (1986), who discusses the concept of critical soil density in relation to optimality of various plant-related processes. The key is to identify limiting factors for given crops, soils, and climatic conditions and to establish relationships between soil compactness and the major soil-related limiting factor. In some situations, the limiting factor may be moisture, in others it may be aeration or nutrient status.

Regardless of quantitative definition, loss of structure may be discerned visibly in cases of excessive compaction (Figure 2). The face of the trench of the compacted soil is much smoother, and few if any pores are visible as opposed to the uncompacted profile shown in Figure 3. Excessively compacted soils suffer from poor aeration and low hydraulic conductivity. The manifestation of excessive compaction on growth is exemplified in Figures 4 and 5, which show the contrast in growth of corn early and late in the season on uncompacted and compacted plots. Less than excessive compaction is not necessarily detrimental to crop growth, as has been found in long-term studies of crop response to compaction (Eriksson et al., 1974; Raghavan and McKyes, 1983).

Figure 3. Uncompacted clay profile 1-m depth.

Figure 4. Left: corn on uncompacted (000) plots. Right: corn on plots subjected to 5 passes at contact pressure of 41 kPa (30 days after seeding).

C. Crop Response to Compaction

Plant growth involves physical, chemical, and biological processes. In field production, there are two arenas of activity, the soil and the atmosphere. Although it may be tempting to separate the two, it must be understood that the atmosphere-related and soil-related processes are interdependent. The demands on the root system at any given time are governed by what

Figure 5. Left: corn on plots subjected to 15 passes at contact pressure of 62 kPa. Right: corn on plots subjected to 1 pass at contact pressure of 62 kPa (65 days after seeding).

we will term *photosynthetic potential*, which may be thought of as some combination of available light, CO_2, and plant ability to absorb these by the aboveground portion of the plant. The degree to which that photosynthetic potential is achieved at any time depends on the ability of the root system to supply water, nutrients, and oxygen from the soil and to exhaust respiratory CO_2. Maximum growth is obtained when the root system can meet the demands of the photosynthetic potential given that the soil contains at least an adequate nutrient and water supply as well as an efficient mechanism for gaseous exchange. Maximum yields are obtained when conditions are optimal throughout the growth cycle, i.e., when a temporal coherence between plant development, weather, and soil conditions through the profile is maintained, as can be done in an artificially controlled environment.

The balance can be upset in compacted soils. The higher mechanical impedance of compacted soils, or compacted layers, restricts the depth of root penetration (Figure 6) as well as the overall root density (Figure 7), which implies slower root development. The most obvious effect of restricted root proliferation on the plant is reduced access to water and nutrient supply. Limited penetration to the subsoil can be critical during dry spells. It has also been suggested that restricted root proliferation can reduce the ability of root systems to overcome the harmful effects of topsoil-resident pathogens (Figure 8). In a survey of pea fields in southwestern Quebec, Vigier and Raghavan (1980) found a relationship between soil dry bulk density and root-rot disease index (Figure 9). A regression model

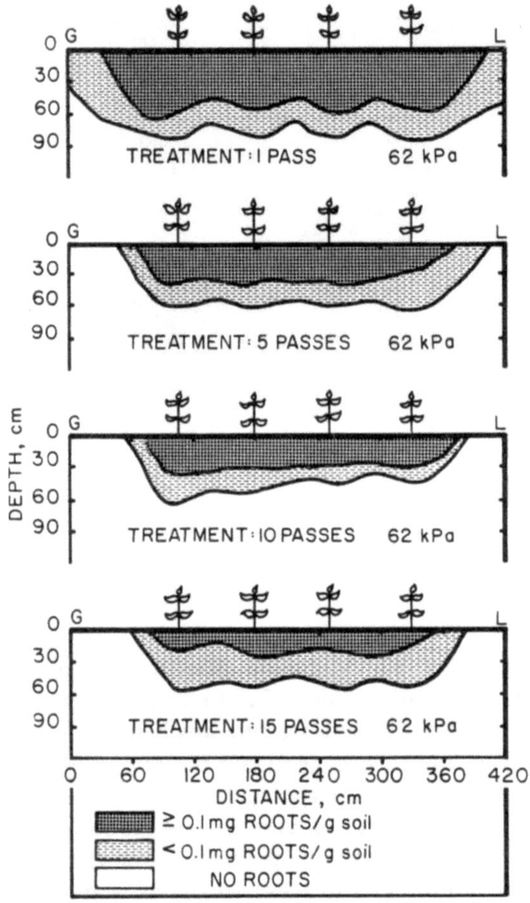

Figure 6. Root density distributions in clay soil subjected to 1, 5, 10, and 15 wheel passes at contact pressure of 62 kPa. [After Raghavan et al., 1979. Reprinted with permission from *J. of Terramechanics*, 16, No. 4, copyright 1979, Pergamon Press plc.]

based on results from later field trials (Raghavan et al., 1982) showed that for given levels of infestation, yield decreases with increases in soil bulk density (Figure 10).

Compaction-induced changes in the air-water regime affect microbial activity such that the nitrogen balance favors ammonium over nitrate nitrogen as compaction levels increase (Voronin, 1982; Sheptukhov et al., 1982) with unfavorable effects on yield. Data from a study on sugar beet grown on plots subjected to various compaction-tillage treatments and then left fallow for six years provided evidence of poorer yields due to the persistent effect of residual compaction on nitrogen balance (Mohammad 1987).

Compaction studies have resulted in the recognition of important general relationships between soil compactness and yield. Figure 11 (Eriksson et al., 1974) is a conceptualization of yield-compactness-weather rela-

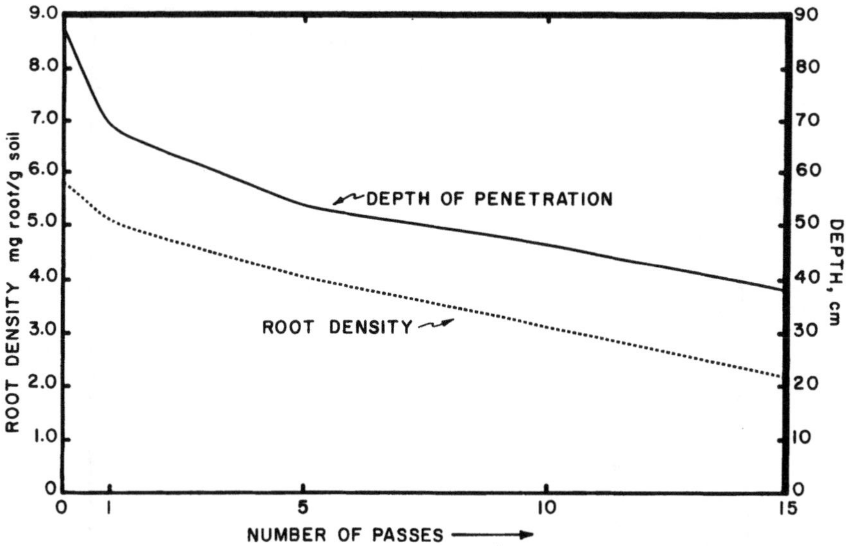

Figure 7. Graph of depth of root penetration and root density of corn versus number of wheel passes on a clay soil. [After Raghavan et al., 1979. Reprinted with permission from *J. of Terramechanics*, 16, No. 4, copyright 1979, Pergamon Press plc.]

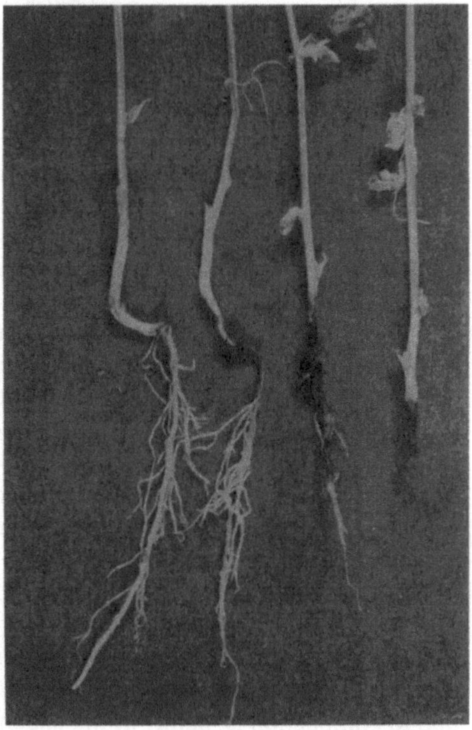

Figure 8. Uninfected pea roots (*Pisum sativum* vr. Rally) and pea roots suffering from root rot. Left to right: disease index 0, 1, 2, and 3. (No yield obtained from disease indexes 2 and 3.)

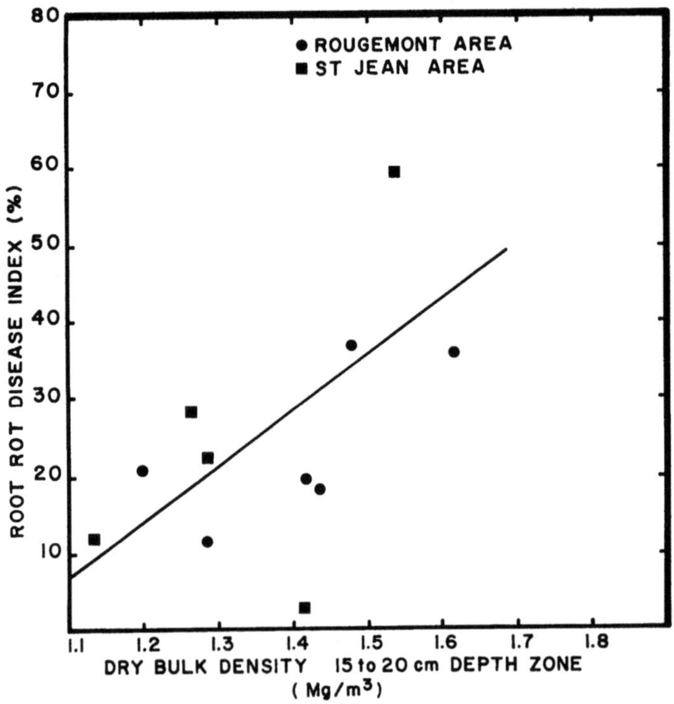

Figure 9. Graph of root rot disease index versus DBD from randomly sampled pea fields in two regions of Quebec. Indicator depth range was 15 to 20 cm. [After Vigier and Raghavan (1980).]

tionships. The essential feature is that the compactness degree (defined relative to a standard compaction level) at which maximum yield is obtainable depends on the weather regime. Data from compaction-weather simulation experiments by Edling and Fergedal (Eriksson et al., 1974) on barley corroborated this view (Figure 12). Raghavan et al. (1979) obtained similar results (Figure 13) in a two-year field study of the response of maize to different levels of compaction. An interesting feature is that in dry years, better yields were obtained on a slightly compacted soil than on loose soil. Raghavan and McKyes (1983) attributed this to differences in available moisture because uncompacted plots had very low moisture due to high evaporative losses whereas highly compacted plots held the water tightly in small pores. The water balance was more favorable at intermediate levels of compaction.

This is an example of bulk definitions of compaction not fully describing the resulting effect without other factors being accounted for. Root-water extraction is more directly affected by pore-size distribution and continuity of pores rather than by total porosity. The change in pore-size distribution due to compaction is mainly at the expense of larger pores associated with

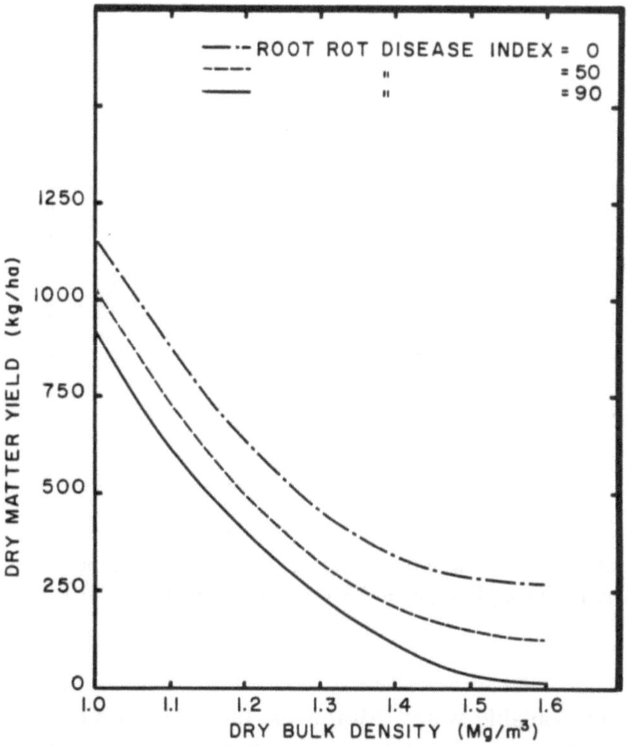

Figure 10. Prediction of dry matter yield of peas versus DBD for various levels of infestation based on model derived from field trials. [After Raghavan et al. (1982).]

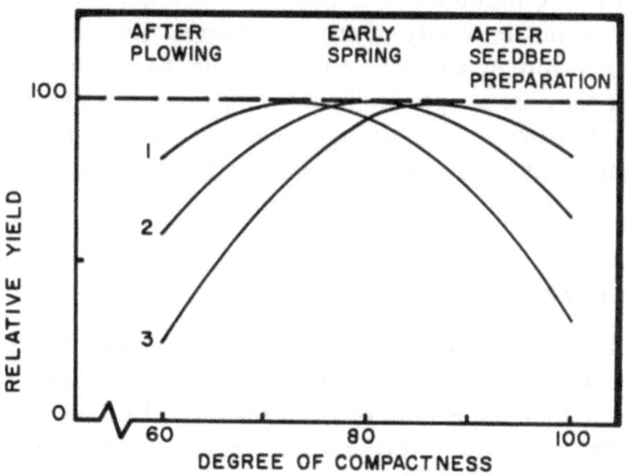

Figure 11. Conceptual representation of compactness-yield-weather relations. Curves 1, 2, and 3 represent wet, normal, and dry years, respectively. [After Eriksson et al. (1974).]

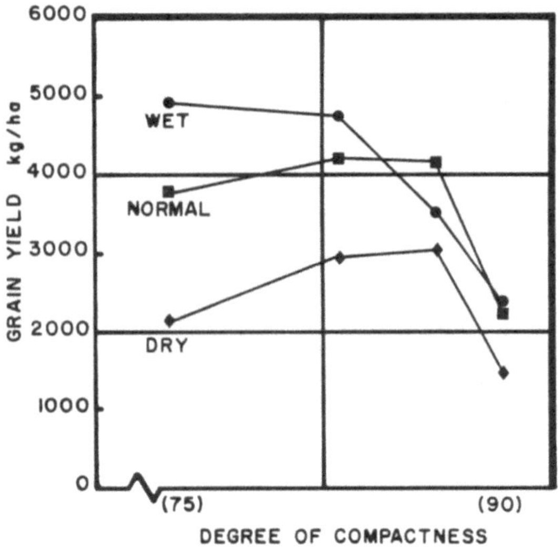

Figure 12. Results from weather-compactness simulation experiment on barley grain yield by Edling and Fergedal. [After Eriksson et al. (1974).]

aeration and available water. Moreover, the relationship between available water and degree of compaction is a function of infiltrating rainfall.

Although optimum curves must be considered to be statistical and vary according to crop, soil type, and soil texture, as discussed by Boone (1986), they provide an interesting integral view of soil-crop relationships as mediated by weather. Optimum curves have been found to vary in peakedness, indicating that the sensitivity of crops to compaction depends on species. One might therefore expect deep-rooted species to be less sensitive to compaction since they are adapted to penetrating typically dense subsoil layers. Other characteristics such as drought resistance or resistance to excessive moisture may be indicative of lower sensitivity to compaction.

D. Crop Response to Tillage

Tillage is practiced for a number of reasons: preparing the seedbed, incorporating residues and manures, controling weeds, controling wind erosion, and so on. General aspects have been reviewed by Larson and Osborne (1982). For the purposes of this paper we view it as an integral part of crop production in most cases and, more specifically, as a practice intended primarily to loosen compacted layers in the soil. The effects of tillage on crop response depend on soil, climate, tillage implement used, and topography. Where topsoil compaction is a problem, tillage certainly has ameliorating effects.

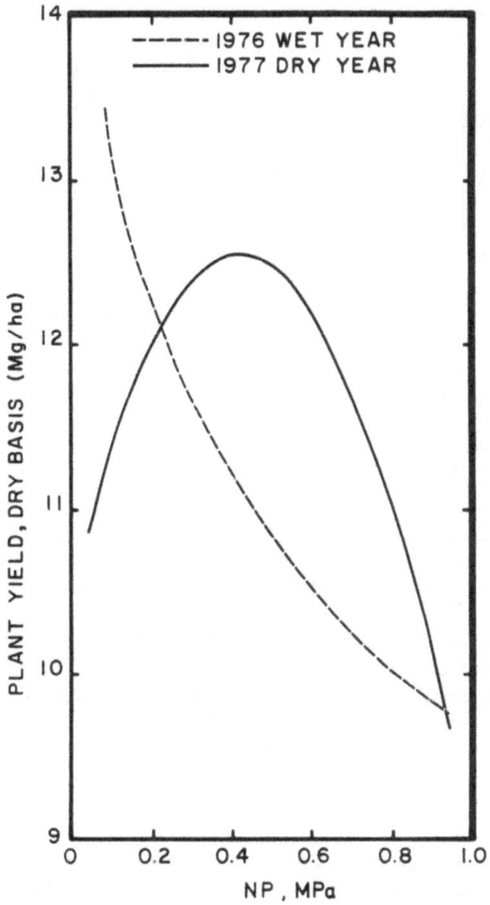

Figure 13. Corn yield versus cumulative ground contact pressure in a wet and a dry year. [After Raghavan et al., 1979. Reprinted with permission from *J. of Terramechanics*, 16, No. 4, copyright 1979, Pergamon Press plc.]

In combined compaction-tillage experiments (Quebec, humid climate), Negi et al. (1981) found that highest yields were obtained on highly compacted but subsequently chiseled or moldboard plowed plots on both a sandy and a clay soil. Uncompacted zero-till plots were comparable. The poorest yields were on compacted zero-till plots. Results obtained on both a clay soil and a sandy loam showed that yield could be expressed as a curvilinear function of soil density, as described in the preceding subsection, even though different tillage implements were used (Figure 14). These conclusions were based on average dry bulk density through the top 20 cm of soil. However, the authors did note that the fit of the regression curve was poor, which may have been due to additional variability introduced by the different effects on soil structure of the implements used.

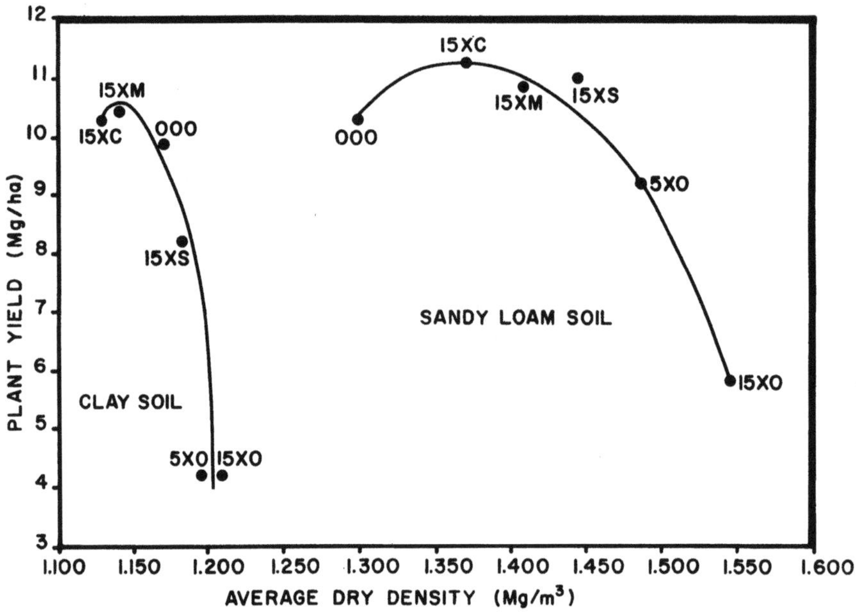

Figure 14. Corn yield versus average dry bulk density on a clay and a sandy loam. First digit(s) represent number of passes, "X" is contact pressure (62 kPa), and last letter represents tillage implement used after compaction treatment; O, none; C, chisel; M, moldboard; S, subsoiler. [After Negi et al., 1981. Reprinted with permission from *J. of Terramechanics*, 18, No. 2, copyright 1981, Pergamon Press plc.]

In drier regions, strategies such as zero-till and conservation tillage are widely accepted as water-conserving and erosion-control measures, and conventional tillage is generally avoided (Larson and Osborne, 1982). Even in such regions, however, tillage may be beneficial in controlling weeds or pathogens or reducing resistance to root penetration (Dann et al., 1987).

E. Soil Response to Machinery Traffic

Agricultural soil is generally subjected to two types of machinery traffic in various sequences, that which compacts (wheel traffic) and that which loosens and redistributes through part of the depth profile, i.e., tillage traffic. Although the tendency is to consider the two separately, their effects on soil properties are not entirely independent in practice because soil response to each is a function of conditions established by the previous operation and the summation of natural forces acting in the interval.

Soil response to compaction is known to be a function of traffic parameters, soil properties, and soil moisture content at time of traffic. Response is usually described in terms of changes in dry bulk density (DBD), porosity (PO), and/or penetration resistance (PR) as functions of applied pressure and soil moisture content (SMC). DBD is the most frequently used parameter in compaction research and can be measured by core sampling or in situ with minimal disturbance to the soil using gamma gauges. PO is easily derived from DBD given the specific gravity of soil solids. PO and DBD are indirect measures of water movement and aeration characteristics. PR is used as a measure of soil strength and mechanical impedance to root penetration, but problems of interpretation of data due to concommitant variables such as moisture content have yet to be fully resolved (Perumpral, 1987). SMC just prior to and after compaction must be known for interpretation of penetrometer data and of changes in bulk density.

Tillage, as it directly affects soil physical properties, depends on type of implement, depth of operation, initial conditions, and soil moisture at time of tillage (Tisdall and Adem, 1986). Tillage tends to raise the soil and change the degree of aggregation with resulting effects on soil physical properties and processes as discussed in a previous monograph (Van Doren et al., 1982). Tillage-induced changes in compaction-related (packing state) parameters are considered transient, particularly in time scales allowing for rainfall effects (Cassel, 1982; Meek et al., 1988), and not uniquely defined in terms of aggregation parameters. This has been considered to be one of the major difficulties in treating compaction and tillage operations sequentially with respect to changes in soil conditions.

F. Modeling Compaction

Wheel traffic generally induces densification of the soil in patterns such as observed by Raghavan et al. (1976a) on a sandy loam soil (Figure 15). The maximum density change ranged from 0.1 to 0.5 Mg/m^3 at a depth of 15 to 20 cm. The magnitude of density changes depends on soil texture, SMC, contact pressure (vehicles of less than 10 tons per axle) or axle load (vehicles greater than 10 tones axle), and number of passes. The region of maximum compaction for lighter vehicles is in the topsoil (0 to 30 cm). Heavy equipment tends to compact the subsoil (30 to 60 cm) and is a more serious problem because alleviating procedures such as subsoiling represent costs over and above normal tillage. Traffic studies have shown that in many conventional cropping systems, the intensity of traffic is such that 90% or more of the field area is subjected to at least one wheel pass per season (e.g., cereals: Eriksson et al., 1974; maize: Voorhees, 1977). These have led to revival of the controlled traffic or zone production concept, which will be discussed later.

Soil response to external loads depends on SMC. For a given energy imparted at the soil surface, the average DBD change is curvilinearly related to SMC as shown in Figure 16. The magnitude of the peak, as determined by the Proctor compaction test described by Lambe (1951), depends

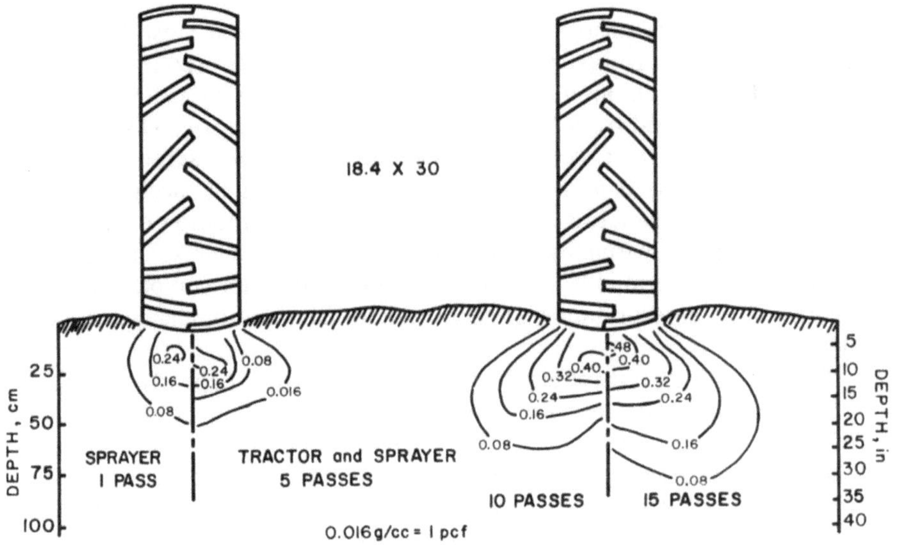

Figure 15. Dry bulk density patterns obtained in a sandy loam orchard soil. [After Raghavan et al. (1976).]

on the energy, while its position along the x-axis depends mainly on soil texture. Raghavan et al. (1977a) observed that the optimum moisture content for maximum compaction also tends to increase with reduced energy in static compaction tests; however, the effect was not apparent in their field data. Finer-textured soils tend to have peaks at higher moisture contents than coarse soils and also tend to have higher moisture-retention capacities. Ohu et al. (1986) showed a shift of the optimum to higher moisture contents with increasing organic matter content as well as lower shear strength for equal compaction levels (Figure 17).

At moisture contents above the Proctor optimum for compaction, wheel slip can contribute to compaction as significantly as loading. Laboratory studies by Raghavan and McKyes (1977) showed that up to 50% of topsoil compaction could be attributed to shear generated by wheel slip (Figure 18). The maximum effect was found to be at slips between 15% and 25%, which included the normal operating range of 20%. This information was corroborated in field studies (Raghavan et al., 1977b, 1978b). At higher slips, topsoil structure is damaged by smearing. Deep ruts and sideways displacement of soil due to greater sinkage are also problems at high moisture content.

Early attempts to model density (or porosity) changes in the soil in terms of load and moisture content were based on laboratory compaction of loosened soil. Soehne's (1958) equation for porosity of a loose soil as a log function of applied static pressure at known moisture content was modified

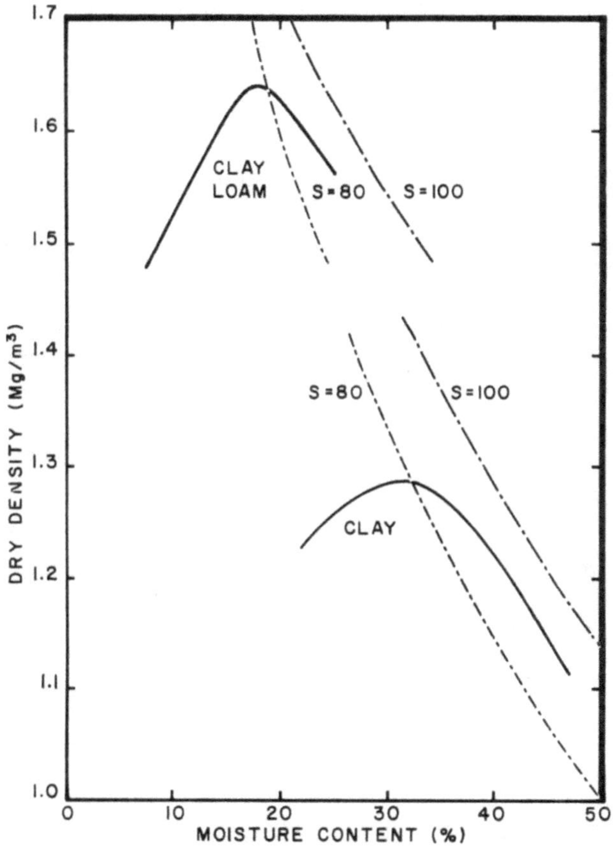

Figure 16. Proctor curves (using gravimetric moisture content) obtained on clay loam and clay soil for imparted energy of 25 Proctor hammer blows applied on each of three layers.

for field soils by Amir et al. (1976), who expressed porosity as a function of soil moisture, applied pressure, and residual pressure (see below). They combined the compaction model with a drainage equation in order to determine the best timing for wheel traffic so as to reduce the risk of compaction. The equation for a change in porosity was

$$dP_0 = B\ln(P_2/P_1) + S^*C\ln(t_2/t_1) \tag{1}$$

where the subscripts 1 and 2 represent pre- and postcompaction, B and C are soil-specific compression characteristics, and S is a soil-specific constant describing influence of time on soil moisture reduction by drainage. P_1 is the residual pressure, or the pressure required to bring a completely loose (virgin) soil to precompacted field porosity. P_2 is the applied pressure. Amir et al. (1976) and Gupta and Allmaras (1987) discuss applications of

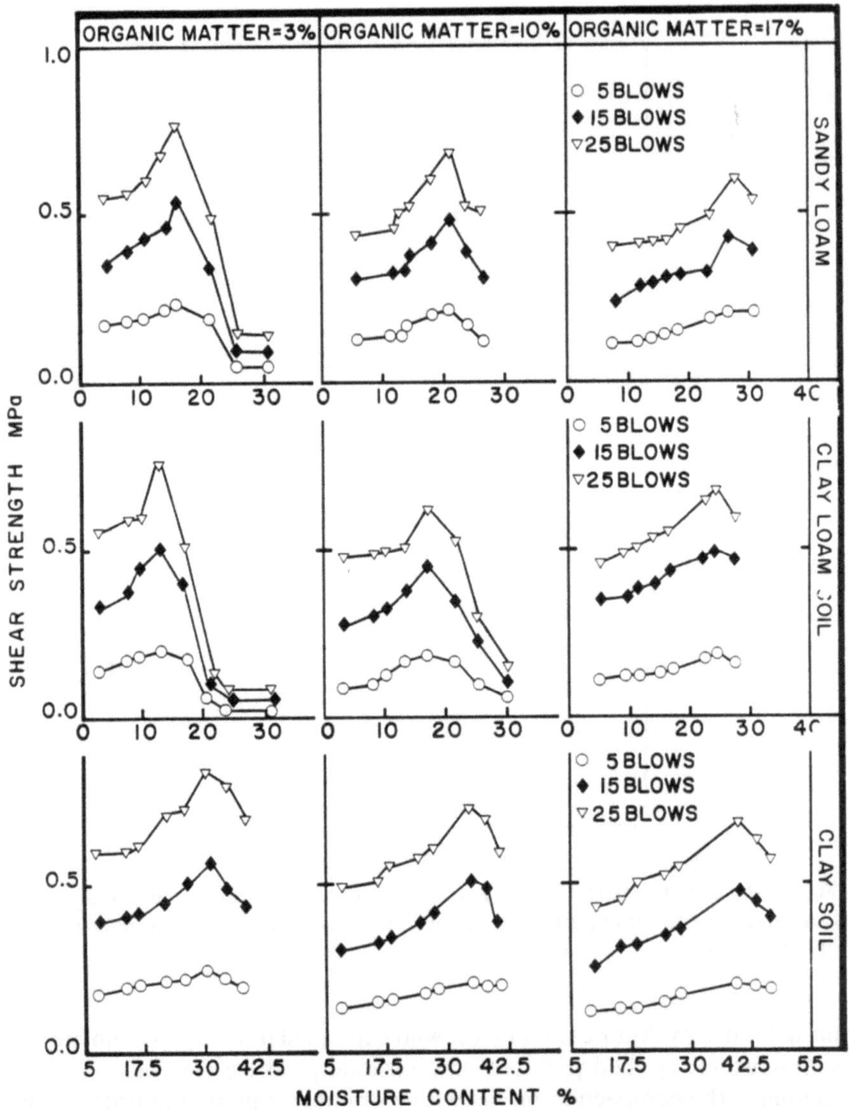

Figure 17. Effect of organic matter on compaction characteristics of three differently textured soils (gravimetric moisture content). [After Ohu et al. (1986).]

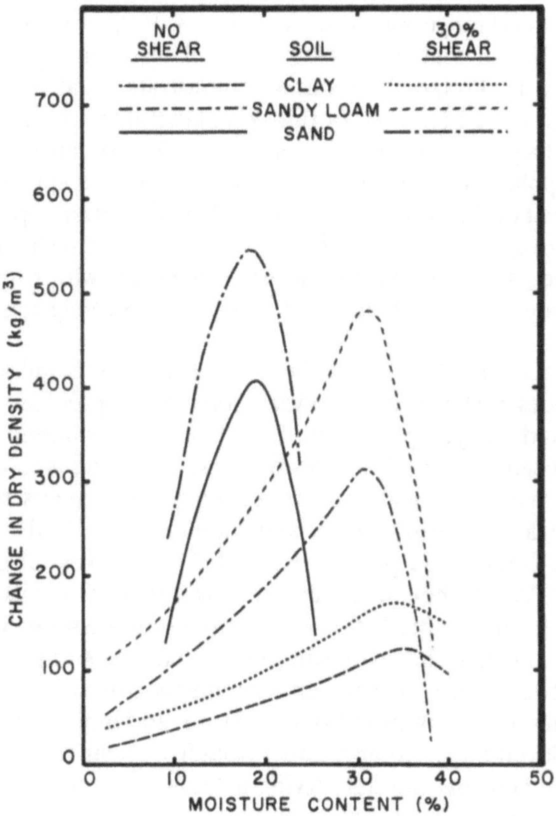

Figure 18. Additional compaction due to slip-induced shear for three differently textured soils (gravimetric moisture content).

this equation with respect to timeliness of operations and reduced compaction levels due to improved drainage and reduced ground pressure.

The practical nature of the applied pressure term has been studied. In field compaction experiments, Raghavan et al. (1976b, 1977a) found that measured depth-averaged dry bulk densities in the topsoil could be reasonably well described by regression techniques with the natural logarithms of applied pressure and soil moisture as independent variables where the applied pressure was expressed as the product of contact pressure and number of passes (nP). Equations for SMC below and above the Proctor optimum were required to provide reasonable fits and consistent coefficients for the independent variables. Regression equations including log functions of nP, SMC, and slip based on field studies (Raghavan and McKyes, 1978) showed reasonable predictive value even though the published equations in this case do not appear consistent with respect to the direction of relationship with nP and depth given the nature of observed compaction patterns and stated ranges of moisture contents. Raghavan and Ohu (1985) have further shown that it is possible to relate contact pressure to the energy applied by Proctor blows and demonstrated that the relationship is independent of soil texture. In this way, laboratory characterization of compaction of a given soil could be applied to a field situation. One practical aspect of applied compactive forces that has not been approached in the literature is that of the time dependence of load for certain operations such as manure spreading or harvesting. The applied load can change considerably from one part of a field to another. However, with some planning, overcompaction of certain field areas, particularly turning zones, could be avoided.

Gupta and Larson (1982) describe a soil compaction and soil breakup model that goes a step further toward solving the problem of series of compacting and tillage operations. Based on the assumption of a relationship between packing bulk density and aggregate-size distribution, they suggest that it is possible to select tillage implements to provide a desired aggregate-size distribution and therefore a desired bulk density. Although the compaction part of the model was successful in field tests (Gupta et al., 1985), they have not yet shown that the combined models could give a reasonable prediction of conditions after a sequence of operations. Success in that endeavor will depend on the strength of the relationships between bulk density and aggregate-size distribution (which probably varies according to soil type, texture, and initial conditions).

Natural alleviation of compaction through freezing/thawing, swelling/shrinking, and cracking through drying is known to contribute to restoration of soil structure after compaction; however, modeling of these is difficult if not impossible. Literature on the subject indicates that the length of time involved for recovery depends on climate, soil type, clay content, organic matter content, and degree of compaction. Estimates of recovery of highly compacted soils on travel routes in the Mojave Desert range upwards of 100 years (Webb et al., 1986). Gameda et al. (1987b) noted that soil bulk density reductions and improvements in crop yield after even a single incidence of high axle loading could take substantially longer than three years on a clay soil in a temperate climate. Moreover, there is evidence that the time scale for improved production is not directly related to that for structural improvement. Hakansson (1982) observed yield improvements over time although high values of DBD and PR persisted and attributed this to developing networks of cracks and earthworm holes that improve water infiltration and root proliferation. Dexter (1986a, 1986b, 1986c) has performed a series of laboratory experiments to lend evidence to the importance of the action of soil organisms in modifying compacted soils to the benefit of root proliferation, thus adding another dimension to natural alleviation. Studies of no-till systems have demonstrated the influence of organic matter accumulation and root action on soil structural improvement (Lal et al., 1979).

A generally applicable soil deformation model has yet to be developed. The limitations of classical mechanics and the drawbacks of empirical equations (Hillel, 1987) have led to implementation of more sophisticated approaches that can take into account the layered nature of soils. Pollock et al. (1986) have adopted the finite-element technique to modeling compaction in a multipass context. Hettiarachi and O'Callaghan (1980) discussed the potential of critical-state mechanical theory as an alternative to classical concepts. Nevertheless, empirical and classical mechanic equations have brought about a firm understanding of soil compaction and provided useful guidelines for reducing the risk of excessive compaction, with respect to both timing of operatons and selection of machinery.

G. Summary

In this section we have presented a conceptual outline of machine-soil and soil-crop relationships, touching upon some research findings relevant to the decision framework to be outlined below. It is clear that the major factors involved in soil response to machinery traffic have been identified. All known soil processes are affected in some way by the change in environment induced by compaction and tillage; however, chemical and biological factors have been largely neglected in the literature, likely due to lack of adequate sampling techniques and to extreme variability. An important point to make here is that although it is possible to take any of these aspects to a high level of complexity, it is doubtful that realistic models of such a dynamic system can be pieced together in the near future. Soils are highly variable spatially and temporally. Plant response to soil condition depends on weather, a stochastic process, as well as on time interactions between weather, plant development, and soil parameters through the profile. There is also evidence that plants respond as much to the nutrient balance created by the physical change in environment as to the physical aspects. The need for integrated studies to determine crop sensitivity to a number of factors simultaneously is evident.

III. Establishing a Decision Framework

A. Basic Considerations

We now suggest, as have Gupta and Allmaras (1987), that it is possible to manage soils to prevent excessive compaction and to maintain optimum physical conditions on the average, based on current knowledge. This implies that we have the ability to control soil physical condition to a certain extent, whether the object is to compact or loosen. Based on the optimum curve concept, it is altogether plausible that soils can be managed to optimize production if climatic and topographic factors are taken into account. It should be reemphasized that compaction is not necessarily the limiting factor, nor is it harmful under all circumstances. In dry regions, the benefits of loosening the soil do not outweigh the additional evaporative losses of moisture associated with surface tillage. Soil and crop-weather models will no doubt be given more attention in compaction control efforts in the future. Gupta and Allmaras (1987) discussed the utilization of a climatic data base for predicting soil water content by horizons in application to delineating critical applied loads at usual times of machinery traffic. Vepraskas (1988) described a method to determine the probability that subsoiling increases tobacco yields, based on rainfall distribution.

The general situation of mechanized crop production is now considered. The basic constraints faced by the farmer are soil type, length of growing season, and climate. These determine the range of crops that may be successfully produced. To this we may add a third set of constraints—economic. The economic constraints depend on the type of enterprise, regional context, and level of government intervention. Here we are suggesting that profit maximization is not necessarily the main goal in crop production. Assuming that maintenance of soil physical condition is the goal, what must the farmer or technical adviser know in order to make decisions?

An outline of the main points to consider is presented in Figure 19. For a given agricultural enterprise, the first priority is to determine the initial soil state, since soil response to traffic depends on it. (It is assumed that fertility levels and pest control are adequate.) Second, the effects of a sequence of operations on conditions over the growing season must be predicted. This implies that soil behavior under mechanical stress is understood. The residual compaction and possibility of alleviating it should also be determined for future crops. A range of probable yields can then be estimated from the optimum curves (yield versus compactness) and likelihood of various weather conditions. Cost-benefit analysis can then be carried out for the season in question or for a sequence of years. Clearly, crop-weather models must play an important role in soil management strategies. The following details are involved:

1. Determining initial state of soil before any operations
   a. Choice of soil structural parameters
   b. Sampling scheme
      i. Choice of relevant space scale
      ii. Variability estimates
2. Predicting conditions through growth period
   a. Estimation of changes due to compaction
      i. Load
      ii. Number of passes
      iii. Slip
      iv. Soil moisture content at time of traffic

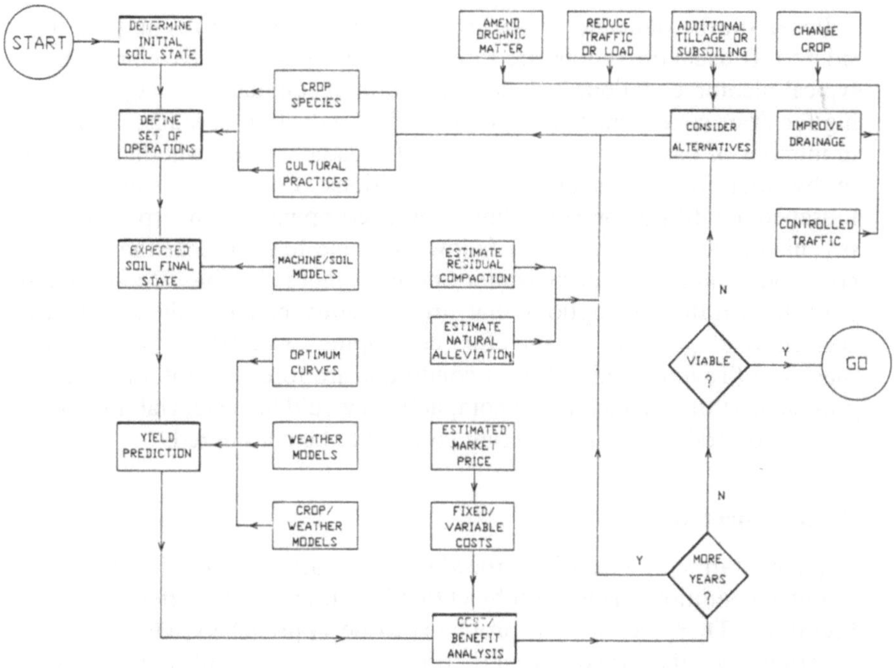

Figure 19. Framework for sustainable soil structure management strategy under regional and economic constraints. Required inputs at various stages are also shown.

  b. Estimation of changes due to tillage
    i. Type of Implement
    ii. Operating depth
3. Estimating yield (and/or economic return)
  a. Prior knowledge of optimum curves
  b. Probability analysis of weather conditions
  c. Crop-weather models
  d. Prediction of market price
  e. Cost estimation
    i. Fixed costs
    ii. Variable costs (e.g., grain drying, transportation)

The initial soil condition may be such that alleviating procedures become immediately necessary, in which case a number of alternatives must be considered. The predicted soil condition may be such that yields may be known to be poor or such that major alleviating steps would have to be taken before the next season. Again, alternatives must be considered. The estimated yield, or economic return, is usually the deciding factor, whether on a one-year or multiyear basis.

Based on the outcome of predictions, various alternatives may be considered. If usual operations are forecast to result in excessive compaction, typical recommendations include installation of adequate drainage or waiting for drier conditions before wheel traffic, reduction of loads, reduction in number of passes, or incorporation of organic matter, whether directly or by adoption of sustainable crop rotations. Compacted soils can be loosened by tillage or subsoiling, cover cropping, or incorporation of organic matter, or may be left fallow for some time. Freeze/thaw effects could be enhanced by early winter flooding where possible. Reduced and controlled traffic are options that are presently being studied, both for conventional machinery and for developing wide-track lateral-move machines (Taylor, 1985). If drier conditions are forecast, optimum curves may suggest that an increase in compactness would be beneficial. Finally, a change to production of a different crop could be considered.

B. Implementation

Implementation of a decision procedure necessarily includes identification of initial conditions. This is a subject that has received little attention in the literature. There are several problems to be approached. The first is to determine whether or not a general problem exists, regardless of degree of compaction. For example, many fields suffer from poor drainage, disease infestation, or toxicity problems, making the question of compaction secondary. Compaction may be a problem only at a certain depth, such as in hardpan formation. Compaction may be limited to the topsoil or extend to the subsoil. In any case, general condition must be determined by sampling.

The second is to determine the relevant parameters. With respect to compactness, dry bulk density and penetration resistance are generally accepted as good indicators. With respect to predicting compaction, soil moisture and organic matter content should also be known, as should expected applied loads. The third is to develop a sampling technique and scheme that is quick and easy to perform and interpret. Presumably, the work involved is to be performed by regional field technicians with access to specialized equipment. Assuming that moisture content, density, and penetration resistance are sufficient descriptors, can be measured simultaneously, and can be rapidly analyzed, the problem is one of variability. Cassel (1982) presented an equation representing the relationship between sample size and variability. Initial studies to determine the relationship between variability and sampling interval should also be performed, since the interval used should be suited to the time constraints of the technician and precision requirements of the farmer. The degree of curvature around the peak of the Proctor curve can also be used to decide a probability range for SMC estimation. The sharper the peak, the more critical the accuracy

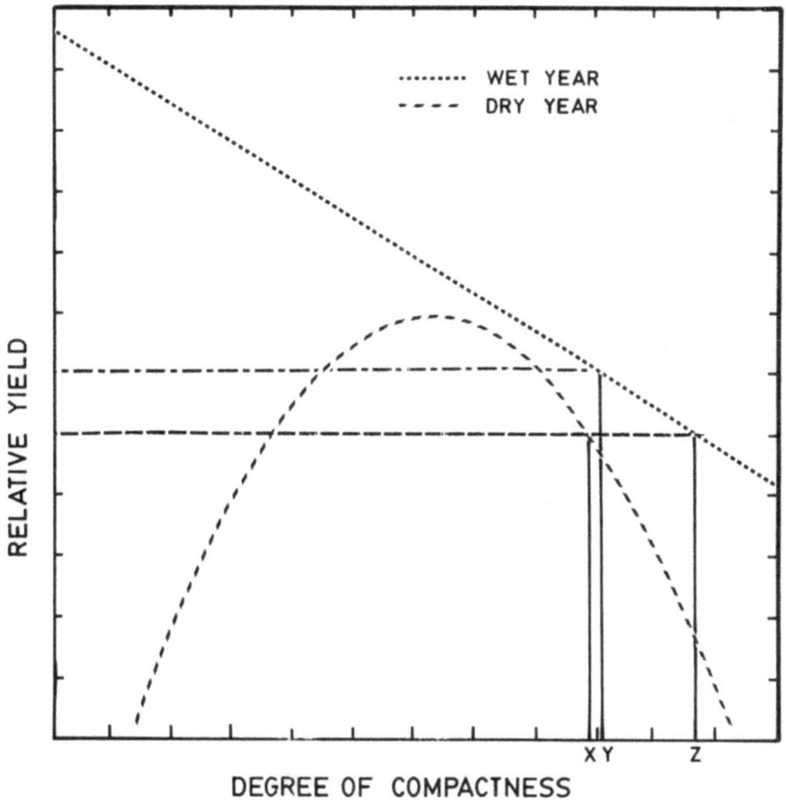

Figure 20. Critical compaction as defined on optimum curves and subject to economic definition of critical compaction levels. (———, economic threshold yield; ———, higher threshold due to costs associated with irrigation system; X, critical compaction level (CCL) for dry year and no irrigation; Y, CCL for dry year with irrigation system; Z, CCL for wet year.)

of the estimate if moisture content tends to be near the optimum at usual times of traffic.

Optimum curves can serve as guidelines for establishing critical soil compaction limits in terms of yield for expected climatic conditions. In Figure 20, yield-compaction curves for a wet and a dry year are presented. Two break-even economic return curves are also plotted, one for a situation with irrigation, the other without. The maximum degree of compactness for various situations is found by dropping perpendiculars from the intersections of return and yield-compaction curves. Here, it is assumed that irrigation results in the same yield-compaction curve as for a wet year. The importance of predictive models of compaction and tillage is clear. Opti-

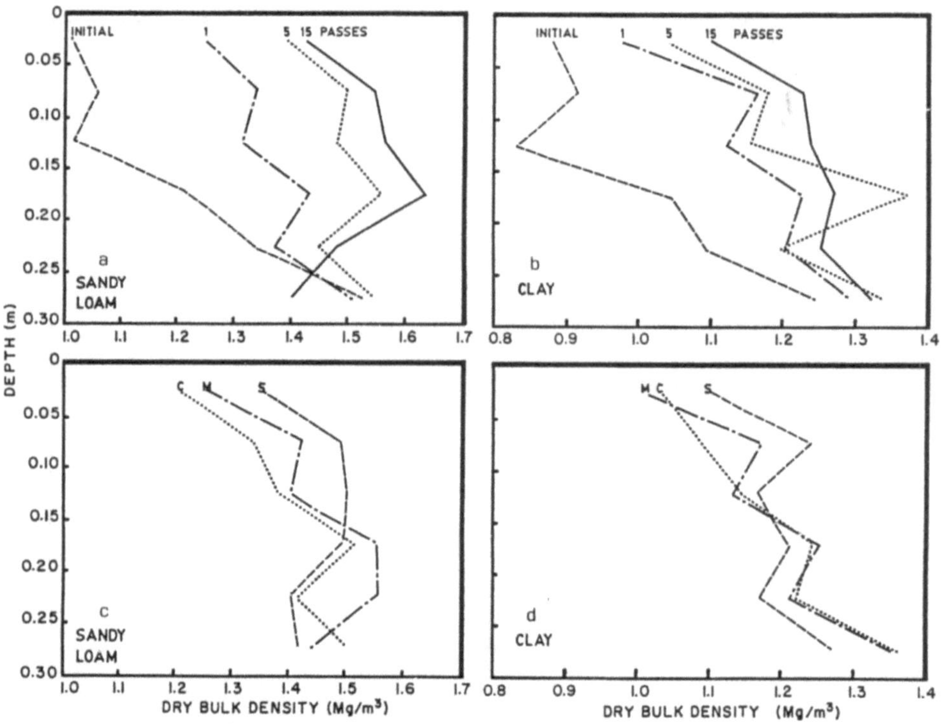

Figure 21. (a, b) Average dry bulk density profiles of sandy loam and clay plots, respectively: ———, initial profile; –·–·–, 1 pass; ----, 5 passes; ———, 15 passes. Contact pressure: 62 kPa. (c, d) Dry bulk density profiles of sandy loam and clay plots subjected to 15 passes with contact pressure of 62 kPa and subsequently loosened with: C, chisel; M, moldboard; S, subsoiler.

mum curves expressed as yield versus degree of compactness could, by way of models, be expressed in terms of yield as a function of sequences of operations, i.e., in terms easily interpreted at the farm level. Although there are fundamental theoretical problems to be resolved, confidence limits on conditions resulting from sequences of operations could initially be established experimentally for typical soils in a given region. Initial bulk density and profiles resulting from different levels of compaction are shown in Figures 21a and 21b for a sandy loam and a clay soil, respectively. The loosening effects of subsequent tillage with various implements on the most heavily compacted plots are shown in Figures 21c and 21d. The loosening and residual compaction could be expressed statistically given a sufficient number of replicates for a range of initial states. In a subsoiling experiment, Bernier et al. (1989) showed the additional lossening possible by two passes of a subsoiler compared to one only. Further studies of this nature, perhaps including combinations of tillage equipment and initial levels of compaction, may give a better idea of how to maintain soil physical structure in a suitable range.

C. Economic Considerations

From the point of view of the typical North American farm enterprise, which is essentially a business that must stand on its own, any decision process associated with soil quality is subject to economic constraints as well as natural constraints pertaining to soil type, topography, and climate. Thus, an orientation toward soil-conserving practices, including crop rotations that maintain organic matter, fertility, and soil structure, depends greatly on the producer's willingness to undertake certain financial risks over the short term in order to reap the benefits of a more stable productive potential over the long term. This risk may also be spread nationally through incentive programs.

An idealized view can also be presented, where the choices are in fact cost-independent, i.e., in a system that would be concerned only with quantities produced as defined by the needs of a specified population. In such a system, maximization of yields would be crucial in the case of undersupply, whereas stabilization of yields would be the goal in a zero-growth context. The threshold yield would therefore be subject to a different set of constraints. This juxtaposition has been made with the intention of emphasizing a possible impending global change of context, one in which agriculture is seen to supply the raw energy needed to allow pursuit of other endeavors. This scenario could be a consequence of the trends toward larger farming enterprises, automation, and eventual widespread reorganization of agriculture.

In the present context, however, economic considerations require more attention. Relationships among yield, market price, compaction by machinery traffic, and costs were recently investigated for three economically important crops in Quebec: peas (Gunjal and Raghavan, 1986), corn silage, and grain corn (Gunjal et al., 1987). The studies were based on data from compaction experiments where whole plot areas were compacted. Yield data from various years were expressed as functions of nP. Yield losses (difference between compacted plot and uncompacted plot yields) were translated to farm level by computing a compaction area percentage (CAP) for different tractor sizes. The CAP is the ratio of the area under the rear wheels and the area covered by a standard implement operation in one pass. This ratio decreases as tractor size increases. Economic analysis therefore included farm level yield losses due to compaction for different weather conditions and machinery costs.

Total per-acre costs for peas as a function of tractor size and number of passes are presented in Figure 22, and similar data for corn silage and grain corn are presented in Figure 23. The relevant results here are that optimum tractor size depends on crop and weather conditions and that small tractors

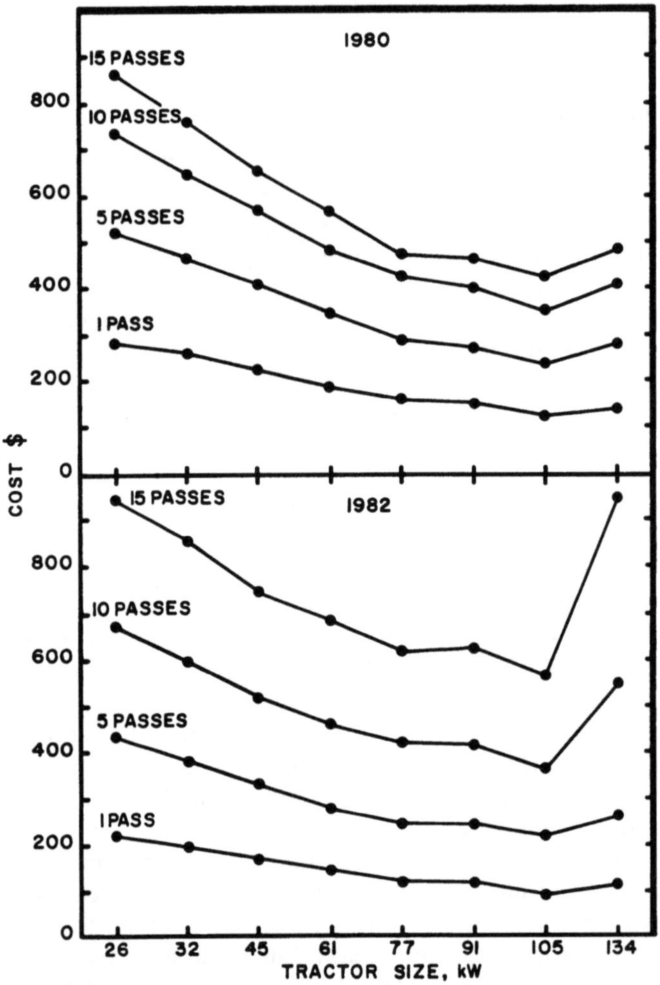

Figure 22. Total per-acre costs for pea production in Quebec as a function of tractor size and number of passes. [After Gunjal and Raghavan (1986).]

are not necessarily better if traffic intensity (percentage of area covered) is taken into account. There is a savings to be drawn from reducing the number of passes (Figure 24). An omission from both of these studies was that residual compaction effects from the stated least-cost tractor sizes were not considered; however, work is under way to apply linear programming procedures to economic studies of compaction (personal communication—G. Lavoie) on a multiyear monocultural and rotational basis that will necessarily include residual effects and possible alleviating effects of direct incorporation of organic matter (Ohu et al., 1985a, 1985b) or adoption of specialized crop rotations.

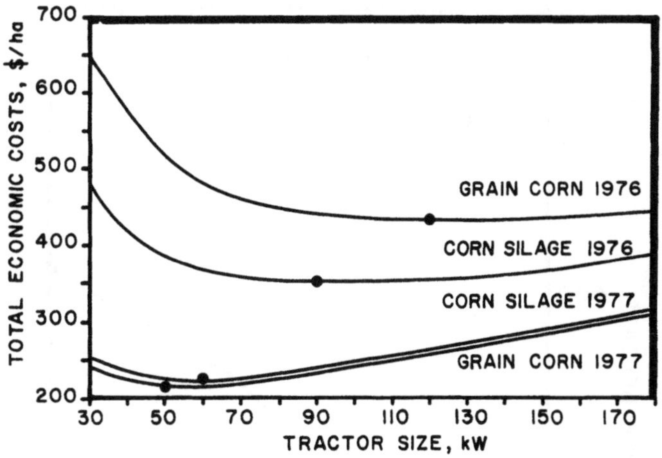

Figure 23. Total per-acre costs for corn silage and grain corn as a function of tractor size and number of passes for a wet year (1976) and a dry year (1977). [After Gunjal et al. (1987).]

An alternative that is presently receiving more attention is that of controlled traffic. The concept behind this approach is that a small percentage of the total field area is devoted to machinery traffic, with energy advantages from the point of view of traction, while the rest of the field is never compacted. This is perhaps a revival of Halkett's guideway system (Figure 25), which was considered too expensive when first developed in 1858. One approach to implementation has been through the development of specialized wide-frame and power units with lateral-move capabilities such as that shown in Figure 26. The possibility of using conventional machinery in this context is also being investigated. Assuming that conventional traffic can typically be associated with a potential yield reduction of 15%, a controlled traffic situation where only 10% of the surface area is compacted would result in a total yield reduction of only 10% if nothing is grown in the traffic lanes. Assuming yield reductions of 50% on traffic lanes and no reduction on the rest of the land, the total reduction should amount to only 5%. Lateral-move machines provide the added advantage that turning zones are not needed and that, therefore, more of the total field area can be used.

IV. Summary and Conclusions

We have attempted to give a broad perspective of the problem of soil degradation due to machinery traffic. Machine-soil-crop relationships are well understood in principle, but quantification is and will likely remain difficult due to the interaction of temporal and spatial variability of soils

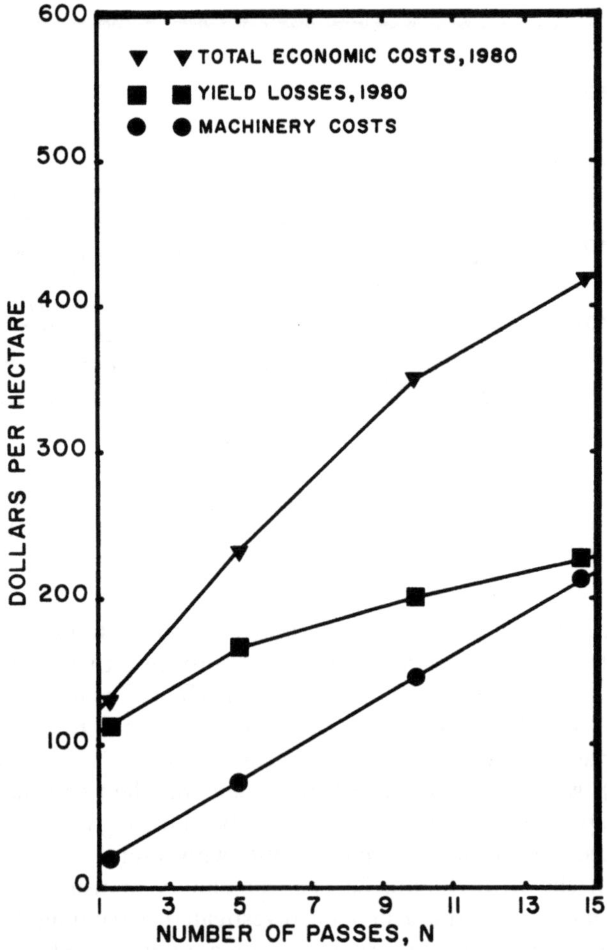

Figure 24. Breakdown of total economic costs of corn production as a function of number of passes using optimum tractor size. [After Gunjal and Raghavan (1986).]

and weather with crop growth. Research to date has identified the major factors involved in soil compaction and has led to recommendations for reducing the risk of compaction, as well as methods for alleviating it. These include timing with respect to moisture conditions, traffic patterns, and intensity as well as wheel, tire, and load characteristics, subsoiling, incorporation of organic matter, and special rotations designed to maintain organic matter and fertility levels.

Implementation of strategies to control soil physical condition is possible but depends on cost-benefit evaluations, possible policy and market changes to ensure the economic feasibility of alleviating procedures and beneficial crop rotations, and the development of appropriate techniques

Figure 25. Controlled traffic using twin steam engines on guideway system as conceived by Halkett in 1858. [From Partridge (1973); sketch by S. Tinker.]

Figure 26. "Modern version": wide-frame vehicle used in controlled traffic research by USDA. [From Taylor (1985); NTML Photo No. P10 347a, sketched by S. Tinker.]

and inexpensive equipment for monitoring compaction levels. Statistical descriptions of compaction, tillage, and natural effects on soil properties, and probability analysis of weather conditions, should be adopted and used to serve as intermediate and long-term decision guidelines.

The problem is as much in adopting and implementing them as it is in developing them. Culver and Seecharan (1986) pointed out that there is a general awareness that soil degradation must be dealt with, however; Girt

(1986) showed that, in Canada, the net returns per hectare for certain conservation cropping systems are economically viable compared to degrading systems only on a time scale of 10 years or more. To marginally successful agricultural enterprises, conservation systems are therefore unattractive. Among the factors involved in the degree to which soil conservation strategies are adopted by the farming community, Culver and Seecharan (1986) noted personal (farmer age and education), economic (farm size, net return), government-level (financial assistance, technical assistance, research), and physical ones (topography, climate). We therefore reach the conclusion that it is now a question of who will pay for a sustainable soil structure.

An aspect that has been alluded to in this paper is that of tactic. In a sense, agricultural research tends to be oriented toward productivity, where optimization is subject to short-term economic constraints. Soil degradation is a long-term problem, and more emphasis should be placed on sustainable rather than maximum productivity unless we believe that soils will eventually become obsolete as a growth medium for food.

Acknowledgments

The authors wish to acknowledge the many contributors around the globe to the field of soil management in general. Funding extended to soil compaction research through an NSERC strategic grant, through a number of Agricultural Engineering Research and Development contracts, and by the Quebec Ministry of Agriculture is gratefully acknowledged. Our thanks also to our associate, Samuel Gameda, for his help in preparing the manuscript and to Steven Tinker for drafting the figures.

References

Amir, I., G.S.V. Raghavan, E. McKyes, and R.S. Broughton. 1976. Soil compaction as a function of contact pressure and soil moisture content. *Can. Agric. Eng.* 18(1):54–57.

Barnes, K.K., W.M. Carleton, H.M. Taylor, R.I. Throckmorton, and G.E. Vanden Berg. 1971. Compaction of agricultural soils. ASAE Monograph, St. Joseph, MI.

Bernier, H., G. Bostock, G.S.V. Raghavan, and R.S. Broughton. 1989. Subsoiling effects on moisture content and bulk density in the soil profile. *Appl. Eng. Agric.* 5(1):24–28.

Blake, G.R., W.W. Nelson, and R.R. Allmaras. 1976. Persistence of sub-soil compaction in a Mollisol. *Soil Sci. Soc. Am. J.* 40:943–948.

Boone, F.R. 1986. Towards soil compaction limits for crop growth. *Neth. J. Agric. Sci.* 34:349–360.

Bowen, H.D. 1981. Alleviating mechanical impedance. In: G.F. Arkin and H.M. Taylor (eds.), *Modifying the root environment to reduce crop stress.* ASAE, St. Joseph, MI.

Cassel, D.K. 1982. Tillage effects on soil bulk density and mechanical impedance.

pp. 45–67. In: *Predicting tillage effects on soil physical properties and processes.* ASA Special Publ. No. 44. Madison, WI.

Chancellor, W.J. 1976. Compaction of soil by agricultural equipment. Bull. 1881, Div. Agr. Sci., Univ. California, Davis.

Coote, F.R. 1980. The deterioration of agricultural land: How sick are our soils? *Agrologist* 9(4):12–14.

Culver, D., and R. Seecharan. 1986. Factors that influence the adoption of soil conservation technologies. *Can. Farm Econ.* 20(2):9–13.

Dann, P.R., A.G. Thomas, R.B. Cunningham, and P.H.R. Moore. 1987. Response by wheat, rape and field peas to pre-sowing herbicides and deep tillage. *Aust. J. Exp. Agric.* 27:431–437.

Dexter, A.R. 1986a. Model experiments on the behaviour of roots at the interface between a tilled seed-bed and a compacted sub-soil. I. Effects of seed-bed aggregate size and sub-soil strength on wheat roots. *Plant and Soil* 95:123–133.

Dexter, A.R. 1986b. Model experiments on the behavior of roots at the interface between a tilled seed-bed and a compacted sub-soil. II. Entry of pea and wheat roots into sub-soil cracks. *Plant and Soil* 95:135–147.

Dexter, A.R., 1986c. Model experiments on the behavior of roots at the interface between a tilled seed-bed and a compacted sub-soil. III. Entry of pea and wheat roots into cylindrical biopores. *Plant and Soil* 95:149–161.

Dumanski, J. 1980. The agricultural land resource: Locating the loss. *Agrologist* 9(4):15–17.

Eriksson, J., I. Hakansson, and B. Danfors. 1974. The effect of soil compaction on soil structure and crop yields. Bull. 354, Swed. Inst. Agr. Eng., Uppsala, Sweden. (Eng. Trans. by J.K. Aase).

Gameda, S., G.S.V. Raghavan, E. McKyes and R. Theriault. 1985. A reivew of subsoil compaction. *Int. Conf. Soil Soil Dynamics Proc.* 5:970–978.

Gameda, S., G.S.V. Raghavan, E. McKyes, and R. Theriault. 1987a. Sub-soil compaction in a clay soil. I. Cumulative effects. *Soil and Tillage Res.* 10:113–122.

Gameda, S., G.S.V. Raghavan, E. McKyes, and R. Theriault, 1987b. Sub-soil compaction in a clay soil. II. Natural alleviation. *Soil and Tillage Res.* 10:123–130.

Gill, W.R. 1971. Economic assessment of soil compaction. ASAE Monograph, St. Joseph, MI.

Girt, J. 1986. The on-farm economics of sustainability and public intervention. *Can. Farm Econ.* 20(2):3–8.

Gunjal, K.R., and G.S.V. Raghavan, 1986. Economic analysis of soil compaction due to machinery traffic. *App. Eng. Agric.* 2(2):85–88.

Gunjal, K.R., G. Lavoie, and G.S.V. Raghavan. 1987. Economics of soil compaction due to machinery traffic and implications for machinery selection. *Can. J. Agric. Econ.* 35:591–603.

Gupta, S.C., and R.R. Allmaras. 1987. Models to assess the susceptibility of soils to excessive compaction. *Adv. Soil Sci.* 6:65–100.

Gupta, S.C., A. Hadas, W.B. Voorhees, D. Wolf, W.E. Larson, and E.C. Schneider. 1985. Field testing of a soil compaction model. *Int. Conf. Soil Dynamics Proc.* 5:979–994.

Gupta, S.C., and W.E. Larson. 1982. Modelling soil mechanical behavior during tillage. pp. 151–178. In: *Predicting tillage effects on soil physical properties and processes.* ASA Spec. Pub. 44, Madison, WI.

Hadas, A., W.E. Larson, and R.R. Allmaras. 1988. Advances in modeling machine-soil-plant interactions. *Soil and Tillage Res.* 11(4):349–372.

Hakansson, I. 1982. Long-term effects of vehicles with high axle load on subsoil compaction and crop response. 9th Conf., Int. Soil Tillage Res. Org. (ISTRO) (Osijek, Yugoslavia), pp. 213–218.

Hakansson, I., W.B. Voorhees, and H. Riley. 1988. Vehicle and wheel factors influencing soil compaction and crop response in different traffic regimes. *Soil and Tillage Res.* 11(3):239–282.

Hettiaratchi, D.R.P., and J.R. O'Callaghan. 1980. Mechanical behaviour of agricultural soils. *J. Agric. Eng. Res.* 25:239–259.

Hillel, D. 1987. Modeling in soil physics: A critical review. In: L.L. Boersma (ed.), *Future developments in soil science research.* Socil Sci. Soc. Am., Madison, WI.

Lal, R., G.F. Wilson, and B.N. Okigbo. 1979. Changes in properties of an alfisol produced by various crop covers. *Soil Sci.* 127:377–382.

Lambe, T.W. 1951. *Soil testing for engineers.* p. 165. Wiley, NY.

Larson, W.E., and G.J. Osborne. 1982. Tillage accomplishments and potential. In: *Predicting tillage effects on soil physical properties and processes.* ASA Spec. Pub. 44, Madison, WI.

Lindstrom, M.J., W.B. Voorhees, and G.W. Randall. 1981. Long-term tillage effects on interrow runoff and infiltration. *Soil Sci. Sco. Am. J.* 45:945–948.

Lovejoy, S.B., and T.L. Napier (eds.). 1986. *Conserving soil: Insights from socioeconomic research.* Soil Conservation Society of America, Ankeney, IA.

Meek, B.D., E.A. Rechel, L.M. Carter, and W.R. DeTar. 1988. Soil compaction and its effects on alfalfa in zone production systems. *Soil Sci. Soc. Am. J.* 52:232–236.

Mehuys, G.R. 1984. *Soil degradation of agricultural land in Quebec. A review and impact assessment.* Science Council of Canada, Ottawa.

Mohammad, F. 1987. Sugarbeet development under conservation tillage. M.Sc. thesis, McGill University, Montreal, Canada.

Negi, S.C., E. McKyes, G.S.V. Raghavan, and F. Taylor. 1981. Relationships of field traffic and tillage to corn yields and soil properties. *J. Terramechanics* 18(2):81–90.

Ohu, J.O., G.S.V. Raghavan, and E. McKyes. 1985a. Peatmoss effect on the physical and hydraulic characteristics of compacted soils. *Trans. ASAE* 28(2): 201–424.

Ohu, J.O., G.S.V. Raghavan, E. McKyes, K.A. Stewart, and M.A. Fanous. 1985b. The effects of soil compaction and organic matter on the growth of bush beans. *Trans. ASAE* 28(4):1056–1061.

Ohu, J.O., G.S.V. Raghavan, E. McKyes, and G. Mehuys. 1986. Shear strength prediction of compacted soils with varying organic matter content. *Trans. ASAE* 29(2):351–355.

Partridge, M. 1973. *Farm tools through the ages.* Osprey, Reading, Berk. United Kingdom.

Perumpral, J.V. 1987. Cone penetrometer applications—A review. *Trans. ASAE* 30(4):939–944.

Pollock, D., Jr., J.V. Perumpral, and T. Kuppusamy. 1986. Finite element analysis of multipass effects of vehicles on soil compaction. *Trans. ASAE* 29(1):45–50.

Raghavan, G.S.V. 1985. Methods of alleviating soil compaction. Soil Compaction Conf. (Ohio State Univ., Columbus, OH), April 1985.

Raghavan, G.S.V., and E. McKyes. 1977. Laboratory study to determine the effect of slip-generated shear on soil compaction. *Can. Agric. Eng.* 19(1):40–42.

Raghavan, G.S.V., and E. McKyes. 1978. Statistical models for predicting compaction by off-road vehicular traffic in different soil types. *J. Terramechanics* 15(1):1–14.

Raghavan, G.S.V., and E. McKyes. 1983. Physical and hydraulic characteristics in compacted clay soils. *J. Terramechanics* 19(4):235–242.

Raghavan, G.S.V., E. McKyes, M. Chasse, and F. Merineau. 1976a. Development of compaction patterns due to machinery operation in an orchard soil. *Can. J. Plant Sci.* 56:505–509.

Raghavan, G.S.V., E. McKyes, I. Amir, M. Chasse, and R.S. Broughton. 1976b. Prediction of soil compaction due to off-road vehicle traffic. *Trans. ASAE* 19:610–613.

Raghavan, G.S.V., E. McKyes, and B. Beaulieu. 1977a. Prediction of clay soil compaction. *J. Terramechanics* 14(1):31–38.

Raghavan, G.S.V., E. McKyes, and M. Chasse. 1977b. Effect of wheel slip on soil compaction. *J. Agric. Eng. Res.* 22:79–83.

Raghavan, G.S.V., E. McKyes, G. Gendron, B. Borglum, and H.H. Le. 1978a. Effects of soil compaction on development and yield of corn (maize). *Can. J. Plant Sci.* 58:435–443.

Raghavan, G.S.V., E. McKyes, and B. Beaulieu. 1978b. Clay soil compaction due to wheel slip. *Trans. ASAE* 21(4):646–649, 653.

Raghavan, G.S.V., E. McKyes, F. Taylor, P. Richard, and A. Watson. 1979. The relationship between machinery traffic and corn yield reactions in successive years. *Trans. ASAE* 22:1256–1259.

Raghavan, G.S.V., and J.O. Ohu. 1985. Prediction of equivalent pressure of Proctor compaction blows. *Trans. ASAE* 28(5):1398–1400.

Raghavan, G.S.V., F. Taylor, B. Vigier, L. Gauthier, and E. McKyes. 1982. Effect of compaction and root rot disease on development and yield of peas. *Can. Agric. Eng.* 24(1):31–34.

Sheptukhov, V.N., A.I. Voronin, and M.A. Shipilov. 1982. Bulk density of the soil and its productivity. *Soviet Soil Sci.* 14(5):97–107.

Soane, B.D. 1985. Traction and transport systems as related to cropping systems. *Int. Conf. Soil Dynamics Proc.* 5:863–935.

Soane, B.D., P.S. Blackwell, J.W. Dickson, and D.J. Painter. 1981a. Compaction by agricultural vehicles: A review. I. Soil and wheel characteristics. *Soil and Tillage Res.* 1:207–237.

Soane, B.D., P.S. Blackwell, J.W. Dickson, and D.J. Painter. 1981b. Compaction by agricultural vehicles: A review. II. Compaction under tyres and other running gear. *Soil and Tillage Res.* 1:373–400.

Soane, B.D., and F.R. Bonne, 1986. The effects of tillage and traffic on soil structure. *Soil and Tillage Res.* 8:303–306.

Soane, B.D., J.W. Dickson, and D.J. Campbell. 1982. Compaction by agricultural vehicles: A review. III. Incidence and control of compaction in crop production. *Soil and Tillage Res.* 2:3–36.

Soehne, W. 1958. Fundamentals of pressure distribution and soil compaction under tractor tires. *Agric. Eng.* 39:276–281, 290.

Taylor, J.H. 1985. Controlled traffic: A spin-off of soil dynamics research. *Int. Conf. Soil Dynamics Proc.* 5:1101–1111.

Taylor, J.H., and W.R. Gill. 1984. Soil compaction: State-of-the-art report. *J. Terramechanics* 21(2):195–213.

Tisdall, J.M., and H.H. Adem. 1986. Effect of water content at tillage on size-distribution of aggregates and infiltration. *Aust. J. Exp. Agric.* 26:193–195.

Van Doren, D.M., R.R. Allmaras, D.R. Linden, and F.D. Whisler (Organ. Committee). 1982. Predicting tillage effects on soil physical properties and processes. ASA Pub. 44, Madison, WI.

Vepraskas, M.J. 1988. A method to estimate the probability that sub-soiling will increase tobacco yields. *Soil Sci. Am. J.* 52:229–232.

Vigier, B., and G.S.V. Raghavan. 1980. Soil compaction effect in clay soils on common root rot in canning peas. *Can. Plant Disease Survey* 60(4):43–45.

Voorhees, W.B. 1977. Soil compaction: Our newest natural resource. *Crops Soils Mag.* 29:13–15.

Voorhees, W.B. 1980. Energy aspects of controlled wheel traffic in the Northern Corn Belt of the Unied States. *Proc. Int. Soil Tillage Res. Organization, 8th Conf.* 1979 (Univ. of Hohenheim, Germany) 2:333–338.

Voorhees, W.B., W.W. Nelson, and G.W. Randall. 1985. Soil and crop response to wheel traffic on highly productive soils of the northern corn belt. *Int. Conf. Soil Dynamics Proc.* 5:1120–1131.

Voronin, A.I. 1982. Bulk density of an irrigated Chestnut soil and its productivity. *Soviet Soil Sci.* 14(3):78–84.

Webb, R.H., J.W. Steiger, and H.G. Wilshire. 1986. Recovery of compacted soils in Mojave Desert ghost towns. *Soil Sci. Soc. Am. J.* 50:1341–1344.

Hardsetting Soils: Behavior, Occurrence, and Management

C.E. Mullins,* D.A. MacLeod,† K.H. Northcote,‡
J.M. Tisdall,‰ and I.M. Young¶

*Plant and Soil Science Department, University of Aberdeen, Aberdeen AB9 2UE, Scotland.
†Agronomy and Soil Science Department, University of New England, Armidale, NSW 2351, Australia.
‡Formerly Division of Soils, CSIRO, Private Bag no. 2, Glen Osmond, SA 5064, Australia.
‰Institute for Irrigation and Salinity Research, Tatura, Victoria 3616, Australia.
¶Now at Scottish Crop Research Institute, Dundee DD2 5DA, United Kingdom.

I. Introduction

Hardsetting soils are soils that set to a hard, structureless mass during drying and are thereafter difficult or impossible to cultivate until the profile is rewetted. The term *hardsetting*, originally used when the concept was first defined, will be used here rather than the hyphenated spelling that has sometimes been used. Only in Australia is this form of behavior specifically identified, mapped, and given the name "hardsetting" (Northcote et al., 1975).

About 13% of Australian soils have duplex profiles with a hardsetting A_1 horizon, but considerably more have hardsetting surface properties or become hardsetting under cultivations. Large areas in the Australian wheat belts have a hardsetting surface horizon, and it is clear that there are many other parts of the world where soils with this type of behavior are widespread. For example, at least 12% of the soils of Zambia are in a class that has been identified as hardsetting (Spaagaren, 1986).

In order to germinate and establish plants in a soil with a hard surface, it is first necessary to loosen it. With hardsetting soil, however, irrigation or rainfall after sowing can cause the surface to collapse again, and even when no visible crust is formed, drying may cause the surface to harden sufficiently to prevent seedling emergence (Mullins et al., 1987). Where a plant is established, it may still experience problems because the drying action of its roots causes the soil strength to increase sufficiently to severely impede or halt their further growth. When it is intended to cultivate a hardsetting soil, there can be severe restrictions on timing so that it is difficult to find a time when the soil is trafficable but not too hard to cultivate, and even when cultivated, it is often difficult to produce a seedbed that is neither too cloddy nor too dusty.

Hardsetting soils also experience other physical problems associated with a structure that is not water-stable, such as poor aeration under wet conditions, poor infiltration, runoff, and erosion. Thus hardsetting soils pose a major challenge in terms of evolving management systems that allow for their sustained use, whether under capital-intensive or small-scale farming systems. Surprisingly, however, the problems of managing soils with a hardsetting surface have not received the attention they merit, even in Australia (Northcote, 1982), although there is now a growing interest and recognition of hardsetting as a distinct form of soil behavior that imposes characteristic physical limitations on cultivations and crop production.

Although it is easy to provide a working definition of hardsetting behavior, it is much more difficult to establish boundaries that delimit it from other forms of soil behavior. Furthermore, soils that are not hardsetting in their natural state can become hardsetting as a result of an unsuitable system of management, so that soil boundaries mapped on the basis of the behavior of soils in a virgin state will not include all soils that are now hardsetting. Indeed, it seems probable that all soils with an appropriate particle-size distribution and clay mineralogy are potentially hardsetting in the absence of a sufficient concentration of organic matter or inorganic cementation or stabilization of microaggregates. Therefore, this review starts first with a discussion of the scientific basis for hardsetting behavior before considering classification and occurrence.

Since it is ultimately deterioration of the physical properties of hardsetting soils that can render cropping marginal or uneconomical, a clear understanding of the processes involved in hardsetting is essential. This should provide a sounder basis on which to judge the "portability" of a successful management system from one climate, crop, or soil type to another.

Because hardsetting soils have been classified and mapped only in Australia, and because the United Nations Food and Agriculture Organization (FAO) and the United States soil classifications (Dudal, 1970; Soil Survey Staff, 1975) do not uniquely group together those properties that result in hardsetting, this review of necessity concentrates on the Australian experience. However, it will be clear from Section III that this type of behavior is also widespread outside Australia, especially under mediterranean and tropical conditions; where possible, references to hardsetting behavior elsewhere have also been discussed.

A. What Is Compaction?

Hardsetting of a cultivated soil usually involves slumping, which is a process of compaction (i.e., increase of bulk density) that occurs without the application of an external load. Unfortunately, many experimenters who have observed a high topsoil bulk density at some stage after emergence of a crop refer to such soils as "compacted." This is misleading because it implies the action of some external load (e.g., agricultural traffic or trampling by humans or animals), whereas the forces causing the compaction occur or are a generated within the soil itself. In practice, in many papers on topsoil compaction it is not possible to distinguish between effects due to externally applied loads and the result of wetting structurally weak or unstable soil, although it is common to attribute compaction to the action of an external load. Even in books and reviews on compaction there is rarely any mention of the results of topsoil structural collapse, and a strong and misleading impression has been created that topsoil compaction results only from external loading of the soil. The plow-layer bulk density of 1.7 Mg/m^3 attained some months after plowing an English hardsetting soil (Young et al., 1988) provides a clear demonstration that external loading is not always the cause of high topsoil bulk density. The distinction is of more than academic importance since measures to reduce the area of wheelings on such a soil, but not to otherwise change the management system, may have little influence on the compaction problem. Thus the occurrence and extent of slumping in hardsetting and other structurally unstable topsoils may often have been overlooked.

II. Scientific Basis for Hardsetting Behavior

A. Properties of Soils That Display Hardsetting

Before giving a detailed description of the process of hardsetting, it is appropriate to summarize the envelope of properties that approximately defines the scope of hardsetting soils. Figure 1 is an attempt to indicate the range of textures and clay behavior encompassing potentially hardsetting soils. At the sandy end of the textual range, although Mullins and Panayiotopoulos (1984) have shown that an artifical mixture of sand with as little as 2% kaolinite can behave in a hardsetting manner, the shear and tensile strength of such sands and of some sandy loams is too small for them to present cultivation problems or to be classified as hardsetting (Ley, 1988). Toward the clayey end of the range, shrinkage during drying causes structural cracks, and these soils therefore do not classify as hardsetting. Since a smaller proportion of clay with a high, rather than with a low,

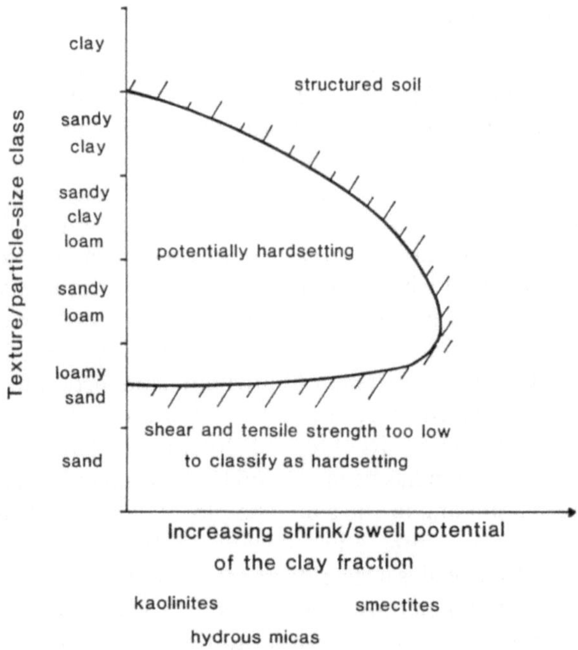

Figure 1. The range of soil textures in which hardsetting is likely to be encountered in relation to clay mineralogy. (Silty soils can also be hardsetting and would be placed somewhere in the middle of this range.)

shrink/swell potential may be needed to cause cracking, it is not surprising that in most reported instances hardsetting soils have a clay mineralogy dominated by hydrous mica (illite) and/or kaolinite (Norrish and Pickering, 1983). In Australian red-brown earths, weak bonding and the tendency to hardsetting have been associated with high contents of silt and fine sand (Cockroft and Martin, 1981; French, 1981).

Where the soil also contains appreciable amounts of cementing or stabilizing materials, such as the oxides and oxyhydroxides of iron and/or aluminum, laterization may occur resulting in a permanently cemented mass, or the soil may consist of water-stable microaggregates (Trapnell and Webster, 1986). Soils that undergo permanent hardening upon drying and do not soften upon rewetting (e.g., as can occur with laterization of cleared forest soils) are not classified as hardsetting. These and other soil horizons that remain hard when moist are classified as fragipans or duripans, depending on the type of cementation (Eck and Unger, 1985). Soils such as sodic soils with a high proportion of dispersible clay are particularly noted for hardsetting behavior. However, the presence of dispersible clay is not a necessary condition for hardsetting, since Young (1987) has observed hardsetting in a soil with no dispersible clay but in which most of the <60 μm size fraction was suspended as 20–60 μm material after immersing a soil sample in water with a minimum of mechanical disturbance. Hardsetting is predominantly observed in soils with low concentrations of organic matter (Cockroft and Martin, 1981), typically <2% (see e.g., Table 8).

B. The Process of Hardsetting

In a previously loosened/cultivated topsoil, hardsetting involves the collapse of some or all of the artificial aggregated structure during and after wetting (Cockroft and Martin, 1981; Murphy et al., 1987; Ley, 1988; Mead and Chan, 1988; Young et al., 1988), and a hardening without restructuring during drying. Thus soils that are hardsetting in their natural state have been unable to develop water-stable aggregates. Figure 2 is a schematic diagram of the processes that can occur during and after wetting of a cultivated layer of hardsetting soil. The processes outlined in this diagram represent the full range of possibilities, but in any given hardsetting event it is not necessary for all these processes to occur. The figure does, however, indicate that hardsetting can be divided into at least two, and sometimes three, physically distinct processes. Attempts to ameliorate these soils must therefore interfere with either the process of slumping or the process of strength development, or both.

1. Slumping

Slumping, which is not unique to hardsetting soils, occurs during and after the wetting of a bed of aggregates that are not water-stable. The aggregates will both soften and swell simultaneously during wetting. The energy re-

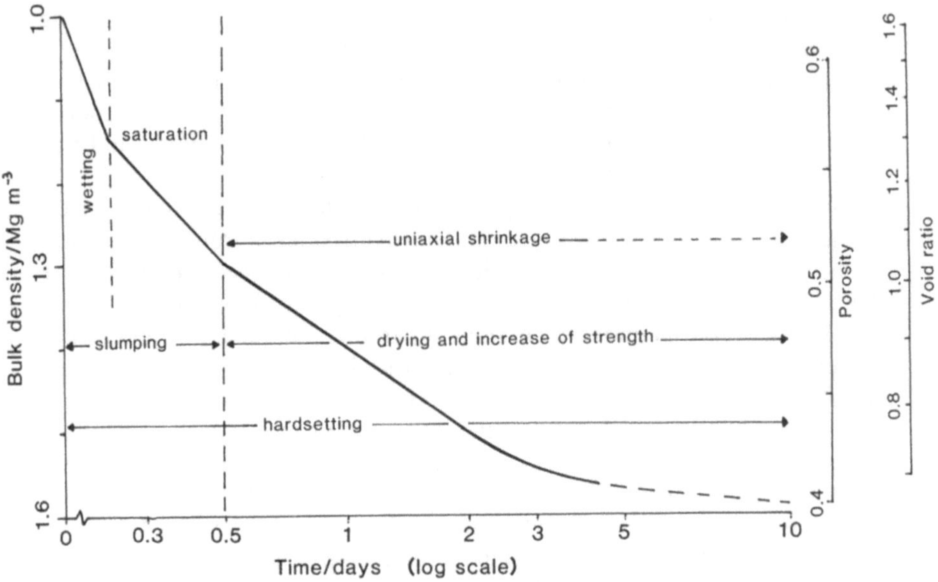

Figure 2. Hypothetical diagram to show the sequence of possible compaction processes operating during hardsetting of an initially aggregated layer of soil at 0.1 m depth.

leased can be sufficient to cause slaking, during which some or all of the silt and clay-sized material becomes suspended (although not necessarily as individual particles) and, under appropriate ionic conditions, some or all of the clay fraction may become dispersed. During wetting, aggregates disintegrate because they have insufficient strength to withstand the stresses set up by rapid water uptake, such as those caused by rapid release of heat of wetting, trapped air, the mechanical action of rapidly moving water (Collis-George and Greene, 1979), or by differential swelling (Emerson, 1977).

Aylmore and Sills (1982) found that the more a soil slakes upon wetting, the greater the tendency for it to set hard on drying. Emerson (1984, 1977) has discussed the influence of clay mineralogy on slaking. The crystalline swelling of a dry aggregate of Ca-montmorillonite is about 25 times that of Ca-kaolinite. The montmorillonite, however, will take up water only slowly because of the extremely small pores between adjoining sheets. In addition, the stresses due to swelling and entrapped air can be relieved by unbending of the montmorillonite sheets, but not of the thicker kaolinite plates. Thus an initially dry montmorillonite aggregate will, upon immersion, slowly swell from the outside, with the aggregate remaining intact; in contrast, the kaolinite aggregate slakes rapidly as air is released.

Many workers have shown that, in structurally unstable soils, saturation at zero matric potential results in a greater loss of porosity in comparison to tension-wetting (e.g., Ghavami et al., 1974; Kemper et al., 1975). Indeed, the contrast between the size distribution of macropores in aggregate beds wetted under these two types of wetting regime is the basis for the Childs test for structural stability (Childs, 1940, 1942; Childs and Youngs, 1958; Rycroft and Thorburn, 1974; Collis-George and Figueroa, 1984).

The precise effects of wetting at different matric potentials are difficult to study because they may combine rate effects that are dependent on the matric potential gradient within the soil and the direct effects of matric potential on soil strength. For tension-wet samples, Ghavami et al. (1974) have obtained a good empirical description of the process of compaction during slumping by:

$$e = e_0 \left[\frac{\Psi_m + c}{\sigma} \right]^b \tag{1}$$

where e is the void ratio of a bed of aggregates with an initial void ratio e_0, which is wet to a matric potential Ψ_m under a total effective stress σ. The latter term is the sum of the weight of the soil overburden, any load applied at the surface, and a proportion of the soil water tension $|\Psi_m|$. Here b is an empirical fitting constant, and c is the cohesion of moist samples. This equation was successfully used to describe slumping of a layer of aggregates wetted at matric potentials between -1.3 and -4.5 kPa and subject to loads of 1.2 to 18.4 kPa (equivalent to the soil overburden pressure at depths >0.05 m). They used these results to successfully predict the soil compaction in a 0.6-m-high column of aggregates wet to a uniform tension by use of a sprinkler, with a suction plate at the base of the column. The void ratio of the uppermost layer decreased by 0.04 but, as a result of the greater overburden, that at the base of the column decreased by 0.23. While their experiment may not exactly simulate the field situation, it does demonstrate the following important points about slumping:

1. Even within the topsoil, overburden pressure can be an important influence increasing the compaction of the lower part of a cultivated and wetted layer.
2. Equation (1) demonstrates the dual role played by matric potential (which occurs in both the top and bottom of the fraction on the right-hand side). This is because matric potential contributes simultaneously to the strength of the system and to the total effective stress that is tending to cause compaction.
3. For aggregates that are not water-stable, the cohesion term in Equation (1) is inapplicable when the soil is saturated at zero potential. In practice, much greater slumping occurred in these samples than in any of the tension-wet samples. This has practical relevance to the situation where the wetting of a soil with an impeding layer at the bottom of the cultivated profile may allow the temporary development of a soil layer at

zero matric potential (i.e., a perched water table). As a consequence, a layer of increased bulk density can be produced (which may subsequently be confused with compaction produced by traffic).

The matric potential of the soil prior to wetting also influences the incidence or severity of slaking. Moist aggregates slake less readily than air-dry ones because they have already completed some or all of their swelling and some pores are already water-filled (Emerson, 1977). The suggestion by Collis-George and Lal (1971) that, a surface mulch may be able to preserve slightly moister conditions at the soil surface that are sufficient to avoid slaking (although not based on the behavior of a hardsetting soil), deserves further investigation.

Eck and Unger (1985) have discussed the literature on plow pans, the structureless dense and/or impeding layer found at the base of some plow layers, and have pointed out that pans are most likely to occur in soils with little or no shrink-swell behavior. Although plow pans are usually attributed to the use of traffic with high axle loads on soil that is wet at depth, the foregoing discussion makes it clear that they can also occur in hardsetting soils as a result of slumping.

2. Uniaxial Shrinkage

Shrinkage is of importance if only because the closer proximity of particles that it entails may make a contribution to the increase in strength observed upon drying hardsetting soils. Laboratory experiments on the behavior of aggregate beds of a hardsetting soil wetted under tension or at zero potential have demonstrated that, at least during the early stages of drying ($\Psi_m > -5$ kPa), uniaxial shrinkage does occur (Blackwell et al., 1988). In contrast, Young (1987) and Ley et al. (1988) have observed little or no shrinkage for small, "undisturbed" cores that were taken from the slumped structureless layer just below the surface (>0.04 m) of some hardsetting soils and equilibrated at a range of matric potentials down to air dryness. These sets of results are not contradictory since the latter samples had already slumped and dried in the field prior to rewetting and equilibration, whereas the former represent cultivated beds of aggregates that have slumped and are in the process of drying for the first time from this new state. It is not necessarily the case that all such hardsetting aggregate beds shrink upon drying, but if they do, the shrinkage must of necessity be uniaxial if the soil is to remain structureless.

Since uniaxial shrinkage is, by definition, anisotropic, it follows that it must be accomplished by realignment of the remanents of the disrupted aggregates and/or within the fabric of the soil. Such a realignment can occur without cracking only if the forces holding the soil together are long-range and nonspecific. This point is important because the only such force is likely to be provided by the matric potential, and therefore, when uniaxial shrinkage occurs, this may be an indication that effective stress contributes a dominant component of soil strength.

3. Hardsetting and the Development of Soil Strength

It is often commented that there is a very sharp increase in the strength of some hardsetting soils during drying, so that they have been referred to by farmers as "lunchtime soils." Such soils can be too wet to cultivate in the morning, yet too dry in afternoon—so they must have been suitable at lunchtime! While this may be a slight exaggeration, it does emphasize their narrow range of suitable working conditions. Young (1987) and Ley et al. (1988) have measured the tensile and unconfined compressive strength of small undisturbed cores of soils displaying hardsetting behavior that were equilibrated at a range of matric potentials. In all cases they observed a characteristic variation of strength with matric potential and water content in which strength increased progressively more steeply with decreasing water content down to a matric potential of −1 MPa. The most pronounced strength increase (by up to a factor of 3) was between potentials of −0.1 and −1 MPa, and occurred over a narrow range of gravimetric water contents (as little as 2%). Mullins and MacLeod (unpublished) have observed similar results for a number of Australian hardsetting soils but with a more gradual increase of strength with decrease in water content).

Mullins et al. (1987) have proposed the following explanation for the development and increase in strength observed in hardsetting soils, starting with a cultivated bed consisting of dry aggregates:

1. Wetting of the system mobilizes some or all of the silt and clay. This may occur through slaking and/or dispersion.
2. During the early stages of drying, the mobilized material is carried behind the retreating water meniscus to occupy concavities on the surface of sand grains and any remaining aggregrates, or to form annular bridges between this larger stable material. This phenomenon has been observed under the microscope by Kemper et al. (1987).
3. As drying proceeds, despite air entry into the soil, the mobilized material remains saturated until a very low potential is reached (reported values for kaolin are <−1 MPa) and consequently the contribution of matric potential to the effective stress provides a major component of the soil strength.

In hardsetting soils in which a large proportion of the soil is mobilized during wetting, reference to annular bridges is misleading since the whole soil becomes a soft, mobile matrix in which sand grains and the vesicular pores containing trapped air are embedded (e.g., Cockroft and Martin, 1981). Nevertheless, stages 1 and 3 remain valid.

The sequence of events above provides a satisfactory explanation for hardsetting, including the sharp increase in strength that is often observed between potentials of −0.1 and −1 MPa, and part 3 of the explanation is equally applicable to describe the effects of subsequent wetting and drying of a hardsetting soil that has already slumped.

Once air entry occurs within the bridging material, the contribution of effective stress to strength must fall to zero and thereafter sample strength must be entirely determined by chemical bonding. It may seem paradoxical that a soil in which the chemical bonding was insufficient to provide water stability can also be a strong soil when dried. However, it must be recalled that wetting soil releases a range of powerful disruptive forces due to double layer swelling, trapped air, and the heat of wetting, which are sufficient to rupture rigid short-range chemical bonds, whereas the flexible polymer bonding such as that provided by polysaccharides in water-stable aggregates may be able to withstand wetting although making a modest contribution to soil strength.

The strength of air-dry samples of hardsetting soil may be several times that of samples equilibrated at a potential of −1 MPa, but there is a general relationship between the two (Young, 1987; Ley et al., 1988) in that soils with high air-dry strengths were also found to have high strength at −1 MPa. This can be understood when it is realized that shrinkage and/or rearrangement of the soil fabric is likely to be involved during the drying process such that the action of the effective stress is to bring particle surfaces sufficiently close together to allow the formation of chemical bonds. At any given value of matric potential, the effective stress depends on the proportion of the area of any potential failure plane that is occupied by the previously mobilized bridging material. However, provided that too many of these bridges do not rupture during air drying, the same area of material will also determine the strength of air-dry samples.

Although no more detailed theory is available to explain the magnitude of the strength of air-dry hardsetting soil, Mullins and Panayiotopoulos (1984) have provided theories to explain the development of tensile and shear strength during drying, up to the point at which air enters the soil matrix. In these theories, both tensile and shear strength are represented as the sum of two terms: the cohesion, which represents the sum total of all strength contributions by chemical bonds; and the effective stress, which represents the contribution of matric potential to strength. The cohesion term is difficult to estimate because, as already explained, even in samples that do not exhibit macroscopic shrinkage, there may be an increase in cohesion due to rearrangement of the fabric during drying.

By estimating the magnitude of cohesion from samples equilibrated at a potential of −3 kPa and assuming that this value does not increase substantially at lower potentials, Ley et al. (1989) have been able to compare the measured strengths of small undisturbed soil cores from a wide range of hardsetting and other structureless soils with theoretical predictions. Given

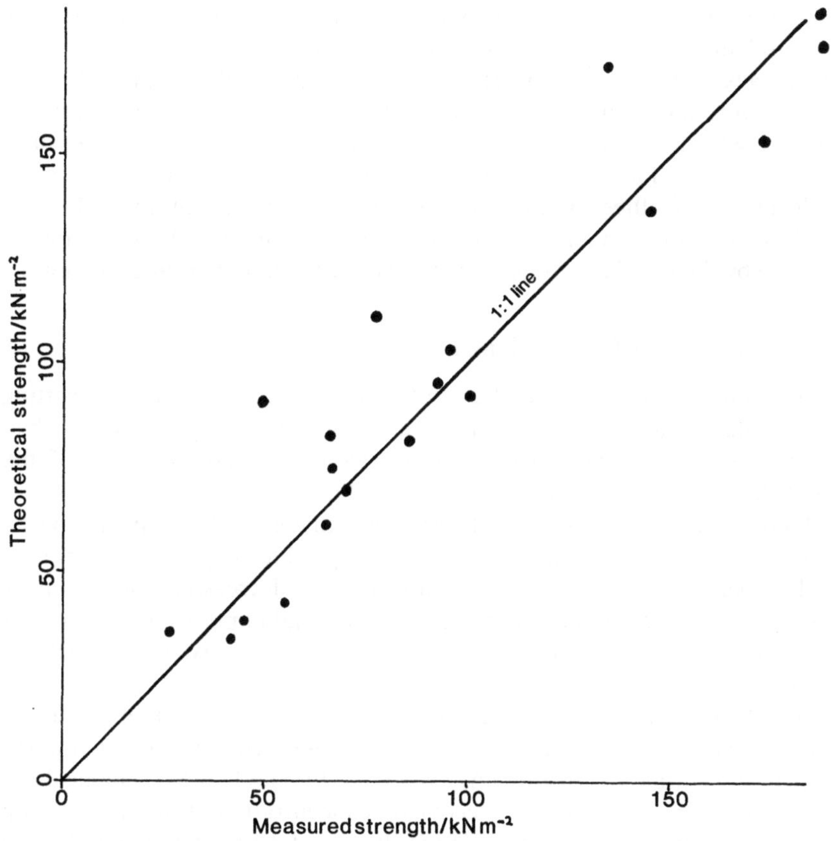

Figure 3. A comparison between measured values of unconfined compressive strength for "undisturbed" cores (0.04–0.08 m depth) from a range of structureless Nigerian soils equilibrated at −100 kPa matric potential, and theoretical predictions. [After Ley et al. (1989).]

the assumptions in the theory and the experimental limitations, there was a suprisingly good agreement between predicted and measured values of crushing strength for samples equilibrated at −100 kPa (Figure 3). A similar degree of agreement was obtained for tensile strength, once the effective stress term had been divided by 2 to allow for the presence of near spherical vesicular pores within the samples [as explained by Snyder and Miller (1985)]. In both cases, the dominant contribution to strength was from the effective stress, so that methods for interrupting the water meniscus, such as the addition of hydrophobic substances, may help the development of microcracks and soil weakening.

For samples equilibrated at −0.3 and −1.0 MPa, theory overestimated measured strengths by progressively greater amounts (up to a factor of 3 times), suggesting that there was progressive air entry into the bridging material or a progressive development of failure .by crack propagation (Braunack et al., 1979) at potentials <-100 kPa. While the preceding explanation is sufficient to explain hardsetting behavior, the possibility that the bridging material may consist of or be reinforced by soluble silica (Daniel et al., 1988, and personal communication) in some hardsetting soils deserves further consideration. In this latter case, continued silica deposition will presumably lead to the development of a fragipan distinguished by the fact that it remains hard and pieces of it are brittle when wet.

4. Crusting and Surface Hardening

Many soils exhibit crusting but not hardsetting. Whereas hardsetting affects the whole of the A_1 horizon, a surface crust is typically less than 10 mm thick and, when dry, separates from and may be lifted off the soil below, which is usually loose (Northcote, 1979).

McIntyre (1958b) found that the crust was made up of two distinct parts: a surface skin, 0.1 mm thick, thought to be produced by raindrop impact; and a "washed-in" zone, up to 3 mm thick, of decreasing porosity attributed to the deposition of mobilized fine material in pores. The "washed-in" zone is often absent and appears to be restricted to soils with unstable structure.

It is clear that crusting involves mobilization and deposition of fine material as in hardsetting. Three factors affecting the surface soil make it particularly susceptible to crusting (Agassi et al., 1981): (1) the mechanical impact of raindrops, (2) the low electrolyte concentration of the soil solution due to leaching by rainwater, and (3) the absence of a soil matrix (which would otherwise slow clay dispersion and mobilization of fine material). A hardsetting soil is distinguished from one that crusts by the fact that the A_1 horizon is so unstable that, without the aid of the above factors, wetting causes the breakdown of aggregates and mobilization of fine material.

A laboratory study of crust formation by Arndt (1985) throws some light on how crusting processes might extend down through the A_1 horizon. He produced a surface crust or seal consisting of a surface layer of slurried soil above a layer of 3 to 10 mm clods cemented together by slurried soil. As a result of reduced infiltration, the clods beneath the crust were slowly wetted by capillary action. This would reduce the amount of entrapped air so that clods were not broken down and were only weakly cemented together. When the clods were wetted by gentle flooding instead of intensive rainfall, the whole layer of clods used in the experiment was saturated and slurried before a protective seal could form. Slumping occurred, and upon drying the whole layer was fused into a crust 20 mm thick, in contrast with the 6-mm seal over 20 mm of intact clods found with the rainfall treatment.

Collis-George and Greene (1979) have studied the infiltration behavior of columns of dry aggregates from a hardsetting soil that were surface ponded to simulate flood irrigation or the situation beneath irrigation furrows, as a function of aggregate size. They observed that slaking caused the development of a surface throttle or restricting layer that subsequently delayed infiltration and therefore the soil beneath will have been tension-wet (as illustrated by Lal et al., 1970). They found that the depth of the zone of slaking increased from 0.02 to about 0.13 m, with an increase in aggregate size from 0.25 to 0.50 mm to 2 to 3 mm.

Both with intense rainfall and with furrow or ponded irrigation, surface slaking and pore clogging are likely to produce a layer capable of restricting further infiltration so that the profile beneath is always tension-wet in the first instance. However, because water movement may be held up by an impeding layer at some greater depth, sufficient rain or irrigation can reduce this tension to zero within a temporary perched water table. An impeding layer can itself be produced by cultivation when soil at the base of the cultivated layer is wet because, even without smearing, disturbing a wet hardsetting soil will mobilize silt and clay and the overburden will aid further compaction at this depth. Therefore the precise water regime and hence the amount of compaction likely to occur at various depths within a hardsetting horizon remain questions that can best be determined from tensiometer measurements made in situ during cultivation, rainfall, or irrigation.

5. Tension Wetting and Flooding

There is a distinction between tension-wet soils and those that are saturated at zero potential since, apart from the different degrees of slumping already discussed, tension wetting will not cause such extensive aggregate disruption (Emerson and Bakker, 1973) and the mobilized material will not be so free to move, although it will still be concentrated at zones of contact between aggregates.

Kemper et al. (1975) found that when flooded, aggregate beds of a structurally unstable silt loam that were subsequently air-dried had a modulus of rupture almost 8 times greater than tension-wet samples. In the field, flood irrigation of this soil gave <50% emergence, while tension wetting gave essentially 100% emergence. However, Hulme (1983) describes an experiment where cotton seeds were planted in columns of aggregates from a hardsetting soil that were capillary-wet by immersing their bases in water for 24 h. The seeds germinated but completely failed to emerge from 0.05-m-high columns; emerged but with mechanical damage to their cotyledons from the 0.1- and 0.2-m-high columns; and had 100% undamaged emergence from the 0.4-m columns. In this case, normal furrow irrigation in the field had resulted in only 50% emergence.

These two experiments represent different degrees of hardsetting behavior, and further work is needed to clarify the influence of the matric potential at wetting in relation to the subsequent strength development. In practical terms, Cockroft and Martin (1981) have obtained more effective tension wetting by slowly running water along the bottom of irrigation furrows (as opposed to filling them quickly), so that most of the furrow is tension-wet.

C. Dispersion and Hardsetting

Because hardsetting is particularly common in soils liable to dispersion, it is necessary to review the factors that control dispersion with particular reference to hardsetting.

The dispersion of soil colloids is controlled by the nature and distribution of the exchangeable cations that counter the permanent and/or variable charges on colloidal surfaces. These counterions, held by electrostatic or coulombic forces, are located in either the Stern layer or the Gouy-Chapmen diffuse part of the electrical double layer (Shanmuganathan and Oades, 1983a).

Dispersions of clay are stable when most of the counterions exist in the diffuse double layer and the thickness of the layer is of the same order of magnitude as the diameter of the individual colloidal particles. Repulsive osmotic forces between double layers prevent the particles from coming sufficiently close to enable short-range attractive forces, mainly van der Waals forces, to bring about the coagulation of the particles. This situation is favored by the presence of monovalent counter cations, particularly sodium.

If the thickness of the double layer is reduced sufficiently, the attractive forces become dominant and cause the coagulation or flocculation of the particles. Compression of the double layer can be brought about by (1) replacing exchangeable monovalent cations with ones of higher charge, and (2) increasing the electrolyte concentration of the soil solution. Altering the surface charge on particles also affects flocculation-dispersion.

1. Exchangeable Cations and Dispersion

a. Exchangeable Sodium

The adverse effect of exchangeable sodium on soil physical properties, through its effect on clay dispersion, is widely recognized. Sodic soils contain sufficient exchangeable sodium to seriously affect the soils' physical behavior. The U.S. Salinity Laboratory Staff (1954) recommend an exchangeable sodium percentage (ESP) of 15 to define the boundary between sodic and nonsodic soils. Colloidal dispersion and aggregate breakdown can, however, occur at considerably lower ESP values. For example, Emerson (1977) reported that for the 0–10 cm samples of 51 soils from southeast Australia, dispersion started at an ESP of 5, and for most samples was complete at an ESP of 10.

McIntyre (1979) has proposed a critical ESP value of 5 as being more relevant for Australian conditions. The onset of adverse physical conditions at this lower value, compared to that adopted in the United States, is attributed to the preponderance of clay soils in Australia and their relatively high ratios of exchangeable magnesium to exchangeable calcium.

b. Exchangeable Magnesium

For a given ESP, aggregates are less stable when magnesium is the complementary exchangeable cation rather than calcium. Emerson and Bakker (1973) found that for illitic red-brown earths in Victoria, the ESP required for dispersion was about halved when magnesium replaced calcium as the complementary ion. In a comparison of the effects of exchangeable magnesium and exchangeable sodium on the dispersion of an otherwise calcium-illite, Emerson and Chi (1977) showed that magnesium was 10% as effective on an equivalent basis as sodium in causing dispersion.

The hydration energy of magnesium is 20% greater than that of calcium. This results in an increase in the percentage of ions that are present in the diffuse double layer compared with calcium ions (Bakker et al., 1973).

For a given ESP level, soils vary in their response to similar percentages of exchangeable magnesium. Clay mineral composition is considered to be the main factor affecting a soil's response (Rahman and Rowell, 1979). The U.S. Salinity Laboratory Staff (1954) groups magnesium and calcium together as similar ions for classifying soils and irrigation waters. It is clear, however, that for many soils the specific effect of magnesium should be taken into account.

c. Exchangeable Aluminum

Due to their high charge, exchangeable aluminum ions exist largely in the Stern layer, so that the diffuse double layer is compressed and repulsive forces are correspondingly decreased. Interparticle bonding may also be brought about by polymeric hydroxylated aluminum cations. The influence of exchangeable aluminum is illustrated by the different dispersive behavior observed by Emerson (1983) of samples from two depths of an acidic clay subsoil. The more acidic sample (pH 5.0, ESP 20) contained three times more exchangeable aluminum than the other sample (pH 6.4, ESP 24). The latter dispersed completely in water, whereas the more acidic sample did not disperse at all.

2. Electrolyte Concentration and Dispersion

Dispersion of soil clays is very sensitive to the electrolyte concentration of the soil solution. As the electrolyte concentration is increased, the osmotic stress between clay particles is reduced to such an extent that short-range forces become dominant and cause flocculation. Quirk and Schofield

(1955) proposed the concept of threshold concentration, this being the electrolyte concentration required to prevent deflocculation.

This suggests that clay dispersion occurs in a soil when the electrolyte concentration of the soil solution falls below this threshold value. Quirk and Schofield (1955) found, however, that dispersion within the matrix of their soils occurred at lower electrolyte concentrations than those required for the reverse process of flocculation of clay suspensions. A likely explanation is that the clay fraction of the soil is in intimate contact with stabilizing agents, such as organic matter, and hence is more sensitive to electrolyte concentration compared to when the clay is suspended in columns used to determine threshold concentrations (Oster, 1982). Dispersion of soil aggregates would therefore tend to occur at a lower concentration than in suspension.

Since dispersion of soil clays is sensitive to both electrolyte concentration and exchangeable sodium, the threshold concentration varies with ESP. Conversely, the critical ESP for dispersion decreases as electrolyte level decreases. Thus Shainberg et al. (1980) showed that when the salt content of the soil was 3.0 $mol_c\ m^{-3}$, dispersion occurred only when the ESP was greater than 12. In distilled water, however, dispersion appeared at ESP values as low as 1 to 2.

The marked effect of a small amount of exchangeable sodium at low electrolyte levels explains, in part, the disagreement over the critical value of ESP used to define sodic soils. In arriving at a value of 15, the U.S. Salinity Laboratory Staff (1954) used a salt concentration greater than 3.0 $mol_c\ m^{-3}$ (Shainberg et al., 1980). McIntyre (1979), on the other hand, used a concentration of 0.7 $mol_c\ m^{-3}$ and found that adverse effects occurred at an ESP of 5.

The marked effect of electrolyte concentration on clay dispersion has two important bearings on the development and amelioration of hardsetting soils. First, leaching of salts from the upper part of the soil profile may reduce the electrolyte concentration to very low levels, and hence induce dispersion leading to crust formation and hardsetting. Thus a soil with a moderate ESP may have reasonable physical properties throughout most of the profile, but still be susceptible to dispersion in the topsoil because of the low concentration of the soil solution.

Second, the susceptibility of a soil to dispersion can be reduced by increasing the electrolyte concentration of its solution. Shainberg et al. (1980) have shown that soils that release salt from the solid phase at a rate sufficient to maintain the concentration of the soil solution above the threshold value for a given ESP will not disperse. Salts are released from soil minerals by weathering and hydrolysis. For example, Rhoades et al. (1968) reported that the release of calcium and magnesium from minerals such as hornblende and plagioclase feldspàr increased the salt concentration of leaching solutions by 3 to 5 $mol_c\ m^{-3}$. In calcareous soils, the solution of carbonate minerals can be effective in preventing dispersion due to exchangeable sodium (Kazman et al. 1983; Shainberg and Gal, 1982). As a management strategy, dispersion may be prevented by applying gypsum to the topsoil to increase its electrolyte concentration. Gypsum has been found to be effective in soils with an ESP as low as 1.0 (Kazman et al., 1983).

The influence of electrolyte concentration on clay dispersion is affected by two physicochemical phenomena observed in soils. In leaching column experiments, Pupisky and Shainberg (1979) found that even though the salt concentration remained constant, the concentration of clay released by dispersion initially increased rapidly with leaching. It reached a maximum at about 1.5 pore volumes of leachate and then gradually decreased with further leaching to a near-constant low level. The clay concentration decreased with leaching even though there was still sufficient clay in the leaching column to maintain a higher concentration.

This pattern of a variation in clay concentration can be related to Emerson and Bakker's (1973) theory of dispersion of aggregates in water. They proposed that, for the dispersion of clay from aggregates containing a small amount of exchangeable sodium, two conditions have to be satisfied: (1) the salt concentration of the soil solution has to be less than the threshold concentration for flocculation; (2) the gradient of salt concentration between the dilute solution in the soil pores and the more concentrated solution in the micropores within aggregates has to exceed a certain value. This salt gradient causes osmotic water movement into, swelling within, and the breakdown of the aggregate (Shainberg et al., 1980). When rainwater percolates into a soil, a large concentration gradient exists at the wetting front between the aggregate micropores and the conducting macropores. This would maximize dispersion through the "osmotic explosion" mechanism and would occur at an early stage in a rainstorm. With further leaching, diffusion reduces the salt gradient between the intra- and interaggregates so that dispersion is lessened.

In a study of the influence of irrigation water quality on infiltration, Oster and Schroeder (1979) found that the lowest infiltration rates were obtained in columns irrigated alternately with water having a sodium adsorption ratio (SAR) of 19.9 and distilled water. If it is assumed that infiltration of the coarse loamy sand used in the study is affected mainly by clay dispersion and clogging of pores, the salt gradient effect on dispersion would account for the observed results. The data also suggest that the potential for dispersion is enhanced where irrigation periodically supplements rainfall (or vice versa).

The second phenomenon affecting electrolyte concentration and dispersion is salt sieving, whereby salt is trapped within the microfabric of aggregates (Blackmore, 1976). Negatively charged clay surfaces are surrounded by a region of anion exclusion so that for wet clay surfaces there is always some tendency for salt to be separated from the water. Within aggregates, regions of anion exclusion, associated with surfaces in the narrower pores, restrict the cross section for diffusion of salt from larger interior pores. Thus, large amounts of salt may be trapped when many large pores are isolated from the exterior by much narrower ones. This might occur when clay skins (cutans) are present on the surface of aggregates. Salt sieving is important where dispersion is affected by the rate of diffusion of salts from soil aggregates to the soil solution (Rengasamy et al., 1984).

3. Surface Charge Density

An increase in the negative charge density on colloidal surfaces will increase the repulsive forces between particles, the thickness of the double layer, and hence promote the dispersion of particles. Changes in pH affect the edge charges of clay minerals, and the surface charges of variable-charge constituents such as iron and aluminum oxides and organic matter. Thus the edge surfaces of kaolinite, where broken primary bonds of silica tetrahedrons and alumina octahedrons are disrupted, are positively charged in acidic solution and negatively charged in alkaline solutions. The pH at which the total negative charge equals the total positive charge is referred to as the zero point of charge (ZPC). The ZPC of iron and aluminum oxides lies in the range 7 to 9, so that soils rich in these constituents tend to have relatively high ZPC values.

The increase in clay dispersion with pH observed in many soils is ascribed to increase in net negative charge (Arora and Coleman, 1979; Suarez et al., 1984). At low pH values, edge-to-face bonding between the positively charged edges of kaolinite and negatively charged surfaces occurs; similarly, positive iron and aluminum oxides form an association with negative clay surfaces. This bonding, referred to as mutual flocculation, would inhibit dispersion. As pH increases toward the ZPC, bonding decreases; eventually the edge charge on kaolinite is reversed and an edge-to-face repulsion is created (Frenkel et al., 1978). Soils with large amounts of variable-charge colloids would be most susceptible to the influence of pH on dispersion.

Wide variation in the effect of exchangeable sodium on kaolinitic soils has been reported (Frenkel et al., 1978). Most studies have been conducted on acidic soils containing appreciable amounts of iron oxides. Under such conditions, edge-to-face bonding would be expected to counteract dispersive forces. However, one of the kaolinitic soils used by Frenkel et al. (1978) was nonacidic, so that edge-to-face bonds would be weaker and more susceptible to disruption by repulsive forces related to adsorbed sodium.

Negative charge density on colloid surfaces can be increased by the specific adsorption of anions. Shanmuganathan and Oades (1983a) showed that the addition of anions to soils in which the clay was flocculated resulted in some dispersion, phosphate and fulvate being the most effective ions. Thus the addition of phosphate fertilizers may change the charge characteristics of surface soils.

4. Bonding Agents

The repulsive forces between clay particles are counteracted by bonding agents and the soil water tension (Childs, 1969). The most important agents are organic matter, and iron and aluminium oxides.

a. Organic Matter

Organic matter acts as both a bonding and a dispersing agent in soils. Natural organic anions are produced by the decomposition of organic matter and by root exudates. The anions are associated with organic acids, which range from simple, well-defined types, such as citric, oxalic, and tartaric, through to fulvic acids (Oades, 1984). In addition to increasing the negative charge on colloidal surfaces, they probably affect dispersion through (1) complexing divalent and trivalent metal ions, thus reducing their concentration in solution; and (2) reacting with positive sites associated with trivalent metals (aluminum and iron) situated at the edges of clay lattices and oxide surfaces (Oades, 1984).

Gillman (1974) has demonstrated the dispersing effect of organic matter in a strongly weathered krasnozem profile developed under virgin rain forest in Queensland. Above a depth of 2 m, the clay fraction is dominated by iron and aluminum oxides with subordinate kaolinite. The subsoil is near its ZPC with no net charge to produce repulsive forces between clay particles so that little or no water-dispersible clay is present. In the topsoil, however, adsorption of negatively charged organic matter on the positive sites of mineral surfaces has lowered the ZPC below the pH of the soil. Consequently, substantial amounts of water-dispersible clay occur.

Organic matter is able to counteract the flocculating effect of exchangeable aluminum. Suspensions of an acid kaolinite subsoil were found by Emerson (1971) to flocculate rapidly in water, whereas suspensions of the topsoil were stable. Adsorbed aluminum, considered to be responsible for the flocculation of the subsoil, was screened by organic matter in the dispersed topsoil.

The dispersing effect of negatively charged organic materials appears, at first sight, to be at variance with the well-known stabilizing effect of organic matter on soil aggregates. An explanation of this apparent anomaly is provided in an experiment conducted by Emerson (1954). Surface soil aggregates were washed with a NaCl solution to replace most of the exchangeable cations with sodium. The aggregates were then equilibrated with lower concentrations of NaCl. For aggregates from a permanent arable field, dispersion occurred when the NaCl concentration was reduced to 3.5×10^{-2} M. In contrast, for aggregates of similar inorganic composition from permanent grassland, organic bonds were able to prevent dispersion even when the aggregates were immersed in distilled water. Despite the much greater stability of the undisturbed aggregates from the grassland, a suspension of these aggregates required a NaCl concentration for flocculation 10 times as high as that required for the suspension of arable aggregates. The explanation is that organic bonds can hold clay particles together against the osmotic stress induced by the adsorbed sodium (Emerson, 1977). If, however, these bonds are broken by shearing, as in the preparation of suspensions, and the clay particles are forced sufficiently far apart, the organic anions act as dispersing rather than as flocculating agents.

To counteract hardsetting, organic matter should be maintained in the soil to increase structural stability. Management systems that involve regular cultivation, along with low inputs of organic residues, not only lower organic matter content but also change the composition of the residual organic matter (Oades, 1981, 1984). The ratio of fulvic to humic acid increases so that the organic materials are more soluble and possess more acidic functional groups than in a corresponding soil with two to three times the organic matter content. Loss of organic matter therefore leads to an increase in dispersible clay as a consequence of an increased proportion of fulvic acids and oxidation of organic binding agents.

Aylmore and Sills (1982) suggest that the strength and longevity of organic structural bonds may be related to the exchangeable cation composition of the soil at the time of incorporation of organic residues into the soil. Ulmic and humic acids are believed to be converted to cementing material through the agency of calcium from the roots of legumes. If sufficient sodium ions were present, the cementing process could be retarded, resulting in reduced bonding.

b. Iron and Aluminum Oxides

The association of stable structure with a high content of iron oxide has led to the view that the iron oxide bonds soil particles together. Thus Velasco-Molina et al. (1971) suggest that the stability of aggregates with an ESP of 22 was due to the high iron content (14% Fe_2O_3) of the soil. Emerson (1983) has pointed out, however, that there is no clear evidence that iron oxides contribute to the bonding between clay particles in aggregates. He found that a red-brown earth, whose clay fraction contained 6% free Fe_2O_3, required an ESP of only 5 to induce dispersion of aggregates in water. Soils rich in iron oxides and with stable structure are generally acidic (e.g., oxisols and krasnozems). It may well be that structural stability is due to the effects of low pH on surface charge and exchangeable aluminum rather than to the bonding of iron oxides. A pH effect is indicated by the fact that the readily dispersible red-brown earth mentioned above was neutral, whereas a more acidic red-brown earth containing appreciable exchangeable aluminum required an ESP of 12 for dispersion (Emerson and Bakker, 1973).

Aluminum oxides are most abundant in highly weathered acidic soils, and, as in the case of iron oxides, it is difficult to separate their possible bonding effects from the effects of pH.

5. Clay Mineral Composition

The rate and amount of dispersion of a soil varies with its clay mineral composition. Arora and Coleman (1979) found that sodium-saturated clay minerals differed in their sensitivity to flocculation by $NaHCO_3$ in the following order: illite > vermiculite > montmorillonite > kaolinite. The critical concentrations of $NaHCO_3$ for these minerals were 185, 58, 28 to 60, and 8 $mol_c\ m^{-3}$, respectively.

Oster et al. (1980) found that the flocculation value for Na-illite was 4.6 times that of Na-montmorillonite. In contrast, Velasco-Molina et al. (1971) reported that soils dispersed in the following order: montmorillonitic > halloysite-kaolinitic > micaceous. It should be noted, however, that the micaceous soil contained carbonate and sulfate minerals, which would release salts into solution and thus promote flocculation.

a. Illite

The repulsive forces between illite and montmorillonite clays in the Na form are similar (Oster et al., 1980). The reason for the greater dispersivity of illite is believed to be due to the smaller edge-to-face attraction forces resulting from the shape of illite particles. Electron micrographs have shown that the planar surfaces of the particles are irregular and terraced (Greene et al., 1978). As the particles approach each other, the irregular surfaces would permit only poor contact between the edges and planar surfaces. This would weaken edge-to-face attraction, and flocculation would require a high electrolyte concentration.

Illite soils would be expected to be susceptible to clay dispersion. Rengasamy (1983) found this to be true of illitic red-brown earths in Australia. The tendency for the clay in these soils to disperse would partly account for their propensity to hardset.

Illite has been shown to be sensitive to the specific effect of exchangeable magnesium on dispersion (Bakker and Emerson, 1973). Rahman and Rowell (1979) also reported a specific magnesium effect for illite, an illitic soil, vermiculite, and a mixed illite-montmorillonite soil, but not for pure montmorillonite.

b. Montmorillonite

Oster et al. (1980) found that the flocculation value for calcium-saturated montmorillonite increased rapidly with small increases in exchangeable sodium. This large effect is believed to result from the phenomenon of demixing, whereby the first additions of sodium to a Ca-montmorillonite are adsorbed on external surfaces rather than on internal surfaces, which remained saturated with calcium. This distribution results in a more diffuse double layer on the external surface and a more stable suspension than if the sodium was distributed evently over all the surfaces.

c. Kaolinite

Schofield and Samson (1954) found that sodium saturated kaolinite was flocculated in water at a pH of 5. The ease of flocculation was ascribed to the attraction between positively charged edges and negatively charged planar surfaces. The clay was completely deflocculated by the addition of a small amount of NaOH, which would remove the positive edge charges. Flocculation of clay particles may also be due to the binding effect of exchangeable aluminum, which increases markedly as pH decreases below 5.

Differing reports on the effects of exchangeable sodium on kaolinitic soils appear in the literature. Most studies have been conducted on acidic soils, in which edge-to-face and exchangeable aluminum bonding will counteract the osmotic stress. Frenkel et al. (1978) found, however, that a kaolinitic soil of pH 6.7 showed dispersion at ESP 10 when leached with distilled water. At this pH, bonding forces would be weak.

6. Mechanical Stress

Dry soil aggregates that do not disperse when immersed in water can be made to do so by subjecting them to mechanical stress. Even a soil whose exchange complex is saturated with calcium can be made to disperse by mechanically shearing aggregates at a moisture context above a critical level, referred to as the water content for dispersion (Emerson, 1977). In a study of the dispersive behavior of Australian red-brown earths, Rengasamy et al. (1984) found that surface soils with a SAR greater than 3 (equivalent to an ESP of about 6) dispersed spontaneously in distilled water, whereas those with a SAR less than 3 dispersed only after mechanical shaking. However, since the clay fraction of the soils is predominantly illitic, the soils are susceptible to dispersion when even weak mechanical forces are applied. In a previous study, Rengasamy (1983) had shown that in the absence of electrolyte the physical process of dialysis and subsequent resuspension was sufficient to cause dispersion of calcium-saturated red-brown earths.

Rowell et al. (1969) found that the amount of clay dispersed varied with the mechanical stress applied to a soil. When the amount of stress was small, the amount of clay dispersion was determined by the ESP level. Large stresses could disperse most of the clay, even at very low ESP values.

The effect of mechanical stress is probably to break the organic matter bonds holding clay particles together (Emerson, 1968). The more strongly the particles in an aggregate are held together, the greater is the work that has to be done to break the bonds. Rowell et al. (1969) have discussed the effects of mechanical stress in terms of potential energy. When a swollen clay is in equilibrium with an electrolyte solution, the clay particles are in a position of minimum potential energy and are prevented from moving away from each other by a potential-energy barrier. Mechanical stress is able, however, to overcome this barrier and thus bring about dispersion.

As the electrolyte concentration of the solution increases, the magnitude of the barrier increases, and hence the mechanical stress needed to overcome the barrier increases. Once dispersion has occurred, the barrier opposes reflocculation. Drying can provide the pressure necessary to bring the particles together.

Due to the effects of mechanical stress, the range of electrolyte concentration at which dispersion occurs is large, a 100-fold difference in molarity being observed by Rowell et al. (1969). Surface soil is subjected to mechanical disturbance through cultivation, treading by animals, raindrop impact, and the movement of water through the soil, particularly during irrigation. The operation of these stresses has a number of implications:

1. The potential for dispersion and hence for the development of hardsetting is increased, especially if the soil is cultivated under wet conditions and at low electrolyte concentrations. Thus a soil may have a friable consistency when uncultivated, but if it is very susceptible to dispersion by mechanical disturbance it may rapidly show hardsetting behavior upon cultivation.
2. Dispersion may occur at electrolyte concentrations considerably higher than the threshold concentrations obtained under stationary laboratory conditions. If threshold values are to be used as safety guidelines in the use of irrigation water, the mechanical disturbance of the soil must be kept to a minimum (Rowell et al., 1969).
3. A soil does not need to be sodic for dispersion to occur. Rengasamy et al. (1984) refer to their red-brown earths with a SAR ≤ 3 (equivalent to an ESP of about 6), which disperse when mechanical stress is applied, as low sodic rather than nonsodic.
4. It is not possible to obtain a universal relationship between ESP and threshold electrolyte concentration, as the latter varies with the amount of stress applied.
5. Tests to predict the probability of dispersion occuring under a given management system have to take account of the stresses to which a soil will be subjected in the field. Rengasamy et al. (1984) have proposed a scheme for identifying dispersive behavior by relating SAR and total cation concentration of the soil solution to spontaneous and mechanical dispersion. They suggest that spontaneous dispersion represents the effect of minimum mechanical disturbance, e.g., under zero tillage or when the soil is protected from raindrop impact by a plant cover. The mechanical disturbance obtained by shaking the soil for 1 h is considered to identify the effects of weak to moderate stress. Examples are that caused by the flow of water across irrigation bays or by rainfall of an intensity common in southeast Australia falling on bare soil. To predict the effect of strong mechanical forces, such as occur during the cultivation of wet soils, it may be necessary to use more severe mechanical disturbance, as in Emerson's (1967) test in which the soil is remolded for half a minute when wet.

D. Agronomic Effects of Strength Limitations

Hardsetting soils impose four agronomic limitations due to their high strength. First, the high strength of the dry soil can delay seeding or tillage operations until the soil is sufficiently wetted by rainfall or irrigation. Second, even when there is sufficient power for tillage, it may not be possible to produce a suitable seedbed tilth. Third, crusting after sowing can prevent emergence (see Section II.B.4). And finally, drying of the profile by roots can impede their growth at potentials well above −1.5 MPa (i.e., wilting point).

1. Timing of Cultivations

Many authors have commented on the restrictions to cultivations imposed by hardsetting. In a climate with seasonal rainfall, it may be necessary to wait for the first rains to soften the soil before it is possible to cultivate (Willcocks, 1981), drill, or plant seeds with a jab planter (Ley et al., 1988), and the delay can result in considerable loss in yield. Even where sufficient tractive power makes it possible to cultivate the soil in a dry state, cloddy seedbeds (i.e., with large, hard aggregates) are produced (Willcocks, 1981; Cockroft and Martin, 1981), and where repeated tillage operations are used, the clods break down into a fine, dusty seedbed that is especially prone to structural collapse and impeded infiltration upon wetting (Tisdall and Adem, 1986b; Lenvain and Pauwelyn, 1988).

Where primary tillage has been possible, the difficulty in subsequently reducing the large clods into a suitably sized range of aggregates by further cultivations depends on the friability of the soil. Utomo and Dexter (1981) have devised a test in which friability, k, is defined as the gradient of the line relating the size of aggregate to their tensile strength on a log-log plot. Thus, where there is little or no variation of aggregate strength with size ($k < 0.05$), aggregates are not friable, and at the other extreme, aggregates with $k > 0.40$ are so friable as to be "mechanically unstable." As shown in a discussion of experimental technique (Dexter and Kroesbergen, 1985), there are considerable limitations to the accuracy to which k can be determined, but it is nevertheless a valuable quantitative index of the ease with which an acceptable seedbed can be produced that relates well with actual field experience.

Ley et al. (1988), Haddow (1988), and Chan (1988) have all documented the fall in k value of air-dry aggregates from soils that have become hardsetting (or increasingly hardsetting) under regular tillage in comparison with less intensive management systems (tropical forest or pasture), with k falling from values >0.2 in the less disturbed to <0.05 in the regularly tilled treatments. Friability also varies with water content and tends to be at a maximum in samples equilibrated at a matric potential of about −100 kPa (Utomo and Dexter, 1981; Young, 1987; Ley et al., 1988). Thus hardsetting soil is friable at −100 kPa (which is often comparable with the soil water content at the lower plastic limit), but loses its friability upon drying, in contrast to better soils that do not. This can impose a severe restriction on the timing at which successful secondary cultivations can be achieved, and repeated cultivations are liable to produce an unacceptably high proportion of fine material.

2. Restricted Root Growth

Many workers have noted the restricted depth of rooting in hardsetting soils (Hamblin and Tennant, 1979; Nicou and Chopart, 1979; Ley, 1988; Young *et al.*, 1988), but it is important to distinguish between at least two possible causes. First, in dryland agriculture, where the runoff from hardsetting soils restricts the amount and depth of profile wetting, this effect may simply be due to the limitation imposed on root growth by water stress when the supply of profile available water is exhausted. At this stage, the roots will often have grown through the hardsetting horizon, for example, into the clayey horizon of a duplex soil.

Second, even where rain or irrigation have been able to achieve a profile that is wetted to depth, the increase in soil strength with decreasing matric potential that occurs as roots dry the soil may be sufficient to seriously slow down or even halt root growth when the soil is still at a potential much greater than the wilting point. Indeed, it seems likely that the scenario described by Boone (1988), where the soil can simultaneously have insufficient air porosity for adequate root aeration and too high a strength for root growth, may be not too far from the truth, as evidenced by the lack of weed growth in some hardsetting soils under fallow.

It is therefore of interest to discover at what value of matric potential the penetrometer resistance at any point in a hardsetting horizon exceeds some critical value. A penetrometer resistance of about 3 MPa seems appropriate, since values of this order are associated with severe impedance or a halting of root growth (Greacen et al., 1969). Young (1987) and Ley (1988) have measured penetrometer resistance in "undisturbed" cores of a hardsetting soil equilibrated at various matric potentials but have found that these values can seriously overestimate compared to values measured in the field at the same water content. They concluded that such laboratory measurements are unreliable, especially at lower potentials, because of the effects of wetting of the soil prior to equilibration and the freedom of core samples to shrink isotropically. Using field penetrometer values, Young (1987) and Ley (1988) found values of matric potential of ca. 0.1 and ca. 0.2 MPa, respectively, which corresponded to a penetrometer resistance of 3 MPa in hardsetting soils that showed serious restrictions to crop root growth in the field.

While it is not particularly unusual for soils, even in a relatively moist state, to show such high values of penetrometer resistance, it is by now clearly established that in a structureless soil, penetrometer resistance correlates well with the rate of root growth (Greacen et al., 1969), and more work is needed involving simultaneous field measurements of penetration resistance and matric potential in relation to the limitations imposed on root growth in hardsetting soils. This phenomenon may have particular importance when there is a dry period during the early stages of plant establishment when the plant has a shallow and limited root system, since the restrictions on nutrient and water uptake can significantly delay the development of the whole plant (Hamblin, 1984; Young et al., 1988), although the final effects on yield may not always be evident.

Young et al. (1988) report a sharp contrast in root growth between cabbages grown on two similar soils, both of which were structurally unstable and slumped after cultivation into a structureless mass. Five weeks after transplanting, over half the roots were found to be growing down fine structural cracks in the less degraded soil, which had a greater surface rooting density and rooting depth (>0.5 m) in comparison to the hardsetting soil (roots to <0.3 m), in which no crack formation was observed. The soils had closely similar clay mineralogies and particle-size distributions, and the reason for the contrast in soil structural development remains obscure. There is a long-documented history of systematic differences in yields between these two sites (Costigan et al., 1983), the lower-yielding and less rootable of which had been under continuous cereal cropping and had a lower organic matter content (16 compared to 23 g kg^{-1}) to a depth of 0.34 m, so that organic matter may be implicated.

It is clear that more work on the soil physical and chemical properties that control the onset of structural development within the rooting zone in the field is an important priority, since this defines the boundary between hardsetting and structured soils and should indicate how the soil may be managed to increase the probability of restructuring.

III. Soil Classification and Occurrence of Hardsetting Soils

A. The Australian Experience

1. Historical Perspectives and Definition

The concept of "hardsetting in surface soil" arose from studies made in Australia during the late 1950s to develop a soil classification suitable for the broad-scale mapping of Australian soils consistently over the entire continent, Tasmania, Kangaroo Island, and the other islands in the Australian system. The existing Great Soil Group classification could not be applied consistently across Australia, nor did it provide for some soils that were being found at that time, and its agronomic content at the Great Soil Group level was very limited. The transference of agronomic data from one area to another was unreliable. For example, the class red-brown earth, Urrbrae fine sandy loam, of the land surrounding much of the Waite

Agricultural Research Institute is a nonsodic soil easily adapted to various agricultural and horticultural practices. Many superfically similar soils in northern Victoria and southern New South Wales also classified as red-brown earths presented many severe agricultural and horicultural problems due in part to their sodic nature; see Northcote (1981) also. It was in this general intellectual environment that a new soil classification was sought. The outcome, *A Factual Key for the Recognition of Australian Soils*, was first published in 1960, and was used by several pedologists to map the soils of Australia published as the *Atlas of Australian Soils* during the period 1960–1968 (Northcote et al., 1960–1968). *A Description of Australian Soils* followed and was published in 1975 (Northcote et al., 1975).

The development of the Factual Key, hereafter referred to as the Key, followed a study of about 500 soil profiles from widely distributed areas in Australia. It is necessary to remember this and to understand some details about the Key before the pedological significance of "hardsetting" can be fully appreciated. The basis of the key is the concept of profile form. *Profile form* is the term used to express the overall visual impact of the physical soil properties in their intimate association one with the other, and within the framework of the solum. Thus, the profile is regarded as a physical system, and those physical properties capable of observation and record that may carry along other features or properties, be they physical, chemical, or biological, are used as the critical ones to distinguish between groups at each step in the key. The only exception is the use of Soil Reaction Trend. The properties were chosen after studying about 500 soil profiles, involving simple trial-and-error methods to determine which properties ordered the soils with most simplicity into groups with the same (like) profile form. It should be noted that these properties may not have any other merit than that they make for the easy formulation and operation of a key. But it *should be understood that the object identified by the key is the entire solum and not merely the properties used in key formulation.* In other words, once the profile form is identified, it becomes the object of study and use. It is a soil with a particular profile form (solum) that has a particular physics, chemistry, and biology, and that is used either as a medium for plant growth, or as a material for construction purposes, such as earth dams, and not individual properties of that solum. After all, *individual soil properties cannot exist as* things *outside the body of the soil* (*solum*) and therefore are useful only as a means to identify and should not be regarded as the entire object (solum) they help to identify. Thus it is the profile form (solum) identified by the key that should be studied in relation to plant growth or construction and not the keying properties alone. The properties used to develop the key were the usual physical, morphological soil properties that were familiar to all field pedologists—that were used by them in defining soil types for detailed soil mapping for irrigation and other forms of land culture at the time. They still are. Of all these morphological properties, texture seemed to be outstanding, as it relates to the behavior of soil, including its likely moisture content, a most important consideration in Australia. Again, soil texture was, and is, one of the first things that any self-respecting soil scientist determined. The next question concerned the relations between texture and profile form. Obviously, the use of surface texture alone, or subsoil texture alone, could not be very helpful. For example, it was well known that there are many kinds of soils with sand surface soil texture. What if the profile form sequence of surface soil texture plus subsoil texture is considered as one entity or thing? From detailed soil studies in various parts of Australia and from studies in the United States and Western Europe during 1956, it was clear that in some soils texture remained sensibly constant as between surface and subsoils, in others texture became more clayey with depth, while in others again there was a big change from a less clayey surface soil to a pronouncedly clayey subsoil. Data for the 500 soil profile confirmed this idea. All mineral soils could be divided into three primary profile forms, U, G, and D, where U means a sensibly constant or uniform texture profile for the solum, G means a gradual increase in apparent clayiness (i.e., a gradational texture profile for the solum), and D means a sudden increase in apparent clayiness from the surface soil to the subsoil (i.e., a duplex texture profile for the solum). (The solum equals surface soil or A horizons plus subsoil or B horizons.) An organic primary profile form was added in which the profile is dominated by plant remains, thus dividing the total soil population into four recognizable and definable divisions or primary profile forms. Definitions are given in the Key (Northcote, 1960, 1971, 1979). This result immediately lessened the load of the total mass of soil data.

The 500 profiles of mineral soils were divided into three piles of profile descriptions representing the U, G, and D primary profile forms. Classification was clearly possible within each, *separately*. There was no reason for thinking that the soil properties of most use to classify within the U division, for example, would prove useful in either the G or D division. Quite the contrary, in fact, there seemed sound a priori grounds for assuming that the soil properties most significant for obtaining subdivisions of U soils would not be so for G and/or D soils. For example, the next most important property of U soils is concerned with the particle size of their component materials. Are these coarse or fine? Clearly, this property cannot function in duplex (D) soils, which have both coarse and fine material in the solum. Experience during detailed soil studies had taught that some D soils are better drained and aerated than others and that B horizon color provided a good general guide. This conclusion was supported and refined by trial-and-error groupings of the D profiles using coded punch cards. Five subdivisions resulted and were termed Dr, Db, Dy, Dd, and Dg. Each of these subdivisions represents a range of drainage characteristics that become progressively more intense (less permeable) from Dr→Dg (Dg includes the real blue gray and green-gray colors of gleyed soil materials). Furthermore, because each subdivision was actually a drainage sequence, it was possible to divide each sequence using other indicative soil properties such as the presence or absence of a bleached subsurface, or A_2 horizon, which indicates periodic waterlogging. By this process the classification moved from level to level, based entirely on physical properties used within the framework of profile form, and involving simple trial-and-error methods to determine which properties ordered the soils with most simplicity into sensibly similar groupings. The objective was to arrive at what could be termed the *principal profile form*, a category that seemed suitable for mapping soils and landscapes at the continental scale of 1 : 2 000 000; and this proved to be so. However, a problem arose before that desirable end could be achieved. No matter how the properties of the Dr soils were arranged, the resulting groups were always too variable when judged by any standards apart from the morphological properties used in making up the groups. Reexamination of all Dr soils in the collection showed that although they did have like profile form to a degree, there were surface soil differences in salt contents, organic matter, cation exchange, and so on. It was a case of returning to field notes, as it seemed that not all properties were recorded on the coded punch cards. Thus the properties of thin crusted surface soil, hardsetting surface soil, and surface soil that does not set hard came to recognized. When these surface soil properties were inserted into the Key at the section level, that is, immediately following the Dr subdivision, the whole scheme for the Dr soils fell into place like a jigsaw puzzle. These properties worked equally well for the other subdivisions of D soils, and subsequently also in the field on the *Atlas* mapping. In other words, the nature of the surface soil is an important property for purposes of soil classification as well as land use. A not-altogether-unexpected result, but certainly one that had not been fully appreciated previously, and moreover, one that focused on the hardsetting character of many Australian soils. The hardsetting property of surface soils was defined by Northcote (1960, 1971, 1979) as follows: "A horizons are considered to be hardsetting when a compact, hard, and apparently apedal condition prevails on the drying-out of the soil periodically." Indeed, the hardsetting character may cause the A_1 horizon of some soils to become their most prominent feature after prolonged exposure in cuttings and erosion gullies. In cultivated soils, clods usually retain the hardsetting condition until completely broken down by repeated cultivations.

Moreover, this definition was presented within the context of the "condition of surface soil"; that is, it refers to the natural condition of the A horizon and its reaction to the usual wetting and drying cycle. Since the wet-dry cycle is variable both across the Australian region and between individual years, it seemed sensible to include the word *periodically* within the definition. Another point made within this context was that "cultivation will often alter the condition of the surface soils but the conditions listed (including hardsetting) will reform when the soil is left undisturbed." Indeed, it is now known that inappropriate and/or excessive cultivation aggravates the hardsetting condition so that it may become progressively worse. Two further relevant points made were as follows:

1. "Soil on which a surface soil seal develops may or may not be hardsetting, that is, a surface seal is not a criterion for the hardsetting condition."
2. "The majority of soils throughout the wet-dry climatic zones of Australia set hard in the dry season. Soils which do not set hard are pedal in the dry, as well as in the moist state; or are single grained (sands)."

More recently, McDonald et al. (1984) defined hardsetting as "compact, hard, apparently apedal condition forms on drying. Surface not disturbed or indented by pressure of forefinger. Surface seal is not necessarily associated with hardsetting." Their definition accepts the notion of "hardsetting" as previously defined by Northcote (1960, 1971, 1979), leaves out the reference to the periodicity of the phenomena, and tends to emphasize "hard," that is, "surface not disturbed or indented by pressure of forefinger." Although this is reasonable from some points of view, it does shift the emphasis away from the "setting" part of the phenomena, which may well be fundamental. All soils, and those that set hard are no exception, are friable at their optimum moisture content, namely, at that moisture content present shortly after the wetted soil has drained naturally. This fact supplies an easy solution to digging (working) hardsetting soils under home garden conditions. Other soils retain their structure and thus remain friable even when dry. It is this contrast between the two conditions of surface soil that allow for their easy recognition in the field.

Pressure tests are really concerned with the *degree* of hardsetting, which can be important agronomically. Omission of the periodicity of the phenomena may make for some difficulties vis-a-vis those soils that are more or less permanently dry and thus remain hard except when wetted occasionally in some years, such as some of the massive earths (Gn2 soils) of the desertic parts of Australia.

2. The Present Position

The nature and distribution of hardsetting soils in Australia are currently available from *A Description of Australian Soils* (Northcote et al., 1975) and the *Atlas of Australian Soils* (Northcote et al., 1960–1968). By far the most important hardsetting soils are the hard duplex soils, Table 1, totaling about 12% to 13% of the land area of Australia. Furthermore, they occur in the more reliable, higher, rainfall districts. The duplex soils with nonhardsetting surface soils, that is, the friable duplex soils and the sandy duplex soils, total only about 3% of the land area. They occur under similar rainfall conditions. Thus, the areal importance of the hardsetting surface soil is clear in the Australian context. Moreover, it should be noted that a few other soils also have hardsetting surfaces. This arises from the fact that "hardsetting" was originally recognized as a morphological/

Table 1. Relative extent of various duplex soils in Australia

Duplex soils	Percentage
(a) Hard: Dr2, Dr3, Db1, Dy2, Dy3, Dd1, Dd2, Dg2	12–13
(b) Friable: Dr4, Db3, Dg4	~1
(c) Sandy: Dy5	~2

Source: Northcote et al. (1975).

Table 2. Relative extent of hard soils other than duplex soils in Australia

Soils	Percentage
Pale sands: Uc4 of sandy loam texture	3
Loam soils: All Um soils except for Um4.4, Um6, and Um7 soils	~12
Noncracking clays: Uf6.4, Uf6.5, Uf6.6, Uf6.7	<1
Massive cracking clays: Ug5.4, Ug5.5	<1
Massive earths: Gn2	17

Source: Northcote et al. (1975).

pedological feature of particular significance for classification within the duplex soils. Its agricultural importance was recognized subsequently. Other soils, it was found, could be identified and classified without reference to the hardsetting feature. Some of these are listed in Table 2. Most of the U soils are shallow and are generally significant agriculturally only where they are associated with hard duplex soils. Another highly significant group of soils that may have some hardsetting members is the massive earths, totaling 17% of the land area. The area of these soils that may be hardsetting is not known. Furthermore, it is likely that the hard character found in some of these soils represents a different condition to that under consideration. For example, in the arid to semiarid zones, loam members of the massive earths will support road trains (very large transports) without difficulty until it rains periodically when they become very soft. This condition may well be traceable to their massive, but highly porous, earthy fabric (Northcote et al., 1975). In the semiarid to more humid zones, the massive earths do not seem to hard-set, except for those associated with hard duplex soils, for which some hardsetting may develop after prolonged cultivation. Again, hardsetting tends to develop in the smooth-pedal structured earths (Gn3) only after clearing and continued cultivation, with the brown, yellow, and gray forms being much more prone than are the red Gn3 soils.

Table 3 sets out some of the morphological differences between surface soils that set hard and those that do not. The most readily observed differences are found in structure and consistence, both of which are poorly expressed in the hardsetting soils. Under moist conditions, hardsetting soil

Table 3. Comparison of some morphological properties of the surface soils (A_1 horizon or top 15 cm) of the hard, friable, and sandy duplex soils

Duplex soil	Color	Texture	Structure	Consistence
Hard	Dark gray-brown or dark gray to brown, red-brown, gray-brown or gray	Loamy sands, sandy loams, loams, clay loams	Moist: weak blocky polyhedral or platy[a]	Moist: friable at optimum moisture content
	Munsell value/chroma 3 or > (moist)		Dry: massive or cloddy massive	Dry: hard
Friable	Very dark brown or very dark red-brown to black	Mostly loams and clay loams	Moist: crumb, granular fine polyhedral 10–12 mm or less in size	Moist: friable soft
	Munsell value/chroma <3 (moist)		Dry: pedality evident	Dry: friable firm
Sandy	Light brownish gray to dark gray brown	Sands to loamy sands	Moist: coherent to weakly so	Moist: weak
			Dry: single grain	Dry: weak

[a] Weak crumb in top 2 cm under some natural vegetation.

Source: After Northcote et al. (1975).

may even appear to be pedal and soft with a good tilth, but this desirable state is lost upon drying, and the soil sets down hard. This does not happen in friable duplex soils, which generally exhibit darker soil colors (lower value and chroma on the Munsell scale) than do the hardsetting soils. Surface soil textures range from loamy sands to clay loams, and the degree of hardsetting seems to increase with increasing clay content.

B. Occurrence in Other Parts of the World

One way of gaining some idea of the likely extent of hardsetting soils in other countries is to correlate the hardsetting soils identified by the Factual Key with groups or units of other soil classifications. In doing so, it should be understood that some inconsistencies will occur because although classifiers may use similar soil properties in their classifications, these may be used at differing levels within the different classifications. Therefore correlations between classifications can only be approximate. Table 4 sets out some such correlations. The large collection of great soil groups is to be expected, since the hardsetting property (along with other properties) had not been recognized when these groups were developed.

A much closer relationship exists, however, with soil taxonomy, while the World Soil Map units again show a broader scatter. The reasons may

Table 4. Approximate correlations between Australian hard, friable, and sandy duplex soils, Great Soil Groups, Soil Taxonomy, and World Soil Map units

Factual Key	Great Soil Groups[a]	Soil Taxonomy[b]	World Soil Map[c]
Hard duplex soils: Dr2, Dr3, Db1, Dy2, Dy3, Dd1, Dd2, Dg2	Podzolic soils, non-calcic brown soils, red-brown earths, soloths, solodized solonetz and solodic soils	Alfisols	Luvisols, planosols, solonetz
Friable duplex soils: Dr4, Db3, Dg4	Chocolate soils, euchrozems, red podzolic soils, ?humic gleys	Mollisols, inceptisols, ?ultisols	Luvisols, phaeozems, acrisols, planosols
Sandy duplex soils: Dy5	Podzolic soils, soloths, solodized solonetz and solodic soils	Alfisols	Planosols, solonetz, luvisols

[a] Stace et al., 1968.
[b] Soil Survey Staff, 1975.
[c] Dudal, 1970.

become clear from the definitions used in these classifications. Alfisols, for example, should correlate well with hard duplex soils because clause 2 of the alfisol definition states: "Have an epipedon that is both massive and hard when dry *or have an aquic, udic, ustic or xeric moiture regime.*" Since an epipedon is a special name for the surface soil, the first part of this definition is straightforward; and correlations with soils identified by other classifications in which the condition of the surface soil is spelt out in a similar manner are to be expected. Thus, as shown in Table 4, alfisols equate well with hard duplex soils. However, the second part of clause 2 of the alfisol definition, with its almost all-embracing rider of "moisture regimes," invoking all but one such regime, the "aridic and torric" (both are terms used for one moisture regime), tends to negate the first part of clause 2! This, of course, is the reason why sandy duplex soils correlate also with alfisols. As shown by the Factual Key, there are great similarities between sandy duplex soils, Dy5, and some hard duplex soils, Dy3, the difference being the hardsetting surface soil in the latter. However, both may be natrixeralf or natriustalf, depending on whether the actual soils are in the southern or northern parts of Australia, respectively. The problem with clause 2 is the use of "or"; if an "and" had been used, such confounding of different soils within one great group would not happen. In the case of the units of the World Soil Map, correlations are less sharp; again the reasons are to be found in definitions and the structure of their key. It is only in the definition of chromic luvisols that an "A horizon . . . which hardens when dry" is mentioned specifically. However, these soils have "strong brown or red argilluvic B horizons," which means that hard duplex soils with other than red or brown clay B horizons, that is, the Dy, Dd, and Dg soils, must be placed elsewhere, usually in planosols or solonetz. Really, the World Soil Map units as a soil classification are little better in concept and description than were the older Great Soil Groups. Again, many World Soil Map units are stated as having umbric or ochric A horizons, which means that hardsetting soils may be present in a range of these soil units since hard and massive conditions when the soil is dry may be found in both such epipedons [Soil Survey Staff, 1975]. Nevertheless, it is clear that hardsetting soils are recognized by both Soil Taxonomy and World Soil Map units to some extent. Thus it would be reasonable to say, for example, that all areas described on the World Soil Map as luvisols, planosols, or solonetz would contain some hardsetting soils, or those potentially so, and similarly so for alfisols on maps using Soil Taxonomy.

The presence of luvisols, planosols, and solonetz, as shown by the World Soil Map, is set out very broadly in Table 5. Although specific reference to soil hardening is made in only one area, namely, for the chromic luvisols of South America, Table 5 does suggest that hardsetting in soils may well be extensive throughout the world. This does not necessarily imply that all these soils should be classified together even by World Soil Map units. For example, chronomic luvisols of the Mediterranean region are unlikely to

Table 5. Distribution of the hardsetting soils—luvisols, planosols, and solonetz—as reported by the World Soil Map

W.S.M. areas	Luvisols	Planosols	Solonetz
North America	Common to widespread	Minor—not known in Canada	Minor
Mexico and Central America	Widespread	Minor	Very minor—not dominant in any area
South America[a]	Widespread, particularly in northeastern Brazil, central Chile, Columbia, and Bolivia	Major areas in Argentina and Brazil	Major areas in Argentina
Europe	Widespread, but especially common in Mediterranean areas	Minor but notable in Spain, Yugoslavia, and Romania	Minor but notable in Hungary
Africa	Widespread	Widespread	Common in some areas
South Asia	Widespread, particularly in southern India, Israel, Lebanon, Syria	Minor	Widespread, particularly in northern India and Syria
North and Central Asia	Common to widespread, e.g., eastern China	Minor—many scattered areas	Common, e.g., central Asia
Southeast Asia	Widespread, but in relatively small areas	None	None
Australasia	Widespread, particularly in Australia	Widespread, particuarly in Australia	Common, particularly in Australia

[a]Specific mention is made of the hardening of some soils, the chromic luvisols, upon drying.
Source: FAO-UNESCO (1974).

exhibit the extreme of texture contrast shown by the chromic luvisols (hard red duplex soils) of Australia. This is another issue, however; the main one in the current context is that a greater range of soils than was previously thought may be hardsetting to a greater or lesser degree. Significantly, the World Soil Map reports frequently refer to the prevalence of soil erosion throughout these areas. Soil erosion often accompanies hardsetting soil conditions.

C. Toward an Agronomic Definition of Hardsetting

Although existing definitions of hardsetting have been sufficient to allow mapping of hardsetting soils in Australia, if more attention is now to be focused on the agronomic aspects of hardsetting, it is appropriate to review the definition from the point of view of cultivation and of the growing plant.

1. What Is an "Apparently Apedal" Condition?

Even in soil dominated by a massive structure, it is often possible to find occasional but infrequent vertical cracks. Where the roots of perennial plants (e.g., eucalyptus trees) are growing in a hardsetting horizon, it is also possible to find some cracks that are associated with large roots. However, from the point of view of annual plants, it is clear that if there is hardly any structure visible in the dry profile and no evidence of roots growing down fine cracks, then the soil is structureless from the point of view of root growth. Root growth can then proceed only when the soil is weak (and therefore wet) enough to permit it.

2. How Hard Is "Hard"?

Any definition of hardsetting that concentrates on conditions at the soil surface may fail to distinguish between the hardness of a surface crust in a soil that has not set hard beneath a crust and genuine hardsetting. It is therefore more appropriate to concentrate on soil conditions at and below seed depth (e.g., 0.04–0.1 m), where hardsetting may cause additional problems to those associated with surface crusting. It is desirable to know the matric potential at which the soil attains a value of penetrometer resistance that is critical to root growth (say, 3 MPa), but this cannot be easily obtained in practice.

It is therefore appropriate to concentrate on a simple field estimate of tensile or shear strength that can be made by breaking a sample of known geometry between the fingers, underfoot, or in a simple portable unconfined "compression" testing apparatus (e.g., Chandler and Stafford, 1987). To be meaningful, such a test should not be performed on clods obtained after cultivation, since these can weaken as a result of wetting and drying in their unconfined state, but on small samples carved from the subsurface soil. Crushing a cubic block of soil is also unsuitable, because the sample is sometimes liable to fail in a tensile mode and, where shear failure does occur, the failure surface will pass through the ends of the sample. However, it is acceptable either to fail a cylinder of soil between its ends in order to measure unconfined compressive strength (of a cylinder whose diameter must be $\geq$2 times its length), or on its side to measure tensile strength (e.g., Ley et al., 1988).

In time, the results from such tests should provide a quantitative scale on which to rank the relative magnitudes of the strengths of air-dry, hard-set soils. In the meantime, the definition, soil "not disturbed or indented by pressure of forefinger" (McDonald et al., 1984), when applied to the air-dry profile face 0.1 m below the soil surface, should provide an acceptable strength criterion for hardsetting.

IV. Management and Amelioration of Hardsetting Soils

A. The Need for Sustainable Soil Management Systems

A sustainable agricultural system is a pattern of cropping and soil management that is capable of maintaining steady or increasing yields (with improved varieties, greater fertilizer input, etc.) for the foreseeable future (i.e., 50 years plus).

It is implicit in this definition that levels of soil erosion significantly greater than those under virgin vegetation or pasture are not consistent with sustainable agriculture, although the deleterious effects of erosion on yield may take a considerable time to become apparent. This point may best be understood with reference to Figure 4, which is divided into three phases:

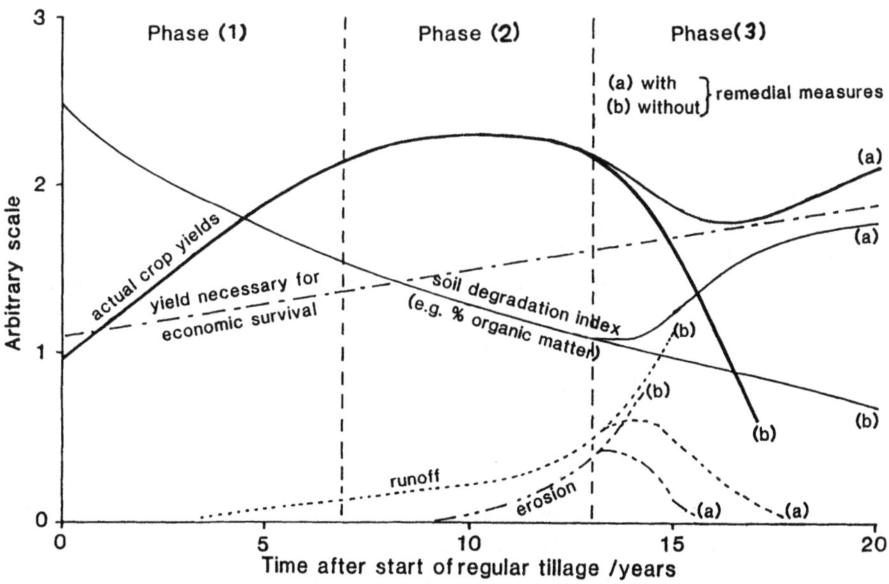

Figure 4. Hypothetical diagram to illustrate the concept of a sustainable agricultural system on soil vulnerable to degradation (time scale based on hardsetting soil behavior in northern New South Wales, Australia).

1. *The learning phase.* In the learning phase, there is a rapid increase in yields as the farmer learns from experience the most effective way to manage the soil for short-term profit. The overall yield necessary for economic survival also increases in line with the costs of any intensification (new varieties, fertilizers, equipment, herbicides, and pesticides).
2. *The warning phase.* In the warning phase, soil degradation has proceeded suffciently to start to seriously reduce the potential yield. However, the effects may become apparent only in certain years, and, if the weather is favorable during this phase, the farmer may be unaware of the progressive decline in soil physical properties. This highlights the importance of monitoring soil physical properties as well as yields. If the farmer has not taken sufficient remedial measures by the end of this phase, it may become impossible or uneconomical to restore the land to a sustainable cropping system.
3. *The equilibrium phase.* Assuming that it is possible to take sufficient remedial action, the equilibrium situation represents the achievement of a sustainable agricultural system.

The time scale in figure 4 is based on experience in northern New South Wales, Australia, but can be shorter or longer depending on the fragility of the soil, the aggressiveness of the climate, and the disparity between the actual management system and that required. Erosion is not necessarily the final cause of the land becoming unusable. Where rainfall erosivity is not high, management difficulties and low and erratic yields associated with hardsetting may be a sufficient cause alone.

B. The Australian Experience

1. Background

In southern Australia, crops are grown traditionally after fallow or in rotation with a legume pasture. With fallow, wheat (*Triticum aestivum*) or other cereals are sown after a long (8 to 10 months) to short (1 to 5 months) tilled fallow. With long fallow, tillage begins when the weather allows, usually after January 1. Farmers use long fallow to conserve water, to mineralize nitrogen, and to control weeds and diseases. Short fallow is used mainly in southeastern Australia, where summer and autumn rain is more common. There is no standard long or short tilled fallow, but the land is usually disked, scarified several times, then tilled lightly before the crop is sown into the bare soil.

With rotations, one or two crops usually follow one or more years of subterranean clover (*Trifolium subterraneum*) or annual *Medicago* species. The only fallow in this rotation lasts for only a few weeks before the next crop is sown.

With rainfed crops on the northwestern slopes of New South Wales, wheat is the main crop, in rotation with barley (*Hordeum vulgare*), sorghum (*Sorghum vulgare*), and sunflowers (*Helianthus annuus*). For maximum storage of water, the fallow begins right after the previous crop has been harvested.

Because tillage damages soil structure and increases the risk of erosion, and new herbicides are available to control weeds, reduced tillage or direct drilling is gradually replacing traditional tilled fallow. Reduced tillage is any system with far fewer operations than traditional tilled fallow. This may be as little as one tillage, with herbicides before and after the crop is sown. In direct drilling, the implement (e.g., a disk drill, slot seeder, modified combine, or rotary hoe) drills the seed into previously undisturbed soil. Disturbance of the soil by these implements ranges from shallow slots to a fine tilth around the seed; the stubble may or may not be burned.

With such a wide range of systems it is not surprising that responses in soil and crop have been inconsistent, and often apply only to a particular district or even year.

2. Effect of Soil Management on Plant Growth and Yield

Research in southeastern Australia has often shown higher yields with long fallow than with short fallow, but the value of either depends on the rainfall that year. Where the seasonal rainfall was less than 440 mm, and enough extra water was stored, long fallow usually increased yields (French, 1978). In red-brown earths in South Australia, long fallow stored on average an extra 55 mm of water, and at Lockhart, in New South Wales, an extra 35 mm of water (French, 1978; Mason and Fischer, 1986). Higher yields from fallow have also been attributed to better preparation of the seedbed, better control of weeds or nematodes, or higher soil nitrogen.

In a rotation trial that ran for many years in the Victorian Mallee, the yield of all rotations increased over the first 20 years from 1.2 t ha^{-1} in year 1 (Elliott and Jardine, 1972). This was attributed to increased mineralization of nitrogen due to tillage. In this trial, however, eventually a long fallow every second year led to lower yields than a long fallow every third

Table 6. Effect of management of Urrbrae fine sandy loam on soil physical properties and on growth of wheat

Property	Depth (mm)	Days after crop sown	Tillage	Direct drilling
Water content (kg/kg dry soil)	50–100	8–10	0.157	0.158
Penetrometer resistance (MPa)	50–100	8–10	0.3	2.4
Bulk density (g cm^{-3})	20–100	25–35	1.2	1.6
Total root length (mm $plant^{-1}$)	—	15–18	488	330
Total dry matter (g m^{-2})	—	At harvest	975	1,004

Source: Whiteley and Dexter (1982).

Table 7. Effect of soil management on yield of wheat

Age of experiment (yr)		Yield (t ha^{-1})				Region	Reference
		Traditional tillage	Reduced tillage	Direct drilling[a]	Direct drilling[b]		
1		2.4	—	2.6	—	Rutherglen, Vic.	Reeves and Ellington (1974)
3		1.9	—	2.9	—		
8	−N[c]	0.6	0.7	0.8	0.8	Wagga Wagga, NSW	Osborne et al. (1978)
	+N	1.5	1.7	1.5	1.6		
1		1.7	—	1.8	1.8	Avondale, WA	Hamblin and Tennant (1979)
1		3.0	2.3	—	2.6	Canberra, ACT	Gates et al. (1981)
1		2.8	—	2.5	2.5	Lockhart, NSW	Burch et al. (1986)
3		4.3	—	3.7	3.7		

[a] Soil fully disturbed 0–5 cm depth.

[b] Soil partly disturbed 0–5 cm depth.

[c] −N = no N fertilizer added; +N = 80 kg ha^{-1} N fertilizer added each year.

or fourth year. With long fallow every second year, the yield reached a peak (1.6 t ha^{-1}) at year 20 and by year 28 had dropped back to 1.2 t ha^{-1}, possibly due to loss of organic matter and structural decline of the soil. With a long fallow every third or fourth year, the yield at year 20 was 1.9 to 2.0 t ha^{-1}, and by year 28 was 1.9 to 2.3 t ha^{-1}.

Often, direct-drilled crops grew more slowly early in the season than those tilled traditionally (Table 6; Reeves and Ellington, 1974; Chan et al., 1987; Murphy et al., 1987). The reasons given for this included lower N mineralization and nutrition, lower soil temperature, less water available, mechanical impedence of roots or emerging shoots, waterlogged soil, poor control of weeds, toxins from decomposing organic residues, soil-borne diseases, length of experiment, or rainfall. Even various cultivars responded differently to different systems. Sowing was sometimes delayed under tillage because the soil was too wet, or under direct drilling because the soil was too dry, reducing the yield in either case. However, in spite of poor early growth, grain yields have differed little from traditional systems (Table 7).

3. Effect of Soil Management on Soil Properties

a. Organic Matter and Aggregate Stability

During the fallow phase, few plant residues are added and the soil is tilled often, so more organic materials are oxidized than in nontilled soil (Rovira and Greacen, 1957). In nontilled pasture soils, plant materials are added continually to the soil, leading to a high percentage of organic carbon (Tisdall and Oades, 1980). So after a few years the levels of organic carbon were usually lower in soils under traditional tillage than in pasture, direct-drilled soils, or those under reduced tillage, especially where stubble was retained (Table 8). At Avondale, West Australia, in the sixth year, soil under traditional tillage contained 1.3% organic carbon, reduced tillage

Table 8. Physical properties of a red-brown earth at Lockhart, New South Wales, after three years of soil management

Treatment	Organic carbon (g/kg^{-1})		Aggregate stability (%)[a]	
	0–50[b]	50–100[b]	0–50[b]	50–100[b]
Traditional tillage	20	15	38	26
Reduced tillage	21	16	42	34
Direct drilling	20	15	46	32
Direct drilling stubble retained	21	16	48	28

[a] >0.2 mm diameter.

[b] Depth (mm).

Source: Burch et al. (1986).

contained 1.5%, and direct-drilled soil contained 1.6% (Hamblin, 1987). At the Permanent Rotation Trial at the Waite Agricultural Research Institute, Adelaide, the soils under pasture contained >2.3% organic carbon, soils with fallow in the rotation contained <1.6%, with other tilled plots between these two extremes (Tisdall and Oades, 1980).

The concentration of organic carbon in hardsetting soils is usually related to the water-stable aggregation (Table 8; Grierson et al., 1972; Hamblin, 1984). In Urrbrae fine sandy loam at the Waite Institute, for every 0.1% increase in organic carbon there was a 2% increase in stable macroaggregates (>250 μm diameter); the activity of roots and hyphae stabilized these macroaggregates (Tisdall and Oades, 1980). In irrigated soils in northern Victoria, with extremes of wetting and drying in summer, the water-stable aggregation (>250μm diameter) for soils under old pasture and tomatoes was 90% and 58%, respectively; organic carbon was 1.8% and 1.5%, respectively (Adem and Tisdall, 1984). When organic residues are added, the soil should be kept moist for long enough to allow microorganisms to produce stabilizing substances (Tisdall et al., 1978). The soil should also suit earthworms and other fauna, which are needed to mix organic residues intimately with the soil (Barley and Kleining, 1964; Tisdall, 1978).

As the structure declines over several years of traditional tillage, the soil often becomes more difficult to work and to produce a good seedbed, leading to poor contact with seed and soil water and to reduced emergence. In a red-brown earth at Merredin, Western Australia, in the sixth year of traditional tillage, the slaking/dispersion index of the soil (0 to 10 cm depth) was 13.7, and under direct drilling was 9.9 (Hamblin, 1984). The cloddy, drier seedbed under traditional tillage delayed the emergence of wheat seedlings, reduced emergence by 40%, and delayed maturity by 3 weeks, compared with direct drilling.

Direct drilling, however, especially with organic residues retained, may increase the incidence of fungal disease or phytotoxicity (Kimber, 1973). Rovira and Venn (1985) showed, in South Australia in a direct-drilled sandy loam with little decomposition of organic residues over summer, that the higher incidence of Rhizoctonia root rot accounted for 70% of the variability in early growth of wheat. In the same way, direct drilling may increase the incidence of disease in hardsetting soils.

b. Soil Water

In excessively tilled soil, crusts with low permeability (Table 9) often formed at the surface due to heavy rain or flood irrigation, and reduced infiltration and seedling emergence severely. The finer the tilth, the more severe was the crust (Braunack and Dexter, 1988). Crusts were also severe if free water accumulated at the surface, because dispersed soil dried more quickly than flocculated soil, and was denser and harder, with no large

Table 9. Hydraulic conductivity (K_s) of soil after simulated rainfall of 33 mm in 30 min

Soil	K_s (m s^{-1})
Clay skin	5×10^{-9}
Disaggregated surface layer	5×10^{-8}
Bulk soil below	1×10^{-5}

Source: D.S. McIntyre, Permeability measurements of soil crusts formed by raindrop impact, *Soil Science*, 85:185–189, © by Williams & Wilkins, 1958.

Table 10. Effect of soil management on ponding and runoff from simulated rainfall at 56 mm h^{-1} for 40 min

Treatment	Time to ponding (s)	Runoff rate (mm h^{-1})
Traditional tillage	141	27.9
Direct drilling[a]	171	9.3
Direct drilling[b]	235	8.2

[a] Soil fully disturbed 0–5 cm depth.

[b] Soil partly disturbed 0–5 cm depth.

Source: Burch et al. (1986).

pores to a depth of about 3 mm (Emerson, 1977). The crusts not only restricted infiltration, but also heavy and persistent rain could keep the soil at the surface saturated for several weeks, and the poorly aerated crust sometimes reduced the germination, emergence, and early growth of wheat (McIntyre, 1958a; Millington, 1959). The more tillage during each cycle of the rotation, the more severe were the effects. For example, permeability of the crusts under wheat-fallow was much lower than under pasture-pasture-fallow-wheat. Surface crusts can also trigger erosion on sloping land; the increased runoff can erode the surface soil, then the subsoil, and then start forming gullies.

In comparisons of different systems of tillage, the infiltration of simulated rainfall was greater (and runoff lower) the less the soil had been disturbed (Table 10; Hamblin, 1984; Murphy et al., 1987; Hamblin, 1987). By the sixth year at Merredin, Western Australia, after rain, the profile wetted more deeply (usually 4–6 cm deeper) and into the top of the B horizon, under direct drilling and reduced tillage, than into the poorly structured soil under traditional tillage (Hamblin, 1984). At Cowra, New South Wales, in the sixth year, more water was stored and used throughout the profile under direct drilling than under traditional tillage (Murphy et al., 1987). Between flowering and harvest under direct drilling, roots extracted water from twice the depth (80 cm) than those under traditional tillage

(40 cm), suggesting deeper roots under direct drilling. The higher growth rate under direct drilling at the end of the season was attributed to more water being available to the plants.

After two years of traditional tillage at Cowra, New South Wales, almost all the simulated rain (19.8 mm in 1 h) that infiltrated was retained in the top 200 mm, whereas in the direct-drilled plots, 28% of simulated rain infiltrated below the 200-mm depth (Mead and Chan 1988). This was related to the number of macropores continuous to the suface. Ponded infiltration with dye water showed that, at 120 mm depth, the soil under traditional tillage contained half the number of macropores (>0.5 mm in diameter), with 8% stained, than soil under direct drilling, with 55% stained (Chan and Mead, 1988, personal communication). In such soil, the matrix may be wetted from the faces of the continuous pores rather than from the soil surface (Smettem, 1986).

Tillage often reduced faunal populations in soil, by removing organic residues and hence destroying the food supply and hastening drying of the soil (Tisdall, 1978). This in turn may reduce the number of macropores continuous to the surface, and hence reduce infiltration and aeration. At the Waite Institute, there were fewer earthworms in tilled soil than in soil under pasture (Barley, 1959). At several sites in Western Australia, tillage almost completely eliminated animals such as termites, ants, and beetles, reducing the number of burrows and infiltration, compared with the nearby virgin soil (Abbott et al., 1979). Rovira et al. (1987) found in a red-brown earth at Kapunda, South Australia that three years of traditional tillage halved the number of earthworms (3.6 per 200 cm^2) compared with direct drilling (6.6 per 200 cm^2), although it did not change the number of biopores (>2 mm diameter; 10 cm depth). This may change with time.

In irrigated soils in northern Victoria, under direct drilling (with added cow manure), there were 51 earthworms per square meter, with four times the number of biopores (2 to 10 mm diameter, 50 mm depth) and four times the rate of infiltration compared with tilled soil (no added cow manure), with 6 earthworms per square meter (Tisdall, 1985).

By compacting the soil, grazing animals also reduced infiltration of traditionally tilled soil and soil under reduced tillage (Burch et al., 1986).

c. Porosity and Mechanical Resistance

Soon after tillage, the porosity and macroporosity were higher and the soil softer than in soils under reduced tillage or in direct-drilled soils (Murphy et al., 1987; Mead and Chan, 1988; Table 6). However, Murphy et al. (1987) showed that soils under traditional tillage slumped during the growing season. At sowing, the bulk density of direct-drilled and traditionally tilled soil was 1.2 t m^{-3} and 1.0 t m^{-3}, respectively. Sixty days later, after a total of 76 mm rainfall, the bulk density of the direct-drilled soil was still

1.2 t m^{-3}, whereas that of the traditional tillage had risen to 1.6 t m^{-3}. Also, at Cowra on a red-brown earth, the bulk density before tillage was 1.6 t m^{-3}, but the 100-mm-deep, freshly tilled soil (bulk density 1.2 t m^{-3}) slumped by 21 mm to a bulk density of 1.7 t m^{-3} after simulated rainfall of 19.8 mm over 1 h (Mead and Chan, 1988).

The increased faunal activity in direct-drilled or pasture soils often increases the number of macropores continuous to the surface (see previous section), leading to higher air-filled porosities soon after rain than in tilled soils. For example, at Cowra, 1 h after simulated rain, the air-filled porosity in the 0 to 50 mm and 50 to 100 mm depths of traditionally tilled soil (14.8% and 9.3%, respectively) was lower than that in direct-drilled soil (19.9% and 17.9%, respectively) or in pasture soil (29.4% and 25.2%, respectively). At 24 h in all treatments, air-filled porosities at 0 to 50 mm depth were similar, with a mean of 25.3%. However, at 24 h the air-filled porosity at 50 to 100 mm depth in the traditionally tilled and direct-drilled soil (mean of 17.8%) was still well below that of the pasture soil (26.1%) (Chan and Mead, 1988).

The increased faunal activity in direct-drilled or pasture soils often increases the number of macropores continuous to the surface (see previous section), and the soils drain more quickly than in tilled soils. For example, at Cowra, the air-filled porosity after 1 h of simulated rain at 0 to 50 mm depth in traditionally tilled soil was 14.8%, in direct-drilled soil was 19.9%, and in pasture soil was 29.5%; at 50 to 100 mm depth it was 9.3%, 17.9%, and 25.2%, respectively. At 24 h the 0 to 50 mm depth of the three soils had drained, but at 50 to 100 mm depth the air-filled porosity in traditionally tilled soil was 17.9%, in direct-drilled soil was 17.6%, and in pasture soil was 26.1% (Chan and Mead, 1988).

Although tilled soils slump under heavy rain, machinery or grazing animals also compact, and increase the strength of, soil under pasture or crops (Willatt and Pullar, 1984; Kelly, 1984; Burch et al., 1986; Makin, 1987). Makin (1987) estimated that during one operation, 15% to 55% (depending on equipment) of a paddock at Werribee, Victoria, was covered by wheeltracks. At Rutherglen, Victoria, the passage of a tractor (load 9200 kg) over a red-brown earth (freshly tilled to 300 mm depth) compacted, and increased the strength of, soil down the profile; for example at 75 mm depth, the bulk density increased from 1.2 t m^{-3} to 1.4 t m^{-3}, and the penetrometer resistance from 0.2 MPa to 1.7 MPa. Subsequent tillage simply filled the wheel rut without loosening the profile.

Some hardsetting soils become much stronger as they dry, whether they are tilled or not. For example, the strength of the red-brown earth at the Waite Institute doubles with every 2.5% decrease in gravimetric water content (Dexter, 1988). This was not so with a red sandy clay loam at Avondale, Western Australia (Hamblin and Tennant, 1979). But even at the same water content, the less disturbed the soil, the stronger was the soil (Table 6). In the soil at Avondale at 10 cm depth, at a matric suction of 0.5 MPa, the penetrometer resistance of direct-drilled soil was 3.0 MPa, reduced tillage was 2.6 MPa, and traditional tillage was 2.0 MPa.

The strength may also vary seasonally. At Cowra, at sowing, the penetrometer resistance for traditional tillage and direct-drilled was 20 kPa and 50 kPa, respectively. Once the soil had slumped after 76 mm rainfall, the penetrometer resistance was 120 and 80 kPa, respectively (Murphy et al., 1987).

d. Temperature

Soil temperature has not often been measured under different systems of management. However, at Murrumbateman, New South Wales, the soil at any depth of the traditionally tilled soil was generally warmer during the day and cooler during the night than the soil at the same depth in the reduced tillage or direct-drilled soil +straw (Aston and Fischer, 1986). For example, before the crop was sown, the maximum soil temperature in direct-drill was at the surface 12°C lower, at 1 cm depth 8°C lower, and at 5 cm depth 4°C lower, than in the traditionally tilled treatment. The higher early growth of wheat in the traditionally tilled treatment was attributed to the differences in temperature.

With irrigated crops in summer at Tatura, Victoria, the maximum temperature recorded in soil at 3 cm depth was 32°C in mulched and 41°C in unmulched soil (Tisdall and Adem, 1986a).

4. The Effect of Gypsum

Gypsum, which is cheap, readily available, and slowly soluble, applied to some hardsetting soils or in the irrigation water, has improved the physical properties of the soil, increased seedling emergence, plant growth, and yields, and improved drainage and leaching (Table 11; Matheson, 1969; Rudd, 1974; Loveday, 1981). Gypsum also increased the number of tillers per plant, and the uptake of nutrients (Table 11).

Gypsum increased emergence by reducing the throttle of strong dispersed soil at the surface, and by slowing down the drying of soil around the seed (Loveday and Scotter, 1966; Grierson, 1978; Loveday, 1981).

Table 11. Effect of gypsum on response of wheat grown in a sodic sandy loam

Gypsum	Emergence (%)	Mean day of emergence	Number of tillers	N ($g\ kg^{-1}$)[a]	P ($g\ kg^{-1}$)[a]
0	76	10.2	4.5	6.2	0.28
$8\ g\ kg^{-1}$	84	8.2	6.7	8.2	0.44

[a] $g\ kg^{-1}$ of N or P in plant tissue.

Source: Shanmuganathan and Oades (1983b).

Gypsum also increased water-stable aggregation, macroporosity, hydraulic conductivity, depth penetrated by water; reduced soil strength; increased percent available water; and made the soil easier to work (Scotter and Loveday, 1966; Loveday, 1981; Chartres et al., 1985). For example, in a sandy loam with exchangeable sodium percentage (ESP) = 9, when compared with no gypsum, gypsum (0.8% w/w) reduced dispersion index from 5 to 1, increased hydraulic conductivity from 3.1 cm h^{-1} to 4.7 cm h^{-1}, and increased friability by 66% (Shanmuganathan and Oades, 1983b).

The direct effects of gypsum do not persist once calcium is leached from the surface layers (Greene and Ford, 1985; Grierson, 1978; Chartres et al., 1985). Micromorphological sections showed that gypsum applied to sodic red-brown earths in northern Victoria at both 5 and 15 t ha^{-1} decreased clay dispersion and formation of crusts, and prevented the collapse of macropores formed by tillage. As rain leached gypsum from the surface layers (at 1 t ha^{-1} per 125 mm to 360 mm of rainfall, depending on the rate of gypsum applied), the macroporosity decreased and crusts formed (Greene and Ford, 1985; Chartres et al., 1985). Hence gypsum was needed periodically.

However, at each irrigation of lucerne in a red-brown earth at Tatura, Victoria, where gypsum had been applied three years earlier, and had been leached from the upper 0.3 m depth, water penetrated deeper into the subsoil than in soil without added gypsum. This was attributed to stabilization of the soil pores by roots and soil organisms, possibly aided by gypsum (Taylor and Olsson, 1987).

Not all hardsetting soils responded to gypsum (Loveday, 1981). Grierson (1978) found that only 4 of 10 such soils responded to gypsum. In strongly sodic soils (ESP > 15), gypsum improved soil structure and yield, whereas with a sodic soil with ESP = 8 and exchangeable magnesium percentage (EMgP) = 28, gypsum improved structure only. High concentrations of organic matter, covered soils, direct drilling, neutral pH, low concentrations of organic molecules adsorbed on the clay particles, severe drying, and calcium carbonate may each reduce the response of a soil to gypsum (Emerson, 1983; Rengasamy et al., 1984). Also, some soils that slake with little dispersion still set hard or form crusts (Rengasamy et al., 1987). The advantage of gypsum may also depend on the season (Loveday, 1981). In years when the rainfall is lower than average, the increased emergence of seedlings due to gypsum may lead to insufficient available water later in the season. For irrigated summer crops, gypsum may increase yield more in drier seasons, when increased infiltration and stored water become more effective.

At first it was believed that gypsum improved the physical properties of the soil by exchanging Ca^{2+} for Na^{+} on colloids, since Ca^{2+}-saturated soils were often more stable and more permeable than Na^{+}-saturated soils (Table 12). In many cases dispersion decreased as ESP was decreased when gypsum was added to sodic hardsetting soils. Also, ESP explained

Table 12. Effect of exchangeable cation on infiltration rate after simulated rain (1 mm s^{-1}) for 16 min

Soil	Mean infiltration rate (mm s^{-1})
Natural	1.6
Ca^{2+}-saturated	1.3
Na^{+}-saturated	0.1

Source: C.W. Rose, Some effects of rainfall, radiant drying, and soil factors on infiltration under rainfall into soils. *J. Soil Sci.* 13:286–298, © Blackwell Scientific Publ., Ltd., 1962.

54% of the variance in the strength of hardsetting soils in West Australia (Aylmore and Sills, 1982).

The "gypsum requirement" of a soil aimed for Ca^{2+} to replace Na^{+} completely or to a given level over a given depth. However, the applied Ca^{2+} did not always exchange completely to the desired depth. With gypsum applied at 12.5 t ha^{-1} to a transitional red-brown earth, only 40% to 50% of the Ca^{2+} exchanged in the top 30 to 40 cm, with 25% of the total Ca^{2+} exchanging for Na^{+} and 25% exchanging for Mg^{2+} (Loveday, 1981). Greene and Ford (1985) showed that the Ca^{2+} exchanged with more of the Na^{+} than with the Mg^{2+}: For example, in the 10 to 25 cm layer, 15 t ha^{-1} gypsum halved the ESP, but the EMgP changed little.

However, even Ca^{2+}-clays, if sheared when wet (e.g., under heavy rain or animals), disperse without electrolyte (Rengasamy, 1982). A threshold concentration of electrolyte, depending on the ESP, is needed to flocculate clays and increase permeability (Loveday, 1981). A moderate dressing of about 5 to 7.5 t ha^{-1} may dissolve enough gypsum to increase the concentration of electrolyte to about 8 to 12 m equiv l^{-1}, enough to keep soils of ESP up to about 20 permeable (Loveday, 1974). After the equivalent of 442 mm rainfall (about the mean annual rainfall in South Australian wheat areas), 0.2% gypsum was needed each year to keep a high enough electrolyte concentration to stabilize a sandy loam; higher rates would have wasted gypsum without improving the soil or crop (Shanmuganathan and Oades, 1983b).

Rengasamy et al. (1984) developed a scheme to distinguish between the effects of ESP and the concentration of electrolyte, and hence the amount of gypsum needed to stabilize a soil. They classified the surface layers of both rain-fed and irrigated red-brown earths, and their management, according to their dispersion, sodium adsorption ratio or SAR (about 50% ESP), and total cation concentration (TCC) in 1 : 5 supensions in water (Figure 5). According to this scheme, soils that disperse spontaneously ($SAR > 3$) disperse even when direct-drilled or sown to pasture. It is mainly ESP that controls the stability of these soils, so Ca^{2+} must reduce the ESP with enough electrolyte to stay flocculated. Soils that are potentially dispersive, with $0 < SAR < 3$, also need enough electrolyte to stay

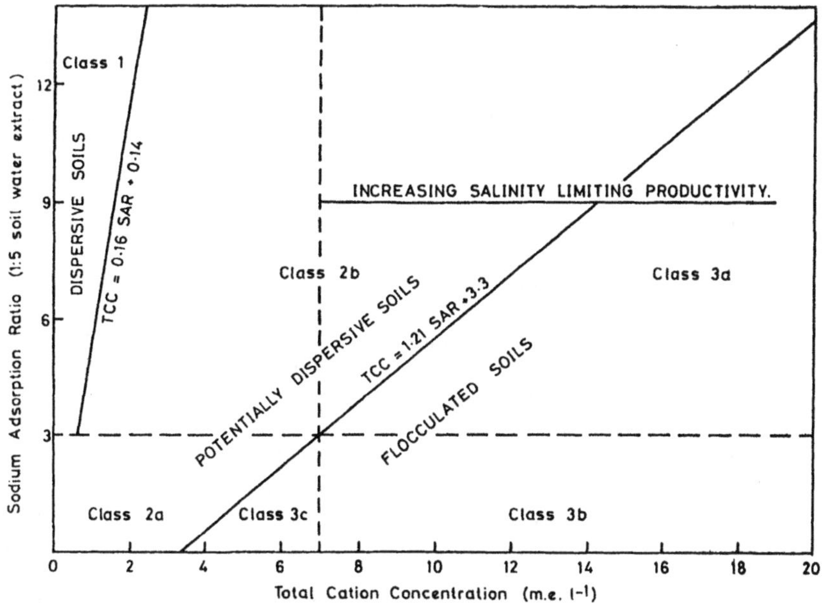

Figure 5. A classification scheme for the prediction of dispersive behavior of horizons of red-brown earths. (After Rengasamy et al., 1984.)

flocculated; however, it would be inefficient and probably not possible to remove all the exchangeable Na^+ from the clay. Potentially dispersive soils with SAR > 3 need enough Ca^{2+} to increase the electrolyte concentration and to reduce the ESP to a feasible level of <6.

5. Integrated System of Soil Management

Many costly experiments have been run, often for many years, comparing the effects of various machines or systems of management on the physical, chemical, and biological properties of soil and on yield. The machines and systems were often developed from those used overseas, with little thought given to the fragile soils in Australia or to the needs of the crop. One of the benefits of a new system may have been that it saved time, but because large replicated trials were set up, operations were sometimes delayed, penalizing the new system. In spite of many years of research and management, productivity is often well below the potential in Australia (French and Schultz, 1984). Results have been inconsistent even for one site and are unlikely to apply universally.

At Tatura, a system was developed by Tisdall and Adem (1988) based on specified physical properties of soil that do not limit infiltration, germination, or root growth (Table 13). These specifications should apply universally. Properties were also specified for unlimited earthworm and microbial

Table 13. Specifications for properties of soils for each purpose, plus levels measured in the surface soil of a red-brown earth managed under the Tatura system

Purpose	Property	Specification	Levels under Tatura system
Infiltration	Aggregate diameter <0.5 mm	≤6%	5%[a]
Germination	Aggregate diameter 0.5–2 mm[b]	100%	21%
	Matric suction	<24 kPa	25 kPa[c]
Root growth	Macroporosity[d]	≥15%	15%[e]
	Penetrometer resistance	≤1 MPa	<1MPa[e]
	Matric suction	<30kPa	≤40kPa[e]
	Aggregate diameter 1–10 mm	100%	23%[a]
Earthworm activity	Matric suction	>100 kPa for <$\frac{1}{3}$ of time	<40 kPa[e]
	Organic residues	10 t ha^{-1} yr^{-1}	10 t ha^{-1} yr^{-1}
Microbial activity	Organic residues	10 t ha^{-1} yr^{-1}	10 t ha^{-1} yr^{-1}

[a] When beds first set up.

[b] 20% diameter of maize seeds; tilth around seeds.

[c] Around maize seeds one day after sown.

[d] % pores >30 μm diameter.

[e] Late summer, after six years under Tatura system.

Source: Tisdall and Adem (1988).

activity in an irrigated red-brown earth in southeastern Australia. Management is not uniform over the entire field, but each operation is tailored to suit a specific zone of soil for a specific purpose at a specific time (e.g., seedling emergence, movement and storage of water, and traffic). The necessary scientific data, machines, and system were defined and/or developed with the aim of producing and maintaining the desired specification in each zone.

In the first autumn of the new system, a hiller sets up land into beds separated by furrows. A spring-tined cultivator tills the prewetted beds at about the lower plastic limit, to produce a coarse tilth for optimum infiltration, aeration, and root growth. A disk seeder drills a cereal into the beds and furrows. In later years, the cereal is drilled into untilled beds. Once the cereal is harvested, a small rotary hoe ahead of the seeder produces a fine tilth on the sowing line for good contact between the water and the seed of the summer crop. This results in good germination and emergence of the seedlings. The mulch from the previous crop conserves water around the seed and keeps the soil at temperatures below lethal levels in summer. The roots and mulch of the previous crop encourage earthworm activity and microbial production of stabilizing substances. The narrow wheels on all equipment travel on the base of the permanent furrows, and several operations are combined in one pass, which saves time and reduces compaction of the beds. Beds are wetted by capillarity from water confined to 5 cm depth in the furrows to minimize slaking on the sides of the furrow.

In one block managed under this system, 10 crops were grown in six years, and the soil remained stable, permeable, and soft, and physical properties of the soil were close to the specified levels (Table 13). The cumulative yield of multiple crops under the Tatura system was far higher than that from crops under traditional systems, which allow no more than three crops in 3 successive years, followed by at least 10 years of pasture to recover.

6. Conclusion

The many trials comparing the effects of soil management on soils and on yields have been costly and laborious. Yet yields were well below their potential, and many results were inconclusive and show that we do not understand many of the mechanisms of productivity. Further research should determine and eliminate from the soil (including the subsoil) the physical, chemical, and biological constraints to high yields. Such research should include physiology and agronomy of the crop as well as soil science. We should then develop new integrative systems of management, based on such results, which would preserve well-structured soil and lead to high yields.

C. Hardsetting in Other Countries

Because aspects of hardsetting behavior (e.g., restricted root growth, cultivation difficulties) are shared by other soil types, it is difficult to be certain from most reports whether the problem soils they refer to are genuinely hardsetting. In particular, many reports on crusting make no reference to topsoil conditions beneath the crust, and references to structure and bulk density of the surface soil are difficult to interpret when they are not given in relation to the date of the last cultivation and the amount and intensity of any intervening rainfall. Even in the many instances where slumping of the cultivated layer is recorded, observations of its subsequent structure (or lack of it) and strength characteristics are often lacking or insufficiently detailed to allow hardsetting to be definitely identified.

El Swaify et al. (1985), in their review of soil management in the semiarid tropics, show that one-third of this area is occupied by alfisols and point out that the present state of the art does not provide technological mangement packages for optimizing productivity of these soils under rainfed conditions. Alfisols in this region are characterized by a lack of structural development, and cultivation is begun only after the start of the rainy

Table 14. References to soils identified or likely to have a hardsetting upper horizon in countries other than Australia

Country	References	Comment/local name
Zambia	Lenvain and Pauwelyn (1988) Lenvain et al. (1987) Spaagaren (1986)	Hardsetting soils appear to occupy at least 12% of land area
Botswana	Willcocks (1981) Sinclair (1985)	Large areas of cultivated land are apparently hardsetting under current management system
Tanzania	Ley (personal communication)	
Senegal	Charreau and Nicou (1971)	"Dek" soil particularly hardsetting
Gambia	Williams (1979)	
Sudan	Razig (1978)	"Gardud" soils in western Sudan
India	Gupta et al. (1984) Reddy et al. (1985) El-Swaify et al. (1985)	"Chalka" soils occupy a large area in Andhra Pradesh. Some alfisols
Nigeria	Ley (1988) Ley et al. (1988)	Degraded alfisols at IITA, Ibadan, become hardsetting
Brazil	Dregne (1976)	Some alfisols
United States	Cassel (1983)	Norfolk loamy sand (typic paleudult)
United Kingdom	Mullins et al. (1987) Young (1987) Young et al. (1988)	Some degraded Salwick and Efford series soils become hardsetting

season, because the dry soil is difficult to handle. Where powerful implements are available, large, hard clods are usually produced. They also make specific reference to the hardsetting type of behavior of some of these alfisols, where "crusting may extend deeper than the immediate soil surface with the result being a consolidation of the soil profile (slumping and hardening). . . ." Dregne (1976), in *Soils of Arid Regions*, also makes specific reference to the massive structure and hardness on drying of African alfisols, and to the surface hardness of alfisols of northern Brazil.

Table 14 provides some specific examples of some of the more clearly defined cases of hardsetting behavior. In India, additions of waste organic matter have ameliorated soil physical conditions and substantially increased yields of groundnuts, millet, maize, and sorghum in a hardsetting "Chalka" soil (Gupta et al., 1984; Reddy et al., 1985), while in Zambia, commercial high-input farming involving incorporation of chopped maize stalks after harvest has also been successful in ameliorating hardsetting behavior, allowing consistent yields of 6 t ha^{-1} to be obtained. However, there are many areas where, for various reasons (stubble burning, grazing, or low yields), the organic matter returned to the soil may fall far below the amount of 6 to 8 t ha^{-1} y^{-1} that Lal (1987a) suggests is sufficient to improve soil physical properties and control erosion.

Charreau and Nicou (1971) reported hardsetting behavior, characterized by a sharp increase in penetrometer resistance with decreasing moisture content, in sandy textured soils in Senegal. Crop root growth and yield responded positively to tillage of these soils (Nicou and Chopart, 1979). Williams (1979) reports rooting problems, high penetration resistance, and high bulk densities (around 1.6 Mg m^{-3} in nonalluvial soils in Gambia. He notes that all the soils (of the continental terminal) are poorly structured and have hard or very hard consistence when dry.

1. Botswana

Sinclair (1985) reports hardsetting behavior of a range of cultivated sandy textured soils in Botswana, and how it is exacerbated by cultivation of the soil well after the start of the rainy season, when the soil is in a wet condition. The delay here is due to the need for draft cattle to recover sufficient strength from grass regrowth to be used for cultivation (Willcocks, 1981). Willcocks has discussed the undoubted benefits to be obtained in terms of increased crop use of the rainfall by the use of tractors to prepare a seedbed before the start of the rainy season, and has demonstrated that tillage that provided a deeper soil loosening allowed deeper crop rooting. In a dry year in which there was only 233 mm of rainfall, he obtained a significant correlation between cultivation depth and yield of sorghum, with yield varying from 0.5 to 1.5 t ha^{-1} for tillage depths from 0.1 to 0.3 m, respectively. Jones (1986) has discussed the socioeconomic contradictions that currently stand in the way of attempts to improve soil productivity and halt soil degradation in this predominantly cattle-rearing country.

2. Zambia

Lenvain and Pauwelyn (1988) have compared soil physical properties of two paleustalfs (near Lusaka) under identical managements systems, in relation to the growth of a wide variety of fertilized crops, with and without irrigation. Consistently lower yields were found on the soil that was positively identified as hardsetting and had a bulk density of >1.7 Mg m^{-3} within the top 0.25 m, in comparison to the better soil, which had a bulk density of <1.5 Mg m^{-3}. Lenvain et al. (1987) have also observed high soil loss (8 t ha^{-1} under maize) and runoff of ca. 20% averaged over a growing season from the hardsetting soil, which was reported to have been cultivated for the last 40 years. Their management recommendations are for soil loosening after harvest, followed by conservation tillage with mulching to improve surface soil physical conditions and reduce runoff and erosion.

3. Nigeria

Ley (1988) and Ley et al. (1988) have studied an oxic paleustalf at IITA, southwestern Nigeria, in which hardsetting behavior has occurred after clearance of the rain forest with a treepusher/root rake and six years of tillage. Bulk density (at 0.04 to 0.08 m) measured just after harvest (but without wheeling) was 1.6 Mg m^{-3}, in comparison to value of 1.2 Mg m^{-3} in a plot that had been manually cleared and cropped to maize for one season. The six-year-cultivated, hardsetting soil had already suffered considerable erosion, and high runoff delayed profile wetting at the start of the rainy season. This, combined with its high strength, meant that several attempts to plant, both with a jab planter and with a seed drill, were unsuccessful, and seeding was delayed by two weeks in comparison to adjacent, less degraded plots (Ley et al., 1988). On the same soil, Lal (1984a, 1985) has documented the degradation of soil physical properties (10 times increased runoff, 42 times increased erosion, 3 times reduced maximum infiltration rate) and reduction in crop yield that occurred after five years of cropping to maize (two crops/year) with conventional moldboard plow/disk harrow cultivations, in comparison to a no-till regime. By the twelfth consecutive crop, yield was 1 t ha^{-1} on the plowed and 3 t ha^{-1} ha on the no-till plot.

Benificial effects from growing mucuna as a mulch immediately after forest clearance, in comparison to maize, include a higher total porosity of the surface (0–0.1 m) soil and greater infiltration. However, only 3.8 t ha^{-1} of mucuna grew on plots cleared with a treepusher/root rake, in comparison to 8.5 t ha^{-1} ha on manually cleared plots, and this had less of a beneficial effect (Hulugalle et al., 1986).

Other measurements on a similar but more sandy textured soil under maize (sandy loam in comparison to the sandy clay loam studied by Ley) at the same site have generally shown less pronounced yield differences between a no-till with mulch treatment in comparison to a variety of tillage treatments, although moisture reserves were greater in the no-till treatment and yields were progressively less with increasing intensity of mechanical tillage (Lal, 1986).

Adeoye (1986) has followed the increase in bulk density at the surface (0 to 30 and 30 to 70 mm) of a loamy fine sand (typic paleustalf) following heavy rainfall, at Samaru, northern Nigeria. Two weeks after disk harrowing to 0.1 m depth, the bulk densities in three treatments were 1.15 to 1.20 Mg m^{-3}. Over the next 14 weeks, bulk density increased progressively to 1.3 Mg m^{-3} on a mulched treatment but to 1.5 to 1.6 Mg m^{-3} on a treatment that was left bare. During this period there were 21 rainfall events that were >20 mm and had an energy of ≥600 J m^{-2}, and even on the treatment that was hoed at two-week intervals, bulk density rose to 1.4 to 1.5 Mg m^{-3}. These results clearly demonstrate the benefit of a surface mulch, although the strength characteristics of this soil suggest that, because of its low silt and clay content, it is only borderline hardsetting (Ley et al., 1988), and Ike (1986) found little difference in crop yields between no-tilled and tilled treatments on this site.

4. United States

Cassel (1983) has made a detailed study of the behavior of a Norfolk loamy sand (typical paleudult) in the Altantic Coastal Plain of the United States. This soil was moldboard plowed to 0.25 m depth one month prior to planting and then disked three times immediately before the planting of a row crop (soybeans). Bulk density was measured within both the disked zone (0 to 0.14 m) and within the plow depth but below that of secondary cultivations (0.14 to 0.28 m). The bulk density of the upper layer was observed to increase by 13%, to 1.63 Mg m^{-3}, in the first three weeks after planting. This slumping was attributed to the effect of 26-and 39-mm rainfall events, and supporting evidence for this hypothesis was provided by the observation that undisturbed cores wetted up for the determination of their water-release characteristic slumped under their own weight.

In the plow zone, below the depth of secondary cultivations, the bulk density was 1.73 Mg m^{-3} and did not increase significantly in a seven-week period after disking. It is improbable that such a high value could exist after plowing unless the soil was unstable and slumped during plowing. Two further possibilities are either that subsequent rainfall caused slumping prior to the disking operation, or that compaction occurred under the tractor wheels during disking and planting. Cassel has given this latter explanation, although the high bulk densities under the wheelings of the last two diskings and the planting operations (which were all confied to the same tracks) were not significantly higher than under the crop rows or in the untrafficked interrows. Thus slumping of the plow layer may well have contributed to this compaction.

Despite its low content of silt and clay (114 and 36 g kg^{-1}, respectively), the soil at 0.14 to 0.28 m depth had a penetrometer resistance of 14.4 MPa seven weeks after planting when the matric potential at 0.2 m depth was only about −60 kPa. Since the soil was also reported as nearly structureless, it is likely to have become impenetrable to roots well before it reached the wilting point and would classify as hardsetting. Restricted rooting has been identified as a problem in this soil in relation to a plow pan, but Cassel's results clearly indicate that the plow zone itself is also likely to present problems.

5. United Kingdom

The comparison between two sandy loam mollisols in the United Kingdom reported by Young et al. (1988) has already been extensively referred to. It is notable that, in terms of their strength characteristics, these soils had greater strengths at any given potential than the worst IITA soil (Ley et al.,

1988) or any of the Australian hardsetting soils studied by Mullins and MacLeod (unpublished results, 1984). However, under United Kingdom conditions, with frequent but low-intensity rainfall before and during the growing season, the agronomic limitations that might be crucial under more aggressive climatic conditions imposed significant but comparatively minor constraints on management and yield.

D. The Influence of Soil Fauna

Lal (1987b) has recently made an extensive review of the occurrence of soil fauna in the tropics, their influence on soil physical and chemical properties, and their relation to soil management. This work contains separate chapters on earthworms, termites, and ants, and except where references are given, this section summarizes the parts of this review that are relevant to physical properties of hardsetting soils.

1. Earthworms

There are three types of earthworm, the litter-dwelling; the soil-dwelling, litter-feeding; and the soil-dwelling, humus-feeding. Not all species of worms have burrows or produce casts, but the main focus of soil research has been on those that do. Burrows are made by either pushing soil aside or ingesting it. Soil casts are subsequently ejected at the surface or into large soil voids.

Earthworm casts are an intimate mixture of soil, organic matter, body wastes, and microorganisms. Not suprisingly, they are higher in organic matter, richer in nutrients, and have greater structural stability than the surrounding soil. They contain no particles larger than can be ingested (ca. 1 mm), and are often higher in silt than the surrounding soil. The soil in casts may be brought from a depth of 150 to 300 mm or more, and burrowing results in a soil layer equivalent to up to 40 mm in thickness being brought to the surface each year. More typically, values around 4 mm yr^{-1} have been observed in soils with a reasonable earthworm population. If this soil were removed from a layer 100 mm thick, it would represent a 4% reduction in bulk density per year.

The effect of casting is to provide a continuous tendency to reduce the bulk density of the surface soil, to increase the surface roughness, and to provide deep vertical channels, often stabilized by lining with body fluids (although such channels are created by only a few species). Many authors have observed the increased infiltration rates that result from the presence of these channels in both tropical and temperate soils. All of these effects should be of great significance in ameliorating hardsetting behavior. Furthermore, since acidic soils are not favorable to earthworms, hardsetting alfisols should represent a more favorable environment for worms than other, more acidic soil types that are common in the topics.

Tillage can greatly reduce earthworm populations by mechanical damage to the worms; exposing the worms to birds and other predators; reducing food supply (where crop residues are removed or burnt); and causing greater extremes of near-surface temperature and moisture regime. Conversely, no-till farming with stubble retention and/or use of mulch or manure, and fallow cropping with grass or legumes, increases the earthworm population and activity. However, some agrochemicals may significantly affect the earthworm population; in particular, carbamate insecticides and the fungicide Benomyl are both toxic to earthworms.

Blackwell (1988) has demonstrated the stability of cylindrical vertical channels of >2 mm diameter under 400 kPa uniaxial loading, and there are many reports of high infiltration rates being maintained in soils under no-till farming as a result of the persistence of earthworm channels, despite the greater topsoil bulk density of such soils in comparison to recently plowed soil. It is therefore surprising that some of the literature on the beneficial effects of no-till farming on soil physical properties makes no comment as to the presence or absence of earthworms.

Since worms may be a major cause of the amelioration of hardsetting soils under no-till management, and since they may be killed as a result of lack of forethought in a pesticide application, it is clear that factors affecting earthworm population and the influence of earthworms on hardsetting behavior merit further study.

2. Termites

Termites occupy two-thirds of the earth's land surface. There are about 2000 recorded species, of which 600 have been recorded in Africa. Many species can live in the same area and compete for food, so that where natural vegetation is cleared for cropping, the balance of species and population changes. Some termite nests consists of galleries in wood and soil, while others are complex above-and below-ground "mounds," which can vary from a few centimeters to 9 m in height and up to 30 m in base circumference. Soil turnover rates due to termite activity lie in the range 0.01 to 0.5 mm yr^{-1}, but with most reported values <0.1 mm yr^{-1}. Thus, over hundreds or thousands of years termites, like earthworms, can be a major factor influencing the form of the soil profile.

Termite mounds are constructed from sand grains that are transported in the termites' mandibles, held together by smaller (usually dominantly clay-sized) material that is ingested and then excreted or regurgitated in the mound. In most soils, especially those of sandier texture, there is an increased clay content in the mound and the substratum is correspondingly higher in gravel and sand. Most mounds have a higher clay : sand ratio than the surrounding soil. The clay is collected from considerable distances from the nest or brought up from the subsoil. Abandoned termite mounds are liable to erosion, which can preferentially wash away the clay from the outside of the mound. Thus the overall effects of termite activity are to recycle clay (and silt) that has been illuviated into the subsoil. Termites have also been held responsible for much of the spatial heterogeneity in surface texture that characterizes many tropical soils.

Soil texture is an important property for mound-building termites, since sandy soils with little silt or clay are unsuitable, as are soils with a high shrink-swell potential such as vertisols. In fact, most hardsetting soils should present an ideal texture for termite mounds, since the hard outer part of the mound is in effect a deliberately constructed hardsetting material and this probably accounts for its strength. Organic matter bonding has sometimes been suggested to be responsible for the high strength of termite mounds. However, the outer layer or crust of the mound usually has an organic matter content that is lower than other parts of the mound and lower than the surrounding soil. Ley (unpublished data, 1987) has measured the compressive strength of undisturbed samples from the outer parts of a mound from Mokwa, Nigeria, that were equilibrated at −100 kPa. These strengths and the very low strength of saturated samples were comparable to the strengths of hardsetting soil of similar particle-size distribution, indicating that this type of mound is really a material that is hardsetting by virtue of its means of construction and is unlikely to be strengthened by organic matter.

The outer layer of termite mounds is closely packed, so there are no macropores and the surface is not permeable. The bulk density of mounds is generally high and greater than that of the surrounding soil. Subterranean galleries that extend up to 50 m from the mound and to 15 m depth have been reported. When in use, these galleries are designed to avoid waterlogging and do not connect directly to the soil surface; but in abandoned mounds that have weathered, these galleries may connect to the soil surface and aid infiltration. However, it is not possible to be categorical about the contrast between the physical or chemical properties of abandoned mounds and those of the suroundling soil. In most regions the conditions over abandoned mounds are more favorable for plant growth, but a number of counterexamples also exist. Williams (1979), in Gambia, for example, found greatly reduced surface infiltration and profile wetting for an old abandoned mound, but when the surface of the mound was made concave instead of convex, to trap water, there was a luxuriant growth of vegetation.

Although termites do not attack native crop and tree species, exotic and introduced crops are particularly vulnerable to termite attack. Other potentially harmful effects include depletion of local vegetation cover, nutrient immobilization, and accelerating soil erosion. Possible beneficial effects include recycling of subsoil and of plant nutrients, litter decomposition, improvement in soil porosity, and modification of the particle-size distribution down the profile. In summary, it is clear that termites can cause a number of marked alterations to soil physical and chemical properties. Depending on termite species, soil type, climate, and vegetation, the balance of these alterations may be beneficial or harmful to crop growth.

Since nest-building termites should do well on hardsetting soils, and since alteration of the particle-size distribution of the surface soil and possible effects on structural stability are likely to have a pronounced effect on hardsetting behavior, it is clear that termites may have a particularly noticeable influence, beneficial or otherwise, on soils liable to hardsetting. The effects of termite activity in relation to hardsetting, and of soil management activities in relation to termite activity, therefore need to be studied with a view to identifying and taking advantage of beneficial effects and minimizing unfavorable ones.

3. Ants

There is little information available on the effects of ants on tropical soil. Ant population decreases from the humid forest region to a low-diversity, low-density population in the savanna. Some species construct nests in trees, but many species are involved in mound building and other soil excavations, and some construct a network of interconnected tunnels and chambers. The net effect of ant activity is to turn over and mix the soil, but there is insufficient information to gain any firm idea of the magnitude of this soil turnover, the source of nest material, and its physical and nutritional characteristics. In comparison to termite nests, there is little difference between the particle-size distribution of the ant nest and that of the surrounding soil, and it seems that ants may have a considerably smaller effect than earthworms or termites on soil properties. However, given Nye's (1955) observation in Nigeria that, where few earthworms are present, the surface layer of soil is formed by ants, it is clear that they may also have an important influence on surface properties of hardsetting soils, and more quantitative data on their influence on soil physical properties and on effects of different management systems on species diversity and population are badly needed.

E. Conclusions

Because the incidence and severity of hardsetting depends on the management system, soil type, climate, and the sequence of rainfall events before and during the growing period, there is no single successful management recipe appropriate to all situations. However, the first stage in treatment should be to diagnose those aspects of hardsetting that are causing problems and identify those features of the current management system that are responsible. Where erosion and runoff are serious problems, remedial measures are overdue and appropriate solutions have already been reviewed by many authors (e.g., Lal, 1984b).

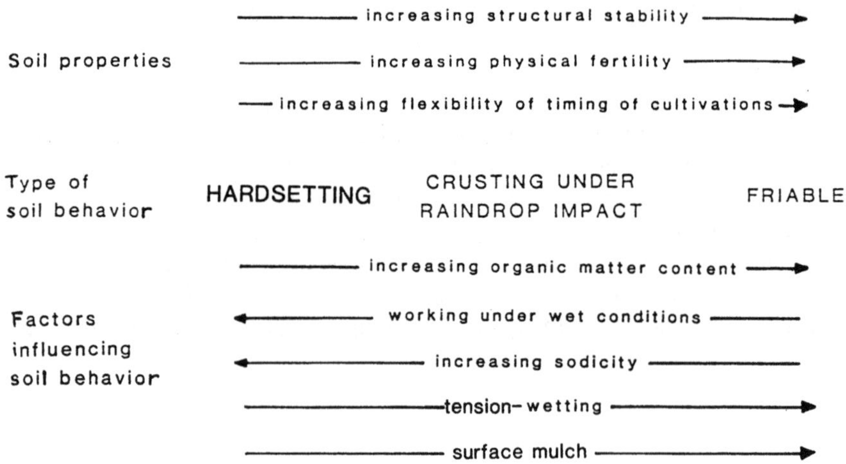

Figure 6. Factors affecting the tendency for hardsetting behavior and its amelioration. [Adapted from Mullins et al. (1987).]

It is not surprising that there is conflicting evidence on the benefits of tillage in hardsetting soils, since those soils that are already in a state where infiltration and/or root growth are suboptimal will respond positively to loosening, while other soils may suffer. The problem with hardsetting soils, even in the absence of traffic-induced compaction, is not how to obtain, but how to maintain sufficient porosity. Figure 6 shows the various factors that are involved.

If sodicity is not a problem, the simplest methods involve increased addition or incorporation of organic matter (or avoidance of its removal). It is clear that, with the exception of any necessary soil-loosening and weed-control operations, tillage should be reduced to a minimum and its timing carefully controlled because of the adverse effects on readily oxidized organic matter, slaking and dispersion, and on earthworm population. Novel approaches that involve cultivation and/or organic matter additions to only a limited portion of the soil surface in the plant's vicinity have the potential to overcome hardsetting problems even where organic residues are in short supply, and such approaches may represent a new way forward.

V. Research and Development Priorities

A. Recognition of the Problem

There is an urgent need for hardsetting to be recognized as a distinct soil physical condition in the same way that crusting already is. As with crusting, the condition must be defined purely in terms of field observations of the behavior of a given soil under a given crop and management system but *without reference* to any pedological classification that has been made of the soil. In this context, current soil classification systems are of use to the extent that they can indicate that a soil is hardsetting in its natural state or is liable to become so after tillage. Clarification of the definition of "hardsetting" is needed so that the term can be used with confidence by farmers, advisers, and agronomists, and a simple scale is needed to indicate the degree of hardsetting (e.g., Section I.V.C.3).

B. The Need for a Diagnostic Approach

Because of the wide variety of potentially limiting physical conditions related to hardsetting, a *diagnostic approach* is needed, in which an attempt is made to identify the particular physical problems that are hampering management and crop growth in any given situation. Problems of subsurface horizons (e.g., the behavior of a sodic clay horizon or restrictions to rooting imposed by aluminum toxicity in acidic subsoil) have also to be considered in such an approach. A diagnostic approach is especially pertinent to hardsetting soils because trials in which different cultivation/management systems are compared are expensive and need to be run for a number of years if reliable recommendations are to be expected. They are also often inconclusive, especially on hardsetting soils, where, depending on the timing, amount, and intensity of rainfall after sowing, the different facets of the problem may or may not become apparent. Thus, for example, regular monitoring of soil physical conditions in relation to the growth of a crop can provide evidence of hardsetting problems (e.g., limited depth of profile wetting, slow establishment) that will not always be evident as a reduction in the yield in years when there is good rainfall soon after establishment. One approach that should be fruitful is to develop and test predictive models for soil bulk density, macroporosity (Gibbs and Reid, 1988), and the strength characteristic. Such models may initally prove quite unreliable, but this in itself should lead to a reappraisal of the influence of the various factors that determine soil physical properties, and hence to a better understanding of how the soil may be manipulated to our advantage.

C. Soil Management Systems

Appropriate management of potentially hardsetting soil depends on the climate and on identification of sodicity problems where they exist.

1. Research is needed on hardsetting in hot, dry regions, particularly where it is difficult to maintain a sufficiently high rate of fresh organic matter additions to ameliorate the hardsetting condition and where there are no earthworms to aid infiltration. Novel approaches to soil amelioration merit further attention, particularly integrated management systems in which an attempt is made to establish and maintain an ameliorated zone in only a part of the soil. These include selective application of organic re-

sidues to a restricted area of the soil surface, permanent beds, and soil slotting (Blackwell, 1988). Soil mixing is also of interest, but the physical and chemical basis for any response needs to be understood before advocating widespread adoption of the technique. Improved management systems must also be consistent, where necessary, with an overall scheme of runoff and erosion control (e.g., contour cultivations and waterways may be a necessary part of the system).

2. Under more humid conditions and where it is feasible to sustain a sufficiently high concentration of soil organic matter to ameliorate hardsetting, a wide range of successful management options are already available. However, where land use is being intensified (e.g., when a new area of land is being brought under mechanized cultivation for the first time), the resilience of the soil to the new regime can only be guessed at. Since the initial decline in soil physical properties may be slow and insidious, there is a need for long-term experiments in which both crop yields and soil physical properties are monitored, and for the development of a set of tests and guidelines that can be used to give an early warning that any given system will be unsustainable.

Although mechanized agriculture has the potential to create spectacular soil degradation more quickly, it is clear that hardsetting can also be produced without the use of powered machinery with inappropriate management. Therefore there is need for a wider awareness of the unsuitability of management systems that involve repeated tillage or reworking of the soil when wet, and do not conserve soil organic matter. The effects of management systems on the populations and activities of soil fauna, particularly earthworms and termites, and of faunal activities on infiltration and hardsetting require more attention.

In countries where land is at a premium and hardsetting results in soils that become so eroded and degraded as to be unusable, the problem may be more related to inappropriate operation of the socioeconomic system rather than the lack of viable alternative management systems. While the solution then lies in the social and political arena, there remains a necessity to identify hardsetting or potentially hardsetting soils.

3. Management of hardsetting soil for intensive irrigated production systems represents a major scientific challenge. The success of the Tatura system (Section III.A.5) demonstrates a possible set of solutions that should have applications elsewhere. The design of this system is based on a diagnostic approach that provides an example against which to model other attempts to adapt or change unsuitable irrigation management systems on hardsetting soils.

4. It is necessary to identify whether or not soil sodicity is a major factor in hardsetting. The scheme presented by Rengasamy et al. (1984) provides appropriate guidelines on which to decide whether gypsum is necessary to reduce clay dispersion.

D. Strategic Research

In many area, limited availability of effective herbicides and specialized machinery makes some form of cultivation of the surface layer a necessity for weed control and, even with direct drilling or jab planting, the disturbed soil around the seed must provide a suitable physical environment. Therefore, further research is needed on the hardsetting process (i.e., structural disintegration, slumping, and hardening) in cultivated soil in order to identify stages where the process can most easily be interfered with so as to reduce strength development and/or promote structural development.

Acknowledgments

One of the authors (C.M.) is grateful to P.S. Blackwell, J.S. Lenvain, and G.D. Smith for their help and advice, and to the Overseas Development Administration (ODA) and the Agriculture and Food Research Council (AFRC) for funding part of the work referred to in this review.

References

Abbott, I., C.A. Parker, and I.D. Sills. 1979. Changes in the abundance of large soil animals and physical properties of soil following cultivation. *Aust. J. Soil Res.* 17:343–353.

Adem, H.H., and J.M. Tisdall. 1984. Management of tillage and crop residues for double-cropping in fragile soils of south-eastern Australia. *Soil Tillage Res.* 4:577–589.

Adeoye, K.B. 1986. Physical changes induced by rainfall in the surface layer of an Alfisol, northern Nigeria. *Geoderma* 39:59–66.

Agassi, M., I. Shainberg, and J. Morin. 1981. Effect of electrolyte concentration and soil sodicity on infiltration rate and crust formation. *Soil Sci. Soc. Am. J.* 45:848–851.

Arndt, W. 1985. Factors affecting the nature and impedance of soil surface seals at Katherine, N.T. CSIRO Aust. Div. Soils Div. Rept. No. 79.

Arora, H.S., and N.T. Coleman. 1979. The influence of electrolyte concentration on flocculation of clay suspensions. *Soil Sci.* 127:134–139.

Aston, A.R., and R.A. Fischer. 1986. The effect of conventional cultivation, direct drilling and crop residues on soil temperatures during the early growth of wheat at Murrumbateman, New South Wales. *Aust. J. Soil Res.* 24:49–60.

Aylmore, L.A.G., and I.D. Sills. 1982. Characterization of soil structure and stability using modulus of rupture–exchangeable sodium percentage relationships. *Aust. J. Soil Res.* 20:213–224.

Bakker, A.C., and W.W. Emerson. 1973. The comparative effects of exchangeable calcium, magnesium, and sodium on some physical properties of red-brown earth subsoils. III. The permeability of Shepparton soil and comparison of methods. *Aust. J. Soil Res.* 11:159–165.

Bakker, A.C., W.W. Emerson, and J.M. Oades. 1973. The comparative effects of exchangeable calcium, magnesium, and sodium on some physical properties of red-brown earth subsoils. I. Exchange reactions and water contents for dispersion of Shepparton soil. *Aust. J. Soil Res.* 11:143–150.

Barley, K.P. 1959. The influence of earthworms on soil fertility: I. Earthworm populations found in agricultural land near Adelaide. *Aust. J. Agric. Res.* 10:171–178.

Barley, K.P., and C.R. Kleinig. 1964. The occupation of newly irrigated lands by earthworms. *Aust. J. Sci.* 26:290–291.

Blackmore, A.V. 1976. Salt sieving within soil clay aggregates. *Aust. J. Soil Res.* 11:149–158.

Blackwell, P.S. 1988. The influence of soil geometry on soil compaction under wheels. *Proc. 11th Int. Soil Tillage Res. Organization Conf., Edinburgh* 1:203–208.

Blackwell, P.S., A. Cass, and S. Gusli. 1988. Personal communication.

Boone, F.R. 1988. Weather and other environmental factors influencing crop responses to tillage and traffic *Soil Tillage Res.* 11:283–324.

Braunack, M.V., and A.R. Dexter. 1988. The effect of aggregate size in the seedbed on surface crusting and growth and yield of wheat (*Triticum aestivum* L. cv. Halberd) under dryland conditions. *Soil Tillage Res.* 11:133–145.

Braunack, M.V., J.S. Hewitt, and A.R. Dexter. 1979. Brittle fracture of soil aggregates and the compaction of aggregate beds. *J. Soil Sci.* 30:653–667.

Burch, G.J., I.B. Mason, R.A. Fischer, and I.D. Moore. 1986. Tillage effects on soils: Physical and hydraulic responses to direct drilling at Lockhart, N.S.W. *Aust. J. Soil Res.* 24:377–391.

Cassel, D.K. 1983. Spatial and temporal variability of soil physical properties following tillage of Norfolk loamy sand. *Soil Sci. Soc. Am. J.* 47:196–201.

Chan, K.Y. 1989. Friability of a hardsetting soil under different tillage and land use practices. *Soil Tillage Res.* 13, 287–298.

Chan, K.Y., and J.A. Mead. 1989. Surface physical properties of a hardsetting sandy loam under different tillage practices. *Aust. J. Soil Res.* (accepted for publication).

Chan, K.Y., J.A. Mead, and W.P. Roberts. 1987. Poor early growth of wheat under direct drilling. *Aust. J. Agric. Res.* 38:791–800.

Chandler, H.W., and J.V. Stafford. 1987. A simple field test for predicting ease of cultivation. *Soil Tillage Res.* 10:269–275.

Charreau, C., and R. Nicou. 1971. L'amelioration du profile cultural dans les sols sableux et sable-argileux de la zone tropicale seche ouest-africaine et ses incidences agronomiques. *L'Agronomie Tropicale, Paris* 26:209–255.

Chartres, C.J., R.S. Greene, G.W. Ford, and P. Rengasamy. 1985. The effects of gypsum on macroporosity and crusting of two red duplex soils. *Aust. J. Soil Res.* 23:467–479.

Childs, E.C. 1940. The use of soil moisture characteristics in soil studies. *Soil Sci.* 50:239–252.

Childs, E.C. 1942. Sability of clay soils. *Soil Sci.* 53:79–92.

Childs, E.C. 1969. *An introduction to the physical basis of soil water phenomena.* Wiley, London.

Childs, E.C., and E.G. Youngs. 1958. The use of soil moisture characteristics in

assessing the soil stability of opencast coal sites. *Medelingen Landbouwhogesch staat gent.* 24:415–421.
Cockroft, B., and F.M. Martin. 1981. Irrigation. pp. 133–149. In: J.M. Oades, D.G. Lewis, and K. Norrish (eds.), *Red-brown earths of Australia.* Waite Agr. Res. Inst. and CSIRO Div. Soils, Adelaide, South Australia.
Collis-George, N., and B.S. Figueroa. 1984. The use of high energy moisture characteristics to assess soil stability. *Aust. J. Soil Res.* 22:349–356.
Collis-George, N., and R.S.B. Greene. 1979. The effect of aggregate size on the infiltration behaviour of a slaking soil and its relevance to ponded irrigation. *Aust. J. Soil Res.* 17:65–73.
Collis-George, N., and Lal, R. 1971. Infiltration and structural changes as influenced by initial moisture content. *Aust. J. Soil Res.* 9:107–116.
Costigan, P.A., D.J. Greenwood, and T. McBurney. 1983. Variation in yield between two similar sandy loam soils. I. Description and measurement of yield differences. *J. Soil Sci.* 34:621–637.
Daniel, H., R.J. Jarvis, and L.A.G. Aylmore. 1988. Hardpan development in loamy sand and its effect upon soil conditions and crop growth. *Proc. 11th Int. Soil Tillage Res. Organization Conf., Edinburgh* 1:233–238.
Dexter, A.R. 1988. Advances in characterization of soil structure. *Soil Tillage Res.* 11:199–238.
Dexter, A.R., and B. Kroesbergen. 1985. Methodology for determination of tensile strength of soil aggregates. *J. Agric. Eng. Res.* 31:139–147.
Dregne, H.E. 1976. *Soils of arid regions.* Elsevier, Amsterdam.
Dudal, R. 1970. *Key to soil units for the Soil Map of the World.* AGL:SM/70/2, FAO, Rome.
Eck, H.V., and P.W. Unger. 1985. Soil profile modification for increasing crop production. *Adv. Soil Sci.* 1:65–122.
Elliott, B.R., and R. Jardine. 1972. The influence of rotation systems on long-term trends in wheat yields. *Aust. J. Agric. Res.* 23:935–944.
El-Swaify, S.A., P. Pathak, T.J. Rego, and S. Singh. 1985. Soil management for optimized productivity under rainfed conditions in the semi-arid tropics. *Adv. Soil Sci.* 1:1–64.
Emerson, W.W. 1954. The determination of the stability of soil crumbs. *J. Soil Sci.* 5:235–250.
Emerson, W.W. 1964. The slaking of soil crumbs as influenced by clay mineral composition. *Aust. J. Soil Res.* 2:211–217.
Emerson, W.W. 1967. A classification of soil aggregates based on their coherence in water. *Aust. J. Soil Res.* 5:47–57.
Emerson, W.W. 1968. The dispersion of clay from soil aggregates. *Trans. 9th Int. Congr. Soil Sci.* 1:617–626.
Emerson, W.W. 1971. Determination of the contents of clay-size particles in soils. *J. Soil Sci.* 22:50–59.
Emerson, W.W. 1977. Physical properties and structure. pp. 78–104. In: J.S. Russell and E.L. Greacen (eds.), *Soil factors in crop production in a semi-arid environment.* Queensland University Press, Brisbane.
Emerson, W.W. 1983. Inter-particle bonding. pp. 477–498. In: *Soils: an Australian viewpoint.* CSIRO, Melbourne/Academic Press, London.
Emerson, W.W., and A.C. Bakker. 1973. The comparative effects of exchangeable

calcium, magnesium, and sodium on some physical properties of red-brown earth subsoils. II. The spontaneous dispersion of aggregates in water. *Aust. J. Soil Res.* 11:151–157.

Emerson, W.W., and C.L. Chi. 1977. Exchangeable calcium, magnesium, and sodium and the dispersion of illites in water. *Aust. J. Soil Res.* 15:255–262.

FAO-UNESCO. 1974. *Soil Map of the World 1 : 5 000 000*. 1-X. Unesco, Paris.

French, R.J. 1978. The effect of fallowing on the yield of wheat. II. The effect on grain yield. *Aust. J. Agric. Res.* 29:669–684.

French, R.J. 1981. Management under low rainfall (South Australia). pp. 97–116. In: J.M. Oades, D.G. Lewis, and K. Norrish (eds.), *Red-brown earths of Australia*. Waite Agr. Res. Inst. and CSIRO Div. Soils, Adelaide. South Australia.

French, R.J., and J.E. Schultz. 1984. Water use efficiency of wheat in a Mediterranean-type environment. II. Some limitations to efficiency. *Aust. J. Agric. Res.* 35:765–775.

Frenkel, H., J.O. Goertzen, and J.D. Rhoades. 1978. Effects of clay type and content, exchangeable sodium percentage, and electrolyte concentration on clay dispersion and soil hydraulic conductivity. *Soil Sci. Soc. Am. J.* 42:32–39.

Gates, C.T., D.B. Jones, W.J. Muller, and J.S. Hicks. 1981. The interaction of nutrients and tillage methods on wheat and weed development. *Aust. J. Agric. Res.* 32:227–241.

Ghavami, M., J. Keller, and I.S. Dunn. 1974. Predicting soil density following irrigation. *Trans. ASAE* 17:166–171.

Gibbs, R.J., and J.B. Reid. 1988. A conceptual model of changes in soil structure under different cropping systems. *Adv. Soil Sci.* 8:123–149.

Gillman, G.P. 1974. The influence of net charge on water dispersible clay and sorbed sulphate. *Aust. J. Soil Res.* 12:173–176.

Greacen, E.L., K.P. Barley, and D.A. Farrell. 1969. The mechanics of root growth in soils with particular reference to the implications for root distribution pp. 265–269. In: W.J. Whittington (ed.), *Root growth*. Butterworths, London.

Greene, R.S.B., and G.W. Ford. 1985. The effect of gypsum on cation exchange in two red duplex soils. *Aust. J. Soil Res.* 23:61–74.

Greene, R.S.B., A.M. Posner, and J.P. Quirk. 1978. A study of the coagulation of montmorillonite and illite suspensions by calcium chloride using the electron microscope. pp. 35–40. In: W.W. Emerson, R.D. Bond, and A.R. Dexter (eds.), *Modification of soil structure*. Wiley, NY.

Grierson, I.T. 1978. Gypsum and red-brown earths. pp. 315–324. In: W.W. Emerson, R.D. Bond, and A.R. Dexter (eds.), *Modification of soil structure*. John Wiley, NY.

Grierson, I.T., J.W. Jijne, and D.J. Greenland. 1972. Changes in some physical properties of red-brown earths of different texture associated with increasing content of organic matter. *Exp. Record* 6:16–22.

Gupta, R.P., S. Kumar, and T. Singh. 1984. Soil management to increase crop production. Report of All India Coordinated Research Projects on Improvement of Soil Physical Conditions to Increase Agricultural Production of Problematic Areas. *Indian Council Agr. Res.*, New Delhi.

Haddow, P. 1988. Friability of hardsetting soils. Honours Project, Soil Science Dept., Aberdeen University.

Hamblin, A.P. 1984. The effect of tillage on soil surface properties and the water balance of a xeralfic alfisol. *Soil Tillage Res.* 4:543–559.

Hamblin, A.P. 1987. The effect of tillage on soil physical conditions pp. 128–170. In: P.S. Cornish and J.E. Pratley (eds.), *Tillage. New directions in Australian agriculture.* Inkata Press, Melbourne.

Hamblin, A.P., and D. Tennant. 1979. Interactions between soil type and tillage level in a dryland situation. *Aust. J. Soil Res.* 17:177–189.

Hulme, P. 1983. Honours Project. Dept. of Agronomy and Soil Science, Univ. of New England, Armidale, Australia.

Hulugalle, N.R., R. Lal, and C.H.H. Ter Kuile. 1986. Amelioration of soil physical properties by mucuna after mechanized land clearing of a tropical rain forest. *Soil Sci.* 141:219–224.

Ike, I.F. 1986. Soil and crop responses to different tillage practices in a ferruginous soil in the Nigerian savanna. *Soil Tillage Res.* 6:261–272.

Jones, M.J. 1986. The dryland farming projects, Botswana, 1971–1985. pp. 62–67. In: *The agricultural dilemma in Africa.* Overseas Development Administration, London.

Kazman, Z., I. Shainberg, and M. Gal. 1983. Effect of low levels of exchangeable sodium and applied phosphogypsum on the infiltration rate of various soils. *Soil Sci.* 135:184–192.

Kelly, K.B. 1984. Root zone limitations in the Goulburn Valley. II Effects of soil modification and treading on pasture growth. pp. 319–328. In: W.A. Muirhead and E. Humphreys (eds.), *Root zone limitations to crop production on clay soils.* CSIRO, Melbourne.

Kemper, W.D., J.S. Olsen, and A. Hodgdon. 1975. Irrigation method as a determinant of large pore persistence and crust strength of cultivated soils. *Soil Sci. Soc. Am. Proc.* 39:519–523.

Kemper, W.D., R.C. Rosenau, and A.R. Dexter. 1987. Cohesion development in disrupted soils as affected by clay and organic matter content and temperature. *Soil Sci. Soc. Am. J.* 51:860–867.

Kimber, R.W.L. 1973. Phytotoxicity from plant residues. II. Effect of time of rotting of straw from some grasses and legumes on the growth of wheat seedlings. *Plant Soil* 38:341–361.

Lal, R. 1984a. Mechanized tillage systems effects on soil erosion from an alfisol in watersheds cropped to maize. *Soil Tillage Res.* 4:349–360.

Lal, R. 1984b. Soil erosion from tropical arable lands and its contol. *Adv. Agron.* 37:183–248.

Lal, R. 1985. Mechanized tillage systems effects on properties of a tropical alfisol in watersheds cropped to maize. *Soil Tillage Res.* 6:149–161.

Lal, R. 1986. Effects of eight tillage treatments on a tropical alfisol: Maize growth and yield. *J. Sci. Food Agric.* 37:1073–1082.

Lal, R. 1987a. Managing the soils of sub-Saharan Africa. *Science* 236:1069–1076.

Lal, R. 1987b. *Tropical ecology and physical edaphology.* Wiley, Chichester.

Lal, R., B.J. Bridge, and N. Collis-George. 1970. The effect of column diameter on the infiltration rate into a swelling soil. *Aust. J. Soil Res.* 8:185–193.

Lenvain, J.S., V.R.N. Chinene, K.S. Gill, and P.L.L. Pauwelyn. 1987. Aspects of compaction, water transmissibility, and erosion on a kaolinitic clay soil. In: Proc. 2nd Regional IBSRAM workshop on land development and management of acid soils in Africa, Lusaka, Zambia.

Lenvain, J.S., and P.L. Pauwelyn. 1988. Comparison of the physical properties of two Zambian soils. pp. 379–389. In: *The red soils of east and southern Africa.*

Proc. Int. Symp., Harare, Zimbabwe, 1986. Manuscript Report, IDRC-MR1 70e. IDRC, Ottawa, Canada.
Ley, G.J. 1988. A study of hard-setting behavior of structurally weak tropical soils. Ph.D. thesis, Aberdeen University, Aberdeen, UK.
Ley, G.J., C.E. Mullins, and R. Lal. 1989a. Hard-setting behavior of some structurally weak tropical soils. *Soil Tillage Res.* 13:365–381.
Ley, G.J., C.E. Mullins, and R. Lal. 1989b. A comparison between theoretical predictions and the measured strengths of undisturbed soil cores. (In preparation.)
Loveday, J. 1974. Recognition of gypsum-responsive soils. *Aust. J. Soil Res.* 25:87–96.
Loveday, J. 1981. Soil management and amelioration. pp. 39–57. In: T.S. Abbot, C.A. Hawkins, and P.G. Searle (eds.), *National Soils Conference 1980 review papers*. Aust. Soc. Soil Sci., Sydney, Australia.
Loveday, J., and D.R. Scotter. 1966. Emergence response of subterranean clover to dissolved gypsum in relation to soil properties and evaporative conditions. *Aust. J. Soil Res.* 4:55–68.
Makin, A.W. 1987. The measurements of some soil properties under tractor wheels. pp 158–161. In: K.J. Coughlan and P.N. Truong (eds.), *Effects of management practices on soil physical properties*. Proc. Natl. Workshop, Toowoomba, Queensland, Australia, September 7–10, 1987.
Mason, I.B., and R.A. Fischer. 1986. Direct-drilling effects on wheat growth and yield in southern New South Wales: I. Lockhart, a 450 mm rainfall site. *Aust. J. Exp. Agric.* 26:457–468.
Matheson, W.E. 1969. Gypsum for hardsetting soils. *J. Agric. South Australia* 72:402–406.
McDonald, R.C., R.F. Isbell, J.G. Speight, J. Walker, and M.S. Hopkins. 1984. *Australian soil and land survey field handbook.* Inkata Press, Melbourne.
McIntyre, D.S. 1958a. Permeability measurements of soil crusts formed by raindrop impact. *Soil Sci.* 85:185–189.
McIntyre, D.S. 1958b. Soil splash and the formation of surface crusts by raindrop impact. *Soil Sci.* 85:261–266.
McIntyre, D.S. 1979. Exchangeable sodium, subplasticity and hydraulic conductivity of some Australian soils. *Aust. J. Soil Res.* 17:115–120.
Mead, J.A., and K.Y. Chan. 1989. The effect of deep tillage and seeded preparation on the growth and yield of wheat on a hard-setting soil. *Aust. J. Exp. Agric.* (accepted for publication).
Millington, R.J. 1959. Establishment of wheat in relation to apparent density of the surface soil. *Aust. J. Agric. Res.* 10:487–494.
Mullins, C.E., and K.P. Panayiotopoulos. 1984. The strength of unsaturated mixtures of sand and kaolin and the concept of effective stress, *J. Soil Sci.* 35:459–468.
Mullins, C.E., I.M. Young, A.G. Bengough, and G.J. Ley. 1987. Hard-setting soils. *Soil Use Management* 3:79–83.
Murphy, B.W., G.J. Hamilton, and I.J. Packer. 1987. Seasonal variation in the soil physical condition of the surface layer (0–10 cm) of a structurally degraded soil and possible effects on the growth of wheat. pp. 52–55. In: K.J. Coughlan and P.N. Truong (eds.), *Effects of management practices on soil physical properties.* Proc. Natl. Workshop, Toowoomba, Queensland, September 7–10, 1987.

Nicou, R. and Chopart, J.L. 1979. Root growth and development in sandy and sandy clay soils of Senegal. pp. 373–384. In: R. Lal and D.J. Greenland (eds.), *Soil physical properties and crop production in the tropics.* Wiley Interscience, Chichester.

Norrish, K., and J.G. Pickering. 1983. Clay minerals. pp. 281–308. In: *Soils: An Australian viewpoint.* CSIRO, Melbourne.

Northcote, K.H. 1960, 1971, 1979. *A factual key for the recognition of Australian soils*, 1st, 3d, and 4th eds. Rellim Tech. Pub., Adelaide, South Australia.

Northcote, K.H. 1981. Morphology, distribution and classification. In: J.M. Oades, D.G. Lewis, and K. Norrish (eds.), *Red-brown earths of Australia.* Monograph, CSIRO Div. Soils, South Australia.

Northcote, K.H. 1982. Soil: A national resource. In: *Man and the Australian environment.* W. Hanley, and M. Cooper (eds.), McGraw-Hill, Australia.

Northcote, K.H., with G.G. Beckman, E. Bettany, H.M. Churchward, D.C. van Dijk, G.M. Dimmock, G.D. Hubble, R.F. Isbell, W.M. McArthur, G.G. Murtha, K.D. Nicolls, T.R. Paton, C.H. Thompson, A.A. Webb, and M.J. Wright. 1960–1968. *Atlas of Australian soils.* Sheets 1–10, with explanatory booklets. CSIRO and Melbourne Univ. Press, Melbourne.

Northcote, K.H., G.D. Hubble, R.F. Isbell, C.H. Thompson, and E. Bettenay. 1975. *A description of Australian soils.* CSIRO Div. Soils, Australia.

Nye, P.H. 1955. Some soil-forming processes in the humid tropics, IV. The action of the soil fauna. *J. Soil Sci.* 6:73–83.

Oades, J.M. 1981. Organic matter in the Urrbrae soil. pp. 63–81. In: J.M. Oades, D.G. Lewis, and K. Norrish (eds.), *Red-brown earths of Australia.* Waite Agric. Res. Inst. and CSIRO Aust. Div. Soils, Adelaide, South Australia.

Oades, J.M. 1984. Soil organic matter and structural stability: Mechanisms and implications for management. *Plant Soil* 76:319–337.

Osborne, G.J., D.L. Rowell, and P.G. Matthews. 1978. Value and measurement of soil structure under systems of reduced and conventional cultivation. pp. 379–385. In: W.W. Emerson, R.D. Bond, and A.R. Dexter (eds.), *Modification of soil structure.* Wiley, Chichester.

Oster, J.D. 1982. Gypsum usage in irrigated agriculture: A review. *Fertilizer Res.* 3:73–89.

Oster, J.D., and F.W. Schroeder. 1979. Infiltration as influenced by irrigation water quality. *Soil Sci. Soc. Am. J.* 43:444–447.

Oster, J.D., I. Shainberg, and J.D. Wood. 1980. Flocculation value and gel structure of sodium/calcium montmorilonite and illite suspensions. *Soil Sci. Soc. Am. J.* 44:955–959.

Pupisky, H., and I. Shainberg. 1979. Salt effects on the hydraulic conductivity of a sandy soil. *Soil Sci. Soc. Am. J.* 43:429–433.

Quirk, J.P., and R.K. Schofield. 1955. The effect of electrolyte concentration on soil permeability. *J. Soil Sci.* 6:163–178.

Rahman, W.A., and D.L. Rowell. 1979. The influence of magnesium in saline and sodic soils: A specific effect or a problem of cation exchange? *J. Soil Sci.* 30:535–546.

Razig, G. 1978. Pedological studies of selected "Gardud" soils from Kordofan Provinces, western Sudan. Ph.D. thesis, Aberdeen University, Aberdeen, UK.

Reddy, D.S., A.P. Raju, and M.S. Reddy. 1985. Management of physically problematic soils of Andhra Pradesh. *Consolidated report of the co-ordinated soil physics project Hyderabad for 1967–1983.* Andhra Pradesh Agricultural Univ. Rajendranagar, Hyderabad.

Reeves, T.G., and A. Ellington. 1974. Direct drilling experiments with wheat. *Aust. J. Exp. Agric. Anim. Husb.* 14:237–240.

Rengasamy, P. 1982. Dispersion of calcium clay. *Aust. J. Soil Res.* 20:153–157.

Rengasamy, P. 1983. Clay dispersion in relation to changes in the electrolyte composition of dialysed red-brown earths. *J. Soil Sci.* 34:723–732.

Rengasamy, P., G.W. Ford, and R.S.B. Greene. 1987. Classification of aggregate stability. pp. 97–101. In: K.J. Coughlan and P.N. Truong (eds.), *Effects of management practices on soil physical properties*, Proc. Natl. Workshop, Toowoomba, Queensland, Australia, September 7–10, 1987.

Rengasamy, P., R.S.B. Greene, G.W. Ford, and A.H. Mehanni. 1984. Identification of dispersive behavior and the management of red-brown earths. *Aust. J. Soil Res.* 22:413–431.

Rhoades, J.D., D.B. Krueger, and M.J. Reed. 1968. The effect of soil-mineral weathering on the sodium hazard of irrigation waters. *Soil Sci. Soc. Am. J.* 32:643–647.

Rose, C.W. 1962. Some effects of rainfall, radiant drying, and soil factors on infiltration under rainfall into soils. *J. Soil Sci.* 13:286–298.

Rovira, A.D., and E.L. Greacen. 1957. The effect of aggregate disruption on the activity of microorganisms in soil. *Aust. J. Agric. Res.* 8:659–673.

Rovira, A.D., K.R.J. Smettem, and K.E. Lee. 1987. Effect of rotation and conservation tillage on earthworms in a red-brown earth under wheat. *Aust. J. Agric. Res.* 38:829–834.

Rovira, A.D., and N.R. Venn. 1985. Effect of rotation and tillage on take-all and Rhizoctonia root rot in wheat. pp. 255–258. In: C.A. Parker, A.D. Rovira, K.J. Moore, P.T.W. Wong, and J.F. Kollmorgen (eds.), *Ecology and management of soilborne plant pathogens.* Am. Phytopathol. Soc., St. Paul, MN.

Rowell, D.L., D. Payne, and N. Ahmad. 1969. The effect of the concentration and movement of solutions on the swelling, dispersion, and movement of clay in saline and alkaline soils. *J. Soil Sci.* 20:176–188.

Rudd, C.L. 1974. More about gypsum. *J. Agric. South Australia* 77:70–72.

Rycroft, D.W., and A.A. Thorburn. 1974. Water stability tests for clay soils in relation to mole draining. *Soil Sci.* 117:306–310.

Schofield, R.K., and H.R. Samson. 1954. Flocculation of kaolinite due to the attraction of oppositely charged crystal faces. *Disc. Faraday Soc.* 18:135–145.

Scotter, D.R., and J. Loveday. 1966. Physical changes in seedbed material resulting from the application of dissolved gypsum. *Aust. J. Soil Res.* 4:69–75.

Shainberg, I., and M. Gal. 1982. The effect of lime on the response of soils to sodic conditions. *J. Soil Sci.* 33:489–498.

Shainberg, I., J.D. Rhoades, and R.J. Prather. 1980. The effect of low electrolyte concentration on clay dispersion and hydraulic conductivity of a sodic soil. *Soil Sci. Soc. Am. J.* 45:273–277.

Shanmuganathan, R.T., and J.M. Oades. 1983a. Influence of anions on dispersion and physical properties of the A horizon of a red-brown earth. *Geoderma* 29:257–277.

Shanmuganathan, R.T., and J.M. Oades. 1983b. Modification of soil physical properties by addition of calcium compounds. *Aust. J. Soil Res.* 21:285–300.

Sinclair, J. 1985. Crusting, soil strength and seedling emergence in Botswana. Ph.D. thesis, Aberdeen University, UK.

Smettem, K.R.J. 1986. Analysis of water flow from cylindrical macropores. *Soil Sci. Soc. Am. J.* 50:1139–1142.

Snyder, V.A., and R.D. Miller. 1985. Tensile strength of unsaturated soils. *Soil Sci. Soc. Am. J.* 49: 58–65.

Soil Survey Staff. 1975. *Soil taxonomy*. U.S. Dept. Agric. Handbook No. 436.

Spaagaren, O. 1986. (Personal communication: Hardsetting soils are mapped as unit no. 9 on: *Soil map of Zambia*, 1983 ed., compiled by W.J. Veldkamp, Soil Survey, Mt. Makulu, Zambia.)

Stace, H.C.T., G.D. Hubble, R. Brewer, K.H. Northcote, J.R. Sleeman, M.J. Mulchay, and E.G. Hallsworth. 1968. *A handbook of Australian soils*. Rellim Tech. Pub., Adelaide, South Australia.

Suarez, D.L., J.D. Rhoades, R. Lavado, and C.M. Grieve. 1984. Effect of pH on saturated hydraulic conductivity and soil dispersion. *Soil Sci. Soc. Am. J.* 48:50–55.

Taylor, A.J., and K.A. Olsson. 1987. Effect of gypsum and deep ripping on lucerne (*Medicago sativa* L.) yields on a red-brown earth under flood and spray irrigation. *Aust. J. Exp. Agric.* 27:841–849.

Tisdall, J.M. 1978. Ecology of earthworms in irrigated orchards. pp. 297–303. In: W.W. Emerson, R.D. Bond, and A.R. Dexter (eds.), *Modification of soil structure*. Wiley, Chichester.

Tisdall, J.M. 1985. Earthworm activity in irrigated red-brown earths used for annual crops in Victoria. *Aust. J. Soil Res.* 23:291–299.

Tisdall, J.M., and H.H. Adem. 1986a. The effect of reduced tillage of an irrigated silty soil and of a mulch on seedling emergence, growth and yield of maize (*Zea mays*) harvested for silage. *Soil Tillage Res.* 6:365–375.

Tisdall, J.M., and H.H. Adem. 1986b. Effect of water content of soil at tillage on size-distribution of aggregates and infiltration. *Aust. J. Exp. Agric.* 26:193–195.

Tisdall, J.M., and H.H. Adem. 1988. An example of custom prescribed tillage in south eastern Australia. *J. Agric. Eng. Res.* 40:23–32.

Tisdall, J.M., B. Cockroft, and N.C. Uren. 1978. The stability of soil aggregates as affected by organic materials, microbial activity and physical disruption. *Aust. J. Soil Res.* 16:9–17.

Tisdall, J.M., and J.M. Oades. 1980. The effect of crop rotation on aggregation in a red-brown earth. *Aust. J. Soil Res.* 18:423–433.

Trapnell, C.G., and R. Webster. 1986. Microaggregates in red earths and related soils in East and Central Africa, their classification and occurrence. *J. Soil Sci.* 37:109–123.

U.S. Salinity Laboratory Staff (L.A. Richards, ed.). 1954. *Diagnosis and improvement of saline and alkaline soils*. Handbook 60, U.S. Salinity Lab., Riverside, CA.

Utomo, W.H., and A.R. Dexter. 1981. Soil friability *J. Soil Sci.* 32:203–213.

Velasco-Molina, H.A., A.R. Swoboda, and C.L. Godfrey. 1971. Dispersion of different mineralogy in relation to sodium adsorption ratio and electrolyte concentration. *Soil Sci.* 111:282–287.

Whiteley, G.M., and A.R. Dexter. 1982. Root development of oilseed, wheat and pea crops on tilled and non-tilled soil. *Soil Tillage Res.* 2:379–393.

Willat, S.T., and D.M. Pullar. 1984. Changes in soil physical properties under grazed pastures. *Aust. J. Soil Res.* 22:343–348.

Willcocks, T.J. 1981. Tillage of clod-forming sandy loam soils in the semi-arid climate of Botswana. *Soil Tillage Res.* 1:323–350.

Williams, J.B. 1979. Soil water investigation in the Gambia. *Tech. Bull.* 3, Land Resource Development Centre, ODA, UK.

Young, I.M. 1987. Soil strength and hard-setting behaviour of some structurally unstable British soils. Ph.D. thesis, Aberdeen University, Aberdeen, UK.

Young, I.M., P.A. Costigan, and C.E. Mullins. 1988. Physical properties of two structurally sensitive soils and their effects on crop growth. *Proc. 11th Int. Soil Tillage Res. Organisation Conf.*, Edinburgh 2:933–938.

Role of Plinthite and Related Forms in Soil Degradation

H. Eswaran,* F. De Coninck,† and T. Varghese‡

I. Introduction

According to Alexander and Cady (1962), laterite is a highly weathered material rich in secondary oxides of iron, aluminum, or both. It is nearly void of bases and primary silicates, but it may contain large amounts of quartz and kaolinite. It is either hard or capable of hardening upon exposure to wetting and drying. This general definition is based on the original work of Buchanan (1807), modified by the work of many others since then. However, the term *laterite* is perhaps one of the most misused terms in earth sciences. Many inconsistencies and confusion exist in the literature

* Soil Management Support Services, Washington, D.C. 20013, U.S.A.
† Geological Institute, University of Gent, 900 Gent, Belgium.
‡ Soil Science Department, Kerala Agricultural University, Trivandrum, Kerala, India.

about the term *laterite* and the process of laterization. The term has been adopted by geologists, mineralogists, mining engineers, and pedologists, and today it encompasses materials that show some kind of sesquioxide accumulation and ranges from weathered rock to hard and cemented ironstone. Further complication was introduced in tropical soils literature, when terms such as *laterite profile* and *lateritic soils* were introduced. Due to this confusion in soils literature, Kellogg (1949) introduced the term *latosol*—LAT derived from *lat*erite—to distinguish them from laterites, but later Maignien (1966) proposed to abandon the term *laterite*. The reader is referred to this excellent review by Maignien (1966) for information on the early work.

With the advent of work on developing a new U.S. soil classification system in 1951, which culminated in *Soil Taxonomy* (Soil Survey Staff, 1975), these older terms were eliminated and new terms with strict definitions were introduced wherever possible. The modern terms include *plinthite*, *petroplinthite* (Sys, 1968), and *petroferric contact*, and the purpose of this paper is to describe these features, explain their genesis, and elaborate the implication of their presence in soils for use and management. The paper does not consider geologic deposits of iron stones and bauxites, which have other modes of formation and which are generally deep-seated.

II. Definition of Plinthite and Associated Forms

Plinthite and related sesquioxide accumulations are not specific to the intertropical areas, as they are the products of specific weathering and soil-forming processes. They are, however, most extensive in the tropics, where the pedological environment is most conducive to their formation.

A. Plinthite

The term *plinthite* was introduced in the early 1960s as a substitute for the term *laterite* and to depict a diagnostic feature in the soil. The term originates from the Greek word *plinthos*, meaning brick. A description of the term is given in *Soil Taxonomy*, and a working definition was provided by Daniels et al. (1978), but as there is no strict definition, users of the system have had difficulties in applying the concept. A provisional definition based on the proposal of Daniels et al. (1978) is given below:

> Plinthite is a discrete body of soil material with less than 25% by volume of rock structure and which is the result of absolute accumulation of iron, is firm to very firm when moist and hard or very hard when dry, and occurs as dark red mottles, which are usually in platy, polygonal or reticulate patterns; the core of the plinthic material may be hard and brittle but is crushed easily by the finger and in the process, smears the fingers. On drying, plinthite hardens irreversibly to petroplinthite.

The absolute amount of iron in plinthite varies with the soil. There is usually a concentration gradient of iron from the core to the diffuse outer part of the plinthite. In the early stages of formation, the plinthite is present as diffuse mottles separated by bleached interconnected zones; through accretion of iron, the individual plinthite is interconnected to form a "continuous phase of plinthite."

The upper boundary of the zone with plinthite may be sharp while the lower boundary is generally diffuse, leading to the bleached and reduced subsoil and then the saprolite. Plinthite may have a range of minerals, reflecting the conditions that prevailed prior to or during plinthite formation. The most elementary form is quartz grains of silt and sand sized in a matrix of clay of kaolinitic composition with some micas. At the other extreme there is accumulation of gibbsite, goethite, and haematite in addition to the quartz and kaolinitic clay (Eswaran et al., 1980). Additions of manganese to the system colors it black; manganese minerals that may be found include birnessite and lithophorite.

Characteristically, plinthite has low organic matter, as it is generally formed at depth in the soil. However, erosion and truncation of the soil may expose the plinthite at the surface, and there may be a secondary enrichment with organic matter.

B. Petroplinthite

Petroplinthite may result from the hardening of plinthite or may be formed directly in the soil. Its genesis is discussed later. Petroplinthite has been referred to as ironstone or lateritic gravel.

> Petroplinthite is nodular or pisolitic material with a hard crust of closely crystallized goethite and/or haematite, enclosing iron enriched soil material; sometimes adjoining nodules may be joined or cemented together or the petroplinthic material may be dendritic in form. The soil material within the petroplinthite is enriched with sesquioxides and may contain gibbsite and manganese minerals.

Petroplinthite frequently occurs in the soil as loose or slightly cemented gravel. The fabric and mineralogy of petroplinthite are provided by Eswaran and Raghumohan (1973) and by Eswaran et al. (1980). The closely crystallizing network of goethite crystals provides the rigidity of the material.

C. Petroferric Contact

Petroferric contact is defined in *Soil Taxonomy* as:

> a boundary between soil and a continuous layer of indurated material in which iron is an important cement and organic matter is absent or is present only in traces. The indurated layer must be continuous within the limits of a pedon but may be fractured if the average lateral distance between fractures is >10 cm.

The processes resulting in a petroferric contact are different from those that lead to plinthite or petroplinthite formation, and are discussed later. The petroferric contact is a root- and water-restricting layer, and when it occurs at shallow depth, it reduces the volume of the soil; when it is exposed at the soil surface, there is no soil for plant establishment.

III. Processes Leading to Accumulation of Sesquioxides

A. Chemistry of Major Elements

Weathering and mineral transformation release cations, and the fate of these elements depends on their solubility and mobility, which is also a function of the pH and redox potential of the system. Alkali and alkaline earth elements (A + AE) form hydrated cations with very low acidity strength (pK_a values >10). As a result they do not form hydroxides in the normal pH range of the soils. Cl, S, P, and C also do not form insoluble hydroxides but may precipitate as salts, e.g., FeS, $Ca_3(PO_4)_2$, $CaCO_3$.

From the point of view of formation of plinthite and related forms, the elements of interest are Si, Al, Fe, and Mn. Silica as a cation, $Si^{+4}H_2O$, constitutes a very strong acid ($pK_a = -1.2$) and is commonly present as a hydroxide, $Si(OH)_4$, with limited molecular solubility (Charlot, 1966; Millot, 1964); when precipitated, it polymerizes by splitting up H_2O molecules. Al, on the other hand, forms a hydrate—$Al^{3+}6H_2O$—with intermediate acidity ($pK_a = 5$), and as a result, below pH of 4.2 it is rather soluble but at higher pH it precipitates as $Al(OH)_3$, which may crystallize as gibbsite (Hem and Roberson, 1967). The solubility of gibbsite is low (K_sp about 32), and so once it is precipitated, it is a rather stable mineral. Fe and Mn are rather special elements, as they can be reduced (Fe^{2+} and Mn^{2+}) or oxidized (Fe^{3+} and Mn^{3+} or Mn^{4+}), as shown by Krauskopf (1967). The reduced compounds form hydrates with low acidity (Pk_a 9.5 and 10.6, respectively) and the presence of the pure $Fe(OH)_2$ or $Mn(OH)_2$ has not been reported. Fe^{3+}, Mn^{3+}, and Mn^{4+} constitute rather strong to strong acids (pK_a less than 3) and therefore are commonly present as insoluble hydroxides or dehydrated oxides at the normal pH range of soils. These fundamental differences in chemical properties of the reduced and oxidized forms cause the solubility to depend on the environmental conditions. The most simple expressions of these differences are the following reactions:

$$Fe_2^{+}6H_2O \rightleftarrows Fe(OH)_3 + 3H^+ + 3H_2O \qquad E_0 = +1.06\ V \qquad (1)$$

$$Mn_2^{+}4H_2O \rightleftarrows Mn(OH)_4 + 4H^+ + 4H_2O \qquad E_0 = +1.23\ V \qquad (2)$$

Besides the redox potential, the presence of H^+ in these reactions indicates the important role of pH in these equilibriums.

Due to its high solubility, Si is easily leached out of the soil system. In the presence of Al^{3+} ions, the solubility of Si is decreased as the entities coprecipitate as amorphous gels, which according to Millot (1964) may age to kaolinite. The solubility and mobility of aluminum are complex and and, as shown by De Coninck (1980), are controlled by concentration and the presence or absence of organic matter. When organic matter is present, chelation of Al occurs and podsolization results; in the absence of organics, gibbsite precipitates. Al and Fe play important roles in the two pedogenetic processes—podzolization and plinthization—and the fundamental difference is the intensity of the role of organic matter. In the formation of plinthite, petroplinthite, or petroferric contact, organic matter has a minimal role or at best an indirect role.

As a result of the low solubility of Fe^{3+} and Mn^{4+}, the mobility is also low. These elements may move as chelates with organic compounds as in podsolization, or they may form coatings on clay minerals and move with the clays. Neither mechanism can result in the accumulations of the elements of the magnitudes required for plinthite or related forms. Consequently, a reduction to convert the iron and manganese to bivalent forms must take place before they can be transported and accumulated.

Although reduction is frequently associated with water saturation, a saturated environment by itself cannot bring large quantities of iron in the bivalent form. Organic matter-free systems, as shown by Schwertmann (1985), do not have constituents to provide electrons to the oxidized compounds. This is evident in soils with aquic soil water regimes, where bleaching is stronger in the surface horizons than in the subsurface horizons. During the microbial decomposition of organic matter, microorganisms utilize the Fe^{3+} (and Mn^{3+}, Mn^{4+}) oxides as final electron acceptors in their oxidative decomposition of organic matter (Brummer, 1973). Consequently, the presence of organic matter accelerates the process of reduction. However, Herbillon and Nahon (1985) have established that reduction may occur in deep-seated situations where the system is free of oxygen, as exemplified by the "pallid zone" in many deep weathering profiles. Figures 1 and 2, based on Equations (1) and (2) presented earlier, explain the reactions taking place. But in actual soil situations, particularly in the organic matter-rich parts of the soil, the role of microorganisms considerably accelerates the processes (Ottow, 1973).

B. Formation of Sesquioxide Accumulations

It was indicated earlier that plinthite, petroplinthite, and petroferric contact are related forms but that each has its own pathway of formation. A very early stage is depicted by the formation of mottles in soil subject to a fluctuating water table. The mottles are of two forms: One represents a zone of iron removal, and the other, a zone of iron enrichment. In *Soil Taxonomy*, the emphasis is on the former and the term, *mottles of chroma 2 or less*, is used to differentiate them. If the zone was one of complete

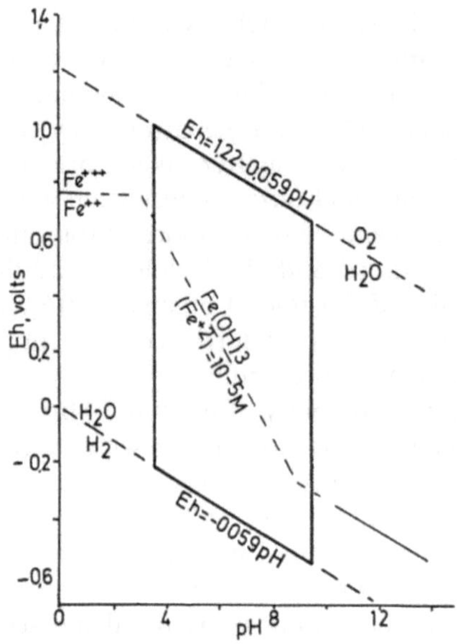

Figure 1. Eh-pH diagram for the reaction

$$Fe^{+} + 3H_2O \rightleftarrows Fe(OH)_3 + 3H^{+} \qquad E_0 = +1.06\ V$$

$$[Fe^{+2}] = 10^{-5}\ M$$

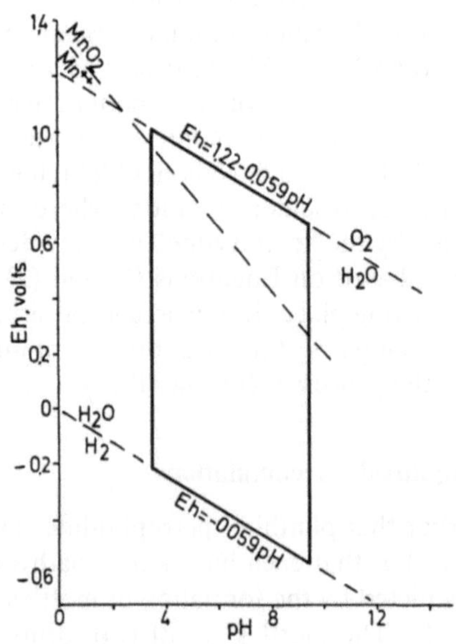

Figure 2. Eh-pH diagram for the reaction

$$Mn^{+2} + 4H_2O \rightleftarrows MnO_2 + 4H^{+} + 2H_2O \qquad E_0 = +1.23\ V$$

$$[Mn^{2+}] = 10^{-5}\ M$$

reduction, the whole zone would have a chroma of 2 or less and the capillary fringe would be where enrichment occurs. In many soils, the low-chroma mottles occur as patches, with the iron-enriched zones, which may be nodules or concretions, interspersed between them. This is a strong argument for the role of microorganisms in the process of Fe reduction.

In the Fe^{2+} form, iron is mobile and is transported to points where accumulation takes place through accretion. Once a nucleus is formed, the nodule grows as the nucleus serves as a template for further deposition. If there is periodicity in the supply of iron, concretions with a concentric fabric result; if there is a continuous supply, nodules with a diffuse fabric result. The low-chroma zones become a permanent feature of the soils, and the morphology may persist even if the water table is lowered. It is for this reason that emphasis is placed on this to define aquic soil water regimes.

The localized redistribution of iron and manganese results in mottles and with time may form nodules and concretions (Stoops, 1970). However, for plinthite formation, studies suggest that a further supply of iron is necessary, and this usually comes from the weathering zone, which may be in a permanently reduced situation. The groundwater table recharges the mottled zone (Gallaher et al., 1974) with iron, and this absolute accumulation of iron is responsible for the considerable enrichment that leads to plinthite. The reticulate pattern is inherited from the processes described earlier, whereby iron is reduced and moves locally.

Plinthite is important because its presence indicates some restrictions to water and root penetration. Daniels et al. (1978) have made extensive observations on this aspect, and they indicate that about 10% platy plinthite is required to perch water. They also add that it is this amount when identification with an auger becomes positive and that the underlying reticulate mottled zone becomes well expressed. Table 1 and Figure 3 show the role of about 15% to 25% plinthite on the perching of water.

Table 1. Percent of measurements[a] that plinthitic and overlying horizons were water-saturated

	Varina			Dothan	
Series site number	1	2	3	1	2
Horizons above plinthite	12	8	11	20	8
Plinthite and reticulate mottled horizon	17	28	24	42	37
Number of measurements	119	119	119	88	86
Percent plinthite (of measurements)	25+	25+	25+	15	15
Depth to plinthite (cm)	107	110	100	100	127
Age of geomorphic surface	—Early Pliocene—			—Mid-Pliocene—	

[a] Measurements were 52 months for Varina and 27 months for Dothan.

Source: After Daniels et al. (1978).

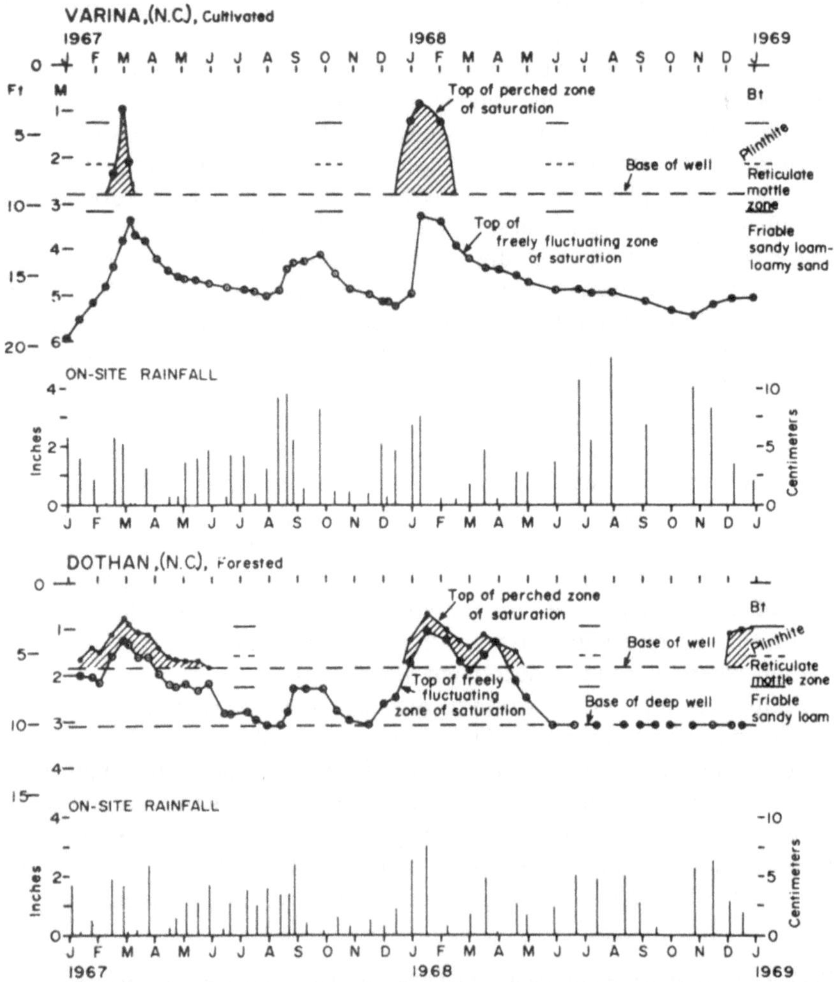

Figure 3. Plot of two-year water table and inferred zone of saturation for Varian and Dothan pedons in North Carolina. The zone of saturation is inferred to be from the base of the well to the water table. Adjacent wells opening in or above the plinthite horizon have similar water table elevations as those opening in the reticulately mottled horizons. Each site had three wells cased with 2.54-cm metallic tubing. Data from only two of the wells are shown because the water table elevations were very nearly the same for both wells in or above the plinthite horizon. All soils have platy plinthite. The rainfall is the amount collected in on-site rain gauges between sampling dates. [*Source*: Daniels et al. (1978). Reprinted with permission from the Soil Science Society of America.]

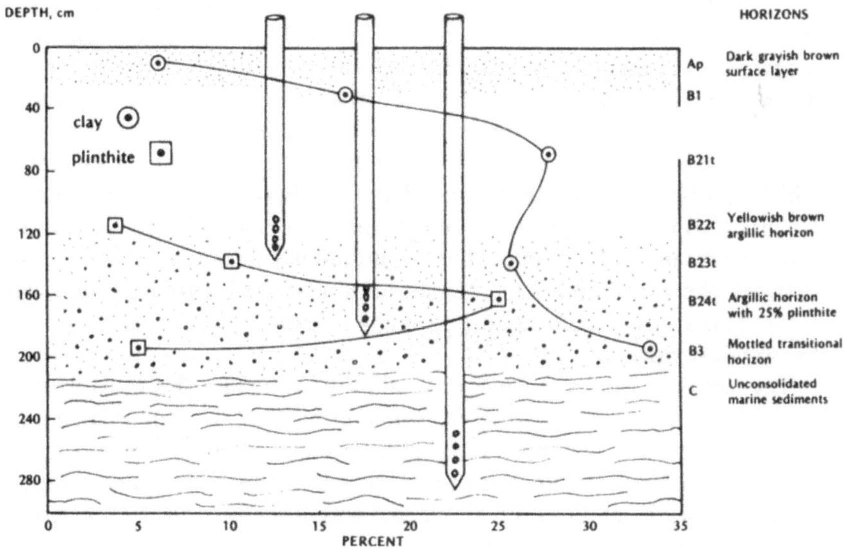

Figure 4. Placement of piezometers and percentages of clay and plinthite in relation to morphology of Dothan soil. [*Source*: Guthrie and Hajek (1979). Reprinted with permission from Soil Science Society of America.]

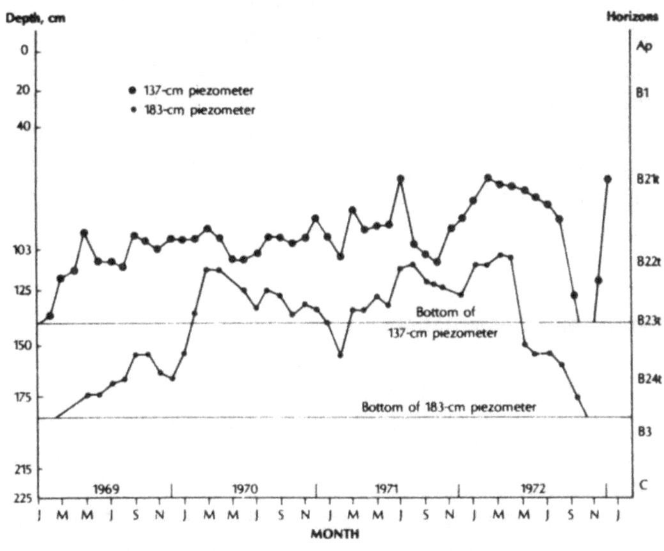

Figure 5. Depth to water table in Dothan soil, January 1969 to December, 1972 (all well pumped dry on May 1, 1972); vegetation at the study site was mainly broomsedge (*Andropogon virginicus* L). [*Source*: Guthrie and Hajek (1979). Reprinted with permission from Soil Science Society of America.]

Guthrie and Hajek (1979) made a more detailed appraisal of the role of the water table in the Dothan soil. Figure 4 shows the placement of the piezometers in their study, and Figure 5 shows the depth of the water table during the study period. They showed that in this soil, a perched water table fluctuates in one continuous zone of saturation above the horizons containing appreciable amounts of plinthite; another zone of saturation, which is discontinuous, is bounded on the bottom by the B3 horizon (Figure 4). Finally, the absence of a water table in the piezometer in the C horizon was used as evidence that a continuously unsaturated zone existed between the continuous zone of saturation and the regional ground water table. This study established the perching role of plinthic material reported by previous workers.

The fact that water perched on the plinthite zone and also moved laterally was established in the study of Blume et al. (1987). They placed a Br^- tracer in a plinthic soil and found greater concentrations above the plinthite than beneath it; they also found that Br^- was carried downslope by lateral water movement both above and within the plinthite. This observation is important for the formation of the petroferric contact. Blume et al. (1987) also showed that significant amounts of Br^- could be detected beneath the plinthite zone, which supports the earlier observations of Daniels et al. (1978) that plinthite is only semipermeable.

Active petroplinthite formation is normally observed at foot slope positions, which suggests a laterally moving groundwater. Petroplinthite also forms in the soil if there is a high and absolute supply of iron by the groundwater table. An initial stage may be plinthite.

In the case of petroferric contact, the processes are similar. In this case, the plinthite or petroplinthite is recemented by a rapid flux of iron; the cement is laminated, indicating that the duration was short and cyclic and that the supply was very high. A petroferric contact may also form on exposed ironstone gravels that were brought to the surface through uplift and erosion.

C. Hardening and Cementation

Hardening and cementation must be distinguished from aggregation and growth by accretion. Aggregation is due to formation of strong chemical bonds between free hydroxides with a positive charge and negatively charged clay minerals or organic matter.

The origin of the cementation is not clearly established. Recently, Chadwick and Nettleton (1988) suggested that cementation in soils generally may be due to two different chemical processes: (1) crystalline cements held together by ionic chemical bonds within the compounds, which occurs in petrogypsic and petrocalcic horizons; (2) cements composed of elements that form covalent bonds within the cementing agents and with other soil constituents. The latter occurs when the accumulating pedogenetic compounds are composed of Si, Al, Fe, Mn, etc. (elements with intermediate electronegativity; Sanderson, 1964). Even in the crystalline form these compounds have outer surfaces composed of OH_2-OH groups, which can always be used to form covalent bonds with the same elements. This process would allow very close approach and therefore a high density of the cement.

Cementation and hardening of petroplinthite, however, result from a close packing of the crystals (Schwertmann, 1985). In the fromation of petroplinthite nodules, the last stage is the formation of the crust. Scanning electron microscopic studies of the crust of the nodules (Stoops, 1970; Eswaran and Raghumohan, 1973; Eswaran et al., 1980) show the closely packed goethite crystals. The acicular crystals are aligned perpendicular to the surface, and this gives the crust its rigidity. In the core of the petroplinthite, however, the goethite crystals are lenticular in form. They occur as rosettes or clumps and are friable or easily crushed between the fingers. The crust with its closely packed acicular goethite crystals was never encountered in plinthite nodules in studies by the authors. This explains why the plinthic material is easily crushed between the fingers and why it stains the fingers.

The lamellar cement binding the petroplinthite nodules in the petroferric contact is also composed of closely packed goethite. Consequently, the cementing due to formations of closely packed macrocrystals is responsible for the hardening (Schwertmann, 1985).

The hardening also requires an alternate wetting and drying situation. It appears that drying should be to the extent that the moisture tension is close to 15 kp. The hardening process normally is subsequent to geomorphic changes in the landscape. Uplift of the land mass, or lowering of the groundwater table, results in drying up of the soil material. In aeas with an ustic soil moisture regime, the hardening is accelerated. Uplift is always accompanied by erosion, which would eventually expose the sesquioxide accumulation zone. This is most evident at the edges of the uplifted zones. The hardened petroplinthite is resistant to further erosion, and escarpments are formed at these edges. These hardened rims of the uplifted zones now control the hydrology of the zone and set the stage for petroferric contact to be formed in the soils within the zone.

On the older continents, such as West Africa and India, uplift is followed by peneplanation. The petroplinthite is broken and transported as gravel and occurs in the soils of the peneplain as a layer a few centimeters to more than 1 m thick. These petroplinthic gravels may have no pedogenetic relationship to the soil in which they occur, and their behavior and roles are similar to stones of quartz or granite in other alluvial/colluvial soils. After deposition, a new set of soil-forming processes commences. If an aquic soil moisture regime prevails, recementation of the petroplinthite may commence again leading to a petroferric contact. The cycle repeats itself.

Table 2. Particle size distribution and Fe_2O_3 content of Plinthite horizons

Series	Material	Sand (%)	Silt (%)	Clay (%)	Fe_2O_3 (%)
Carnegie	White matrix	51.1	10.2	38.7	0.97
	Yellow mottle	57.4	11.8	30.8	4.90
	Red plinthite	64.9	13.9	21.2	14.90
Cowarts	Yellow matrix	57.1	5.9	37.0	0.46
	Yellow mottle	53.6	10.2	36.2	4.33
	Red plinthite	65.7	15.0	19.3	7.10
Fuquay	Light gray matrix	53.0	7.5	39.5	0.26
	Brownish yellow mottle	68.4	11.8	19.8	1.87
	Red plinthite	65.1	15.5	19.4	14.46
Irvington	Light gray matrix	54.0	12.2	37.8	0.46
	Yellowish brown mottles	62.5	16.7	20.8	3.29
	Red plinthite	61.7	19.1	19.2	8.04
Tifton	Brownish yellow matrix	50.0	12.2	37.8	4.90
	Dark red plinthite	71.2	17.6	11.2	12.33

Source: After Wood and Perkins (1976).

D. Properties of Sesquioxide Accumulations

Many studies have shown that associated with the formation of plinthite is a particle-size change. The plinthic material feels sandier and more brittle. Table 2 illustrates this in some soils of Georgia. Wood and Perkins (1976) analyzed the pale soil matrix, mottles, and plinthite and established that there was a decrease in measured clay. They do not indicate the reason for this observation, but it is probably related to aggregation of clay and silt to form silt and sand-sized particles, which has been well established in iron-rich soils. The marked changes are in the free iron content, which shows a 5- to 10-fold increase from the pale matrix to the plinthic materials. Also associated with the increase of iron is the presence of crystalline species of iron such as goethite and haematite.

IV. Occurrence and Distribution

Plinthite and related forms occur in a range of soils. In the literature, many workers have erroneously related them to the deeply weathered soils of the tropics, mainly the Oxisols. Although the process of sesquioxide accumulation is related to weathering, the soil need not be in the ultimate stage of soil formation for these accumulations. The classes provided in *Soil Taxonomy* (Table 3) illustrate the range of soils where they occur. Within any

Table 3. Classes of soils characterized by presence of plinthite[a]

Great groups	Order
Plinthaquepts	Inceptisols
Plinthaquox	Oxisols
Plinthaqualfs	Alfisols
Plinthoxeralfs	Alfisols
Plinthustalfs	Alfisols
Plinthaquults	Ultisols
Plinthohumults	Ultisols
Plinthudults	Ultisols
Plinthustults	Ultisols

[a]Other great groups may have plinthic subgroups.

class, the accumulation may be allochotonous or autochtonous. When the petroplinthite occurs as a band with sharp upper and lower boundaries, the boundaries representing discontinuities in the soil.

Areas with active plinthite formation are associated with relatively young landscapes with a fluctuating water table in the soil. A laterally moving water table may also contribute to the formation of plinthite. The Kerala Coast in South India and some areas in Southeast Asia have soils with current plinthite formation. Characteristically, such soils are associated with other soils, generally of sedimentary origin and with petroplinthite gravel. Figures 6 and 7 show areas in Southeast Asia and in India with soil associations containing plinthite and petroplinthite. In the eastern part

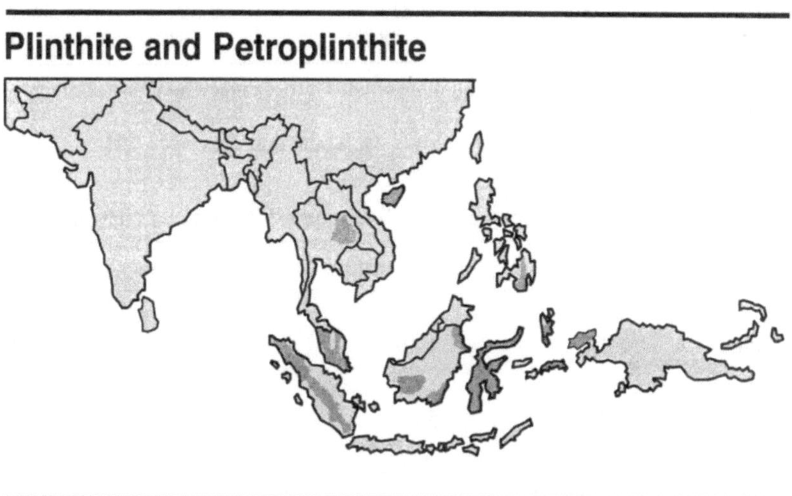

Figure 6. Distribution of plinthite and related forms in Southeast Asia.

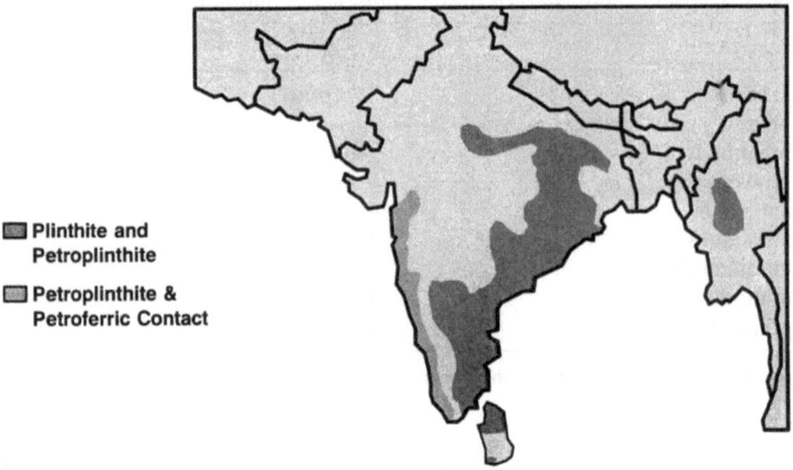

Figure 7. Distribution of plinthite and related forms in South Asia.

Table 4. Estimates of area occupied by soils with plinthite and related forms[a]

Country	Plinthite	Petroplinthite	Petroferric	Total
Indonesia	50	9,553	10,397	20,000
Laos	10	75		85
Malaysia	5	60		65
Philippines	5	25		30
Thailand		100		100
Total	70	9,813	10,397	20,280
Percent	0.35	48.39	51.27	100

[a]Note that some of the soils with petroplinthite or petroferric contacts may have plinthite underneath. Areas in '000 ha.

Source: FAO (1979). *Soil map of the world*, Vol. IX, South East Asia.

Table 5. Estimates of area occupied by soils with plinthite and related forms[a]

Country	Plinthite	Petroplinthite	Petroferric	Total
India	250	4,242	5,676	10,168
Sri Lanka	40	1,107		1,147
Burma		317		317
Total	290	5,666	5,676	11,632
Percent	2.49	48.71	48.80	100

[a]Note that some of the soils with petroplinthite or petroferric contacts may have plinthite underneath. Areas in '000 ha.

Source: FAO (1978). *Soil map of the world*, Vol. VII, South Asia.

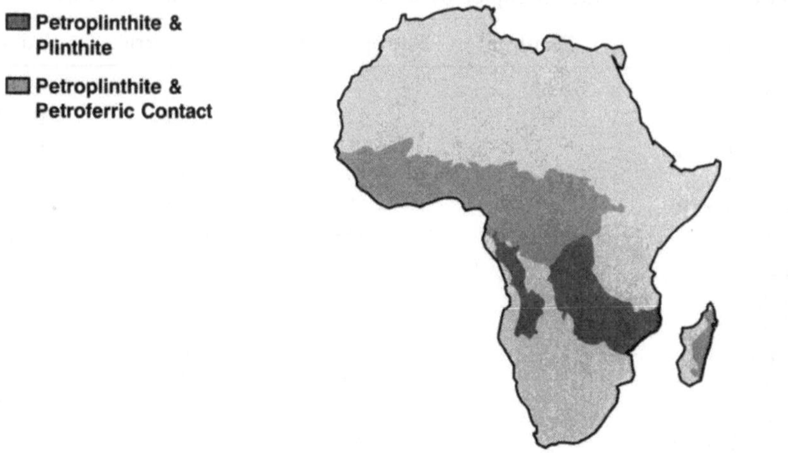

Figure 8. Distribution of plinthite and related forms in Africa.

of India, due to uplift and peneplanation, the dominant alfisols have petroplinthic gravel. A similar situation exists in Northeast Thailand and in Cambodia and Laos. Tables 4 and 5 provide estimates of the distribution of such soils in Southeast Asia and in South Asia.

The largest continguous extent of such soils is in West Africa (Figure 8, Table 6). They are present as a band from the west coast of Gambia in the north to Cameroon in the south and extend to southern Sudan and eastern Zaire. South of this band, soils with petroplinthite are still common, particularly in the soils of the Southern African Plateau. West Africa also has the largest extent of soils with petroferric contact. These generally occur on the mid-Tertiary surfaces of the Ivory Coast, eastern Ghana, Burkina Fasso, and the Central African Republic. In Namibia and southern Africa, the Kalahari sands mask their presence, if any, and they occur sporadically in the eastern part of southern Africa.

V. Management Implications

Formation of plinthite, petroplinthite, or the petroferric contact is a natural soil-degradation process. It is a degradation process due to the fact that their presence reduces the quality of the soil for agriculture and engineering purposes. Their presence reduces the effective soil volume available for moisture and nutrient storage. They, particularly the hardened forms, are obstacles for root penetration.

In general, plinthite or its related forms are formed in the subsoil. When

Table 6. Estimates of area occupied by soils with plinthite and related forms[a]

Country	Plinthite	Petroplinthite	Petroferric	Total
Angola	621			621
Benin	241	7,142	11,888	19,271
Burkina Fasso		6,067	8,421	14,488
Cameroon		5,677	747	6,424
Central Afr. Rep.		24,194	3,309	27,503
Chad		158	6,126	6,284
Congo		5,345		5,345
Ethiopia	859			859
Gabon		2,299		2,299
Gambia		35	2,476	2,511
Ghana		5,605	4,087	9,692
Guinea		2,293	9,858	12,151
Guinea-Bissau		320		320
Ivory Coast		6,067	4,616	10,683
Kenya		307		307
Liberia		9,168		9,168
Madagascar		1,036	1,140	2,176
Malawi	15		450	15
Mauritania				450
Mozambique	48			48
Niger		52	1,278	1,330
Nigeria		15,941	4,176	20,117
Senegal		2,576	5,935	8,511
Sierra Leone		3,476	1,177	4,653
Sudan		15,101	11,962	27,063
Tanzania	907			907
Togo		2,086	1,301	3,387
Uganda	532			532
Zaire	23	63	157	243
Zambia	9			9
Zimbabwe	48			48
Total	3,303	115,008	79,104	197,415
Percent	1.67	58.26	40.07	100

[a]Note that some of the soils with petroplinthite or petroferric contacts may have plinthite underneath. Areas in '000 ha.

Source: FAO (1976). *Soil map of the world*, Vol. VI, Africa.

plinthite is present, some water perches above this layer and potential uses of the soil is restricted to crops requiring water saturation, such as rice. Plinthite by itself poses few constraints to agricultural use of the soil. However, erosion brings the plinthic material close to the surface and even exposes it to atmospheric conditions. This initiates the process of hardening and cementation. In many parts of the world, such as in West Africa and in peninsular India, geologic uplift has exposed the plinthite. This, coupled with global climatic changes—increased aridity—in these parts of the world, has resulted in extensive areas of hardened petroplinthite or petroferric contact at the soil surface or at shallow depths.

Due to population pressures, these soils are being cultivated, resulting in erosion rates of 50 to 100 t ha^{-1} yr^{-1}. In many areas, the petroferric material is exposed and, as it is as hard as concrete, the soil cannot be cultivated and is abandoned. This is turn causes more erosion and soil loss.

The solutions are not easy to come by. The first step is detailed soil surveys to demarcate the areas with plinthite and also the other degraded soils. The soils with plinthite must be carefully managed to reduce soil loss. Intensive soil conservation measures such as terracing must be adopted.

The management of the already barren plateaus where petroplinthite and petroferric contacts outcrop presents a different kind of challenge. The rehabilitation measures are capital-intensive and require digging pits into the crust and planting tree crops. Revegetation of such lands is a slow and time-consuming process but is the only option for the use of the land. Such lands cannot be used intensively for several generations to come, and so reforestation is the only solution. The presence of such barren lands should serve as a warning to all of the long-term repercussions of mismanaging plinthic soils.

In the humid tropics, such as in Malaysia, petroplinthic soils have been used successfully for rubber and oil-palm cultivation. The yields are about 50% lower than in other soils, and in addition, the maturity period for harvesting is about two to four years longer on these petroplinthic soils.

VI. Conclusion

Soils with petroplinthite or petroferric contact, and to a lesser extent with plinthite content, are some of the more problematic soils in the world. There are no good estimates of the area of such soils under cultivation, but in general, they are the last soils to be used by the native population. They present additional problems in the semiarid tropics, where they are the most extensive.

They are most extensive in the tropics, where most of the developing countries are located. Their productivity is low and, specifically for the plinthite soils, they can be easily be mismanaged, resulting in soil degradation. Regretfully, these soils have not received much research attention.

As detailed soil maps are few in the developing countries, their exact distribution is not known. Due to population and socioeconomic pressures, these soils are coming increasingly under cultivation. The challenge of the next decade is to develop the technology to utilize such soils for sustainable agriculture.

References

Alexander, L.T., and J.G. Cady. 1962. Genesis and hardening of laterite in soils. Tech. Bull. 1282, Soil Conservation Service, U.S. Dept. Agric.

Blume, L.J., H.F. Perkins, and R.K. Hubbard. 1987. Subsurface water movement in an Upland Coastal Plain soil as influenced by plinthite. *Soil Sci. Soc. Am. J.* 51:774–779.

Brummer, G. 1973. Redox reaktionen als merkmalspragende Prozesse hydromorpher boden. pp. 17–28. In: E. Schlichting and U. Schwertmann (eds.), *Pseudogley and gley*. Verlag Chemie, Weinheim, West Germany.

Buchanan, F. 1807. A journey from Madras through the countries of Mysore, Canara and Malabar. East India Co., London, 2:440–441.

Chadwick, F., and D. Nettleton. 1988. Micromorphologic evidence of adhesive and cohesive forces in soil cementation. In: Press. Proc. Int. Work. Meet. Soil Micromorphology, San Antonio, Texas.

Charlot, G. 1966. *Les methodes de la chimie analytique*. Masson et Cie, Paris.

Daniels, R.B., H.F. Perkins, B.F. Hajek, and E.E. Gamble. 1978. Morphology of discontinuous phase plinthite and criteria for its field identification in the Southeastern United States. *Soil Sci. Soc. Am. J.* 42:944–949.

De Coninck, F. 1980. Major mechanisms in formation of spodic horizons. *Geoderma* 24:101–128.

Eswaran, H., J. Comerma, and V. Sooryanarayanan. 1980. Scanning electron microcopic observations on the nature and distribution of iron minerals in plinthite and petroplinthite. pp. 335–341. *Lateritisation Processes*, Proc. Int. Sem. Laterisation Processes. Oxford and IBH Publishing Co., New Delhi.

Eswaran, H., and N.G. Raghumohan. 1973. The microfabric of petroplinthite. *Soil Sci. Soc. Am. Proc.* 37:79–81.

FAO. 1976, 1978, 1979. *Soil map of the world.* Vol. VI, Africa (1976); Vol. VII, South Asia (1978); Vol. IX, South East Asia (1979). FAO-UNESCO, Rome.

Gallaher, R.N., H.F. Perkins, and K.H. Tan. 1974. Classification, composition, and mineralogy of iron glaebules in a Southern Coastal Plain soil. *Soil Sci.* 117:155–164.

Guthrie, R.L., and B.F. Hajek. 1979. Morphology and water regime of a Dothan soil. *Soil Sci. Soc. Am. J.* 43:142–144.

Hem, J.D., and C.E. Roberson. 1967. Form and stability of aluminum hydroxide complexes in dilute solution. Geol. Survey, Water Supply Paper 1827. U.S. Govt. Printing Office, Washington, D.C.

Herbillon, A.J., and D. Nahon. 1985. Laterites and laterization processes. pp. 779–796. In: J.W. Stucki, B.A. Goodman, and U. Schwertmann (eds). *Iron in soils and clay minerals.* NATO ASI Series, D. Reidel Publishing Co., Dordrecht, Netherlands.

Kellogg, C.E. 1949. Preliminary suggestions for the classification and nonmenclature of great soil groups in tropical and equatorial regions. Commonwealth Bureau Soil Sci., Tech Commun. 46:76–85.

Krauskopf, K.B. 1967. *Introduction to geochemistry*. McGraw-Hill, NY.

Maignien, R. 1966. Review of research on laterites. UNESCO, Paris.

Millot, G. 1964. *Geologie des argiles*. Masson & Cie, Paris.

Ottow, J.C.G. 1973. Bacterial mechanisms of iron reduction and gley formation. pp. 29–36. In: E. Schlichting and U. Schwertmann (eds.), *Pseudogley and gley*. Verlag Chemie, Weinheim, West Germany.

Sanderson, R.T. 1964. *Chemical periodicity*. Reinhold, NY.

Schwertmann, U. 1985. Some properties of soil and synthetic iron oxides. pp. 203–204. In: J.W. Stucki, B.A. Goodman, and U. Schwertmann (eds.), *Iron in soils and clay minerals*. NATO ASI Series, D. Reidel Publishing Co., Dordrecht, Netherlands.

Soil Survey Staff. 1975. *Soil taxonomy: A basic system of soil classification for making and interpreting soil surveys*. U.S. Dept. Agric. Handbook 436, U.S. Govt. Printing Office, Washington, DC.

Stoops, G. 1970. Scanning electron microscope applied to the micromorphological study of laterite. *Pedologie* 20:268–280.

Sys, C. 1968. Suggestions for the classification of tropical soils with lateritic materials in the American classification. *Pedologie* 18:189–198.

Van Schuylenborgh, J. 1973. Sesquioxide formation and transformation. pp. 93–103. In: E. Schlichting and U. Schwertmann (eds.), *Pseudogley and Gley*. Verlag Chemie, Weinheim, West Germany.

Wood, B.W., and H.F. Perkins. 1976. Plinthite characterization in selected Southern Coastal Plain soils. *Soil Sci. Soc. Am. J.* 40:143–146.

Soil Erosion and Land Degradation: The Global Risks

R. Lal*

I. Introduction

The total land area of the world exceeds 13 billion hectares, but less than half of it can be used for agriculture, including grazing. The world's potentially arable land is estimated at 3031 million hectares, or 22% of the total land area. The potential cultivable land is distributed as follows: 2154 and 877 million hectares, respectively, in developing and developed countries representing 28% and 15% of the land area (Dudal, 1982). About 1461 million hectares or 40% of the world potentially arable land is cultivated (Dregne, 1982; Dudal, 1982), with 784 and 677 million hectares representing 36% and 77% of the potentially cultivable land in developing and

*Department of Agronomy, The Ohio State University, Columbus, OH 43210, U.S.A.

developed countries, respectively. Ironically, the 1461 million hectares of land now being cultivated does not include an estimated 2000 million hectares of once biologically productive land that has been degraded or destroyed.

Productivity of cultivated land may decline from severe soil degradation. For example, the land productivity classes include class I, extremely high; class II, very high; class III, high; class IV medium; class V, low; and class VI, very low. The percentage of land in classes V and VI combined is 57.4 in Africa, 63.7 in Asia, 69.6 in Australia and New Zealand, 57.6 in Europe, 78.5 in North and Central America, 22.0 in South America, and 59.4 in the world (Dregne, 1982). Additionally, percentages of productivity classes V and VI are likely to increase.

A major factor responsible for the degradation of the natural resource is accelerated soil erosion (Kovda, 1977). It is estimated that accelerated soil erosion has irreversibly destroyed some 430 million hectares in different countries. That is about 30% of the present cultivated land area of the world. Worldwide natural erosion is estimated to total some 9.9 billion tons of soil a year, but human-induced accelerated erosion is more than 2.5 times higher—26 billion tons/per year (Holdgate et al., 1982; O'Keefe, 1983; Brown, 1984).

In general, soil erosion is more severe in mountainous than in undulating terrains. Kadomura and Yamamoto (1978) estimated the rate of natural erosion for the world to be on the order of 1.5 to 7 t ha^{-1} yr^{-1} for mountainous regions and 0.1 to 7 t ha^{-1} yr^{-1} for undulating uplands. If global warming trends continue (caused by increases in atmospheric concentrations of CO_2, expected to reach 600 ppm by 2070), global erosion rates may increase considerably (Rind, 1983).

In the context of this volume, *tropic* is defined as regions lying within 23° north to south of the equator. Subtropical regions extend between 23° and 35° north and south of the equator, and temperate regions lie beyond latitude 35°. Tropical regions are further classified by the number of wet months they experience: humid, more than 10; subhumid, 5 to 8; semiarid, 3 to 5; and arid, less than 3. The climax vegetations of these regions are, in order, tropical rain forest, semideciduous rainforest, savanna, and scrub gasses (Lal, 1987). Predominant soils of the humid and subhumid tropics are highly leached oxisols, ultisols, and alfisols, and less leached inceptisols. Soils of the semiarid and arid regions are not leached. Their most predominant soils are vertisols, alfisols, inceptisols, entisols, and aridisols. Characteristics of various ecological regions are discussed in detail by Lal (1987).

II. Technical Data on Soil Erosion

The voluminous technical literature on soil erosion comprises data on rates of soil erosion in relation to landforms, climate, land use, management systems, and natural and artificial factors. But systematic evaluations of

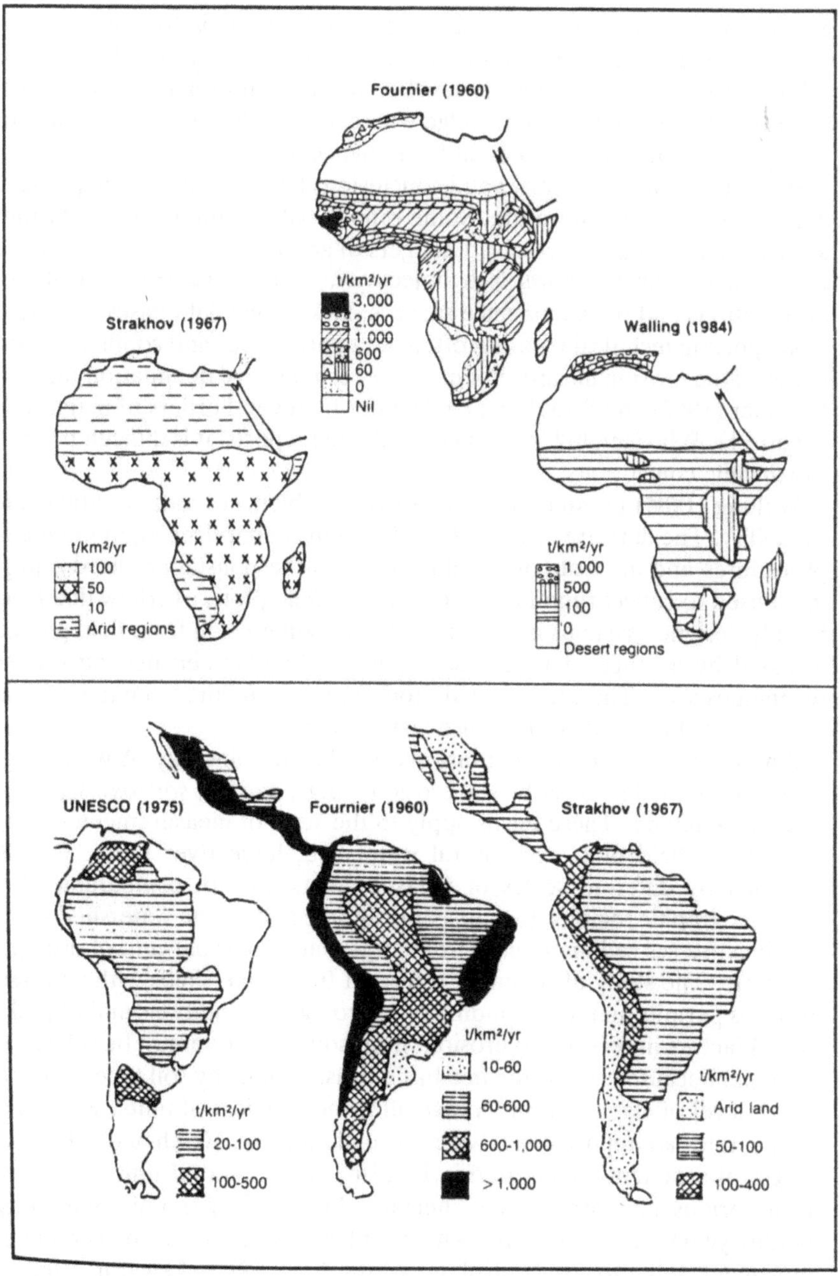

Figure 1. Comparison of suspended sediment yield maps of Africa and South America (Lal, 1988).

erosion processes to establish cause-effect relationships for different climates, ecological regions, geological factors, etc., are few. The literature is full of horror stories. From the level of technical information presented, however, it is often difficult to judge whether an author is "crying wolf" or the threat to natural resources and the environment is genuine.

It is often difficult to draw valid conclusions from the mass of haphazard data available in the literature. Some reports with emotional connotations always indicate real or perceived dangers of accelerated erosion to soil and environments. Such reports lack objectivity. Others, based on qualitative observations and reconnaissance surveys, lack strong data bases. The lack of supporting technical data has often led to erroneous and conflicting estimates of the erosion hazard. For example, estimates of denudation rates by Fournier (1962) and Strakhov (1967) differ by several orders of magnitude (Figure 1). Who is right? With such conflicting reports, it is difficult to draw worthwhile conclusions.

Walling (1984) pointed out some of the problems of data accuracy and reliability. The data are often collected by nonstandardized equipment and procedures and are not comparable. Besides the nonstandardized equipment used to collect runoff and erosion from field plots, overflow due from faultily designed systems is often a problem. Soil erosion from field plots is assessed by a range of techniques. The results obtained depend on the techniques used. The data often are too sparse to support estimates on soil, rainfall, weather records, land-use history, etc.

Another weakness is the confusion caused by terminology. A wide range of terms are used to describe soil erosion—e.g., *erosion, soil loss, sediment yield, denudation.* These terms apply to the scale of measurements—e.g., small area, field plot, agricultural watershed, large river basin. Results obtained by different scales of measurements are not comparable. For large geographic regions, the most appropriate scale is the watershed scale, which expresses sediment yield from large watersheds per unit of area per unit time. The study of sediment transport from watersheds with a variety of scales permits an understanding of hydrological conditions and of pedological factors in relation to erosion. Still, soil erosion cannot be related to soil type, because each watershed basin consists of many soil types, even at a high level of classification. Above all is the problem of units. Results of erosion rates are reported in a wide range of units: soil depth and volume or weight of sediments (measured in English, metric, and SI units), removal from various unit areas (acre, hectare, km^2, mi^2) per unit time (day, month, year). It is not surprising that planners' response to schemes to implement soil erosion-control measures has been lukewarm. Neither should we be surprised that there are so few success stories about controlling erosion. There has, in fact, been little progress in controlling erosion anywhere in the world. Is it difficult to decide who should be blamed—scientists, planners, or farmers.

III. Soil Erosion by Water

The vigor of measurements and the length of time for data collection are relatively less for most tropical regions than for the temperate zone. The literature surveyed below is presented by major geographic regions.

A. Soil Erosion in the Tropics

1. Tropical Africa

Tropical Africa lies between 23° north and south of the equator. The literature regarding Africa's soil erosion, erosion-related degradation, and its perpetual food deficit is voluminous (Fournier, 1967; Ahn, 1975a, 1975b; Golubev, 1983; Nyambok and Ongweny, 1979). Despite all the information, erosion is increasing day by day.

Major ecological zones of Africa and the percentage of area under each zone are shown in Figure 2. Soil erosion by water is most severe in the semiarid region (III), the tropical wet-dry region (IV), East Africa (VI), and the Ethiopian highlands (VIa). With deforestation, soil erosion also becomes severe in the equatorial wet region (V). Walling (1984) estimated sediment yield for regions III, IV, and V from 0 to 1 t ha^{-1} yr^{-1}. The sediment yield for regions VI, VIa, and IX is estimated to range from 1 to 10 t ha^{-1} yr^{-1}. Much higher rates are reported from localized areas.

a. Ethiopian Highlands

In 1975, some 9 to 9.5 million hectares of Ethiopia's nearly 13 million hectares of cultivable land were cultivated. Ethiopia has as many cattle as people: The 1975 estimate was more than 25 million of each (Kuru, 1978). The severe erosion in Ethiopia is a human-created problem. Its drought-triggered famine is merely a symptom of the problem of soil degradation caused by erosion. Data from the Simien Mountains in the Gondar region indicated an average annual soil loss of some 20 metric tons per hectare (Lamb and Milas, 1983). The Central Plateau region of Tigrai, covering about 6000 km^2, suffers from severe soil erosion (Virgo and Munro, 1978). Ethiopian highlands reportedly lose more than 1 billion tons of toposil every year (Brown, 1981). That is equivalent to stripping away 1 m of topsoil from about 80,000 ha. Kuru (1978) has listed many causes of severe erosion in Ethiopia, including continuous deforestation activities, uncontrolled grazing, and unsuitable farming practices.

b. East Africa

The East African region comprises Kenya, Tanzania, and Uganda. Soil erosion is reportedly a serious problem in most of this area, especially the highlands. Finn (1983) estimated that some 4.81×10^8 m^3 of terrigenous

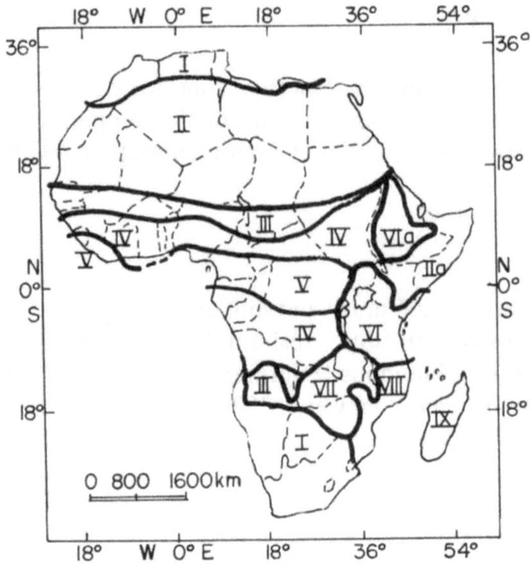

Percentage Area of Africa Covered By The Climatic Zones.

Zone	Region	Percentage Area
I	Mediterranean	2.6
II	Northern Desert	33.0
IIa	Horn of Africa	5.2
III	Semi - Arid	7.2 (6.6 in north, 0.6 in south)
IV	Tropical Wet - Dry	17.5 (11.5 in north, 6.0 in south)
V	Equatorial Wet	6.8
VI	East Africa	6.4
VIa	Ethiopian Highlands	2.3
VII	South Savanna Plateau	5.7
VIII	Mozambique	1.9
IX	Malagasy	2.2
X	Southern Africa	9.2

Figure 2. Ecological map of Africa. [J.F. Griffith, In: *Climates of Africa*, © Elsevier Science Publ., 1972.]

sediments are transported annually from this region to the western Indian Ocean.

In Kenya a popular saying has it that "erosion removes soil corresponding to one lorry load (or about 20 t of soil) from every acre every year." High erosion occurs in regions of steep slopes on the eastern side of Mount Kenya. Biswas and Biswas (1978) reported that the annual rate of soil loss in the catchment area of the Tana River is as high as 15.5 t ha^{-1}. Hove

(1977), Thomas et al. (1981), Barber (1983), and O'Keefe (1983) reported erosion and runoff losses from some unprotected soil surfaces around Nairobi of as much as 146 t ha^{-1} yr^{-1}. Soil erosion is particularly severe in the semiarid region comprising the Machakos Hills. These hills have been occupied since about 1900 (Moore, 1979). Severe surface and gully erosion had already set in by 1930. In this region, erosion rates of 70 to 140 t ha^{-1} yr^{-1} are common. In Machakos, Thomas et al., (1980) measured erosion rates of 1 to 70 t ha^{-1} yr^{-1}. Edwards (1979) discussed the relationship between catchment hydrology and erosion for different regions. Measured sediment yields in some tributaries of the Tana River exceeded 21 t ha^{-1} yr^{-1} (Edwards, 1977). Barber (1983) reported low natural erosion rates of 0.3 to 0.5 t ha^{-1} yr^{-1} for some Kenyan rivers in humid regions with volcanic rock, and 0.6 to 1.7 t ha^{-1} yr^{-1} for the basement complex in semiarid regions. Accelerated erosion rates from arable lands, however, can be as high as 8 to 225 t ha^{-1} yr^{-1} (Dunne and Ongweny, 1976). In this region of dense, high populations, land use is a much more important factor in relation to sediment yield than either catchment area or geology (Dunne, 1979; Edwards and Blackie, 1981).

Areas of Kenya and Tanzania with severe wind and water erosion are discussed by Lundgren (1975). Erosion and sedimentation rates for Tanzania have been reported by Rapp and his colleagues (Rapp, 1972, 1975, 1977a, 1977b). Rapp et al. (1972) and Stromquist (1981) reported rates of 2.4 to 8.4 t ha^{-1} yr^{-1}. Christiansson (1981) estimated denudation rates for five catchments in Tanzania. The data show denudation rates from 1.4 to 11 t ha^{-1} yr^{-1}. Severe soil erosion occurs in the highlands of Rwanda and Burundi, where very steep lands are being used intensively for crop production. In Burundi, Van Lancker and Bruggeman (1982) estimated that in special cases sediment yield may be as high as 6000 t ha^{-1} yr^{-1}. A similarly serious situation exists in neighboring Rwanda.

c. Southern Africa and Madagascar (Regions VII, VIII, and IV)

Severe soil erosion occurs in the intensively used highlands of Malawi (Figure 3). Balek (1977) reported sediment rates of 18 t ha^{-1} yr^{-1} from some catchments in Malawi. Soil erosion is equally severe in Lesotho (Makholibe, 1984). Also in Lesotho, Chaklea (1981) measured sediment rates from 1 to 20 t ha^{-1} yr^{-1}. Severe erosion is also reported from Zimbabwe (Stocking, 1984). Erosion is particularly severe in regions of southern Africa where large-scale farming is practiced without appropriate conservation measures. In Natal province of South Africa, about 200 million tons of topsoil are estimated to be lost each year (Brown, 1981). Whitmore (1972), however, pointed out that this severe soil erosion can be curtailed by correct land management practices. Rooseboom (1978) reported sediment yields of South African rivers from 0.1 to 10 t ha^{-1} yr^{-1}. Watson (1984) observed that high sediment yields were related to burning.

Figure 3. Severe erosion in highlands of Malawi.

In Madagascar, some 80% of the land area is reported as affected by severe erosion. Madagascar's Mangoky River deposits some 19 to 20 million metric tons of sediments each year at its mouth (Randrianarijaona, 1983). In some areas as much as 250 tons of topsoil are lost per hectare per year. Rossi and Salomon (1979) described a severe gully erosion in Madagascar called "Sakasaka" and "Lawaka."

d. Tropical Wet-Dry Regions (Semideciduous Forest and Savannah Regions)

The tropical wet-dry region covers about 17.5% of Africa's total land area and has some intensively used arable lands, predominantly alfisols, which are highly susceptible to water erosion.

In West Africa severe sheet, rill, and gully erosion are reported from arable lands. Severe gully erosion on the Jos Plateau in Nigeria and in semi-arid regions (Figure 4) is caused by excessive grazig and overstocking. Erosion rates exceeding 100 t ha^{-1} yr^{-1} are common (Fournier, 1966; Charreau, 1968; Kowal, 1972a, 1972b; Adu, 1972; Lal, 1976; Roose, 1973; Roose and Godefroy 1977; Uriyo, 1983). Some earlier data of sediment loads in major rivers of the region indicate low annual erosion rates of 0.1 t ha^{-1} for the Niger, 0.21 t ha^{-1} for the Niger-Benue, and 0.50 t ha^{-1} for the Benue River watersheds (Balek, 1977). Sediment loads in rivers from Nigeria have been reported to range from 0.1 to 7.4 t ha^{-1} yr^{-1} (Oyebande, 1981). In Ghana, Akrasi and Ayibotele (1984) reported sedi-

Figure 4. High stocking rate is related in gully erosion in semi-arid Africa.

ment loads of about 3 to 5 t ha^{-1} yr^{-1}. In Zambia, Robinson (1978) observed that erosion of agricultural land has been serious since the 1930s.

e. Equatorial Africa

A major part of the equatorial ecological zone forms the tropical rainforest zone of the Congo Basin region still covered by original forest vegetation, where agricultural activity is generally low. Wherever forest has been removed, however, severe erosion has resulted from unsuitable methods and inappropriate land use. Grove (1972) reported erosion rates of 0.4 to 0.5 t ha^{-1} yr^{-1} for the Congo Basin, where increasing population pressure and intensive land use have since increased erosion rates.

Severe gully erosion occurs in parts of southeastern Nigeria. Osuji (1984) reported that in Imo state, about 13 million metric tons of soil are lost annually from an area of about 9000 km^2. A gully in the Amucha region reportedly increased from about 6 m deep in 1970 to about 48 m deep in 1981. Osuji (1984) observed that another gully advanced 760 m in less than six months.

f. Semiarid and Arid Africa

Both wind and water erosion are severe in the semiarid/arid ecological region. Vast areas of the West African Sahel are characterized by hardened plinthite at the surface, indicating severe sheet erosion that has occurred in the past (Figure 5). These regions are noted for high runoff, flash floods, and severe gully erosion. In the Sahel, severe gully and flash

Figure 5. Hardened plinthite covers vast areas of arid regions in Africa.

flood erosion are commonly observed during the monsoon. Talbot and Williams (1976) observed gullies 150 to 300 m long formed during one short rainy season in the Sahel zone of Niger. Data on river sediment yield taken in the early 1970s indicated low rates of erosion on large watersheds. Grove (1972) reported erosion rates of 0.2 t ha^{-1} yr^{-1} for the Niger, 0.4 t ha^{-1} yr^{-1} for the Niger-Benue, and 0.8 t ha^{-1} yr^{-1} for the Benue River. These rates are lower than those of the Amazon and Mississippi Basins. Severe erosion and high sedimentation rates are reported from semiarid East Africa. Rapp et al. (1972) reported sedimentation rates of 140 to 182 t ha^{-1} yr^{-1} from catchments near the Morogoro and Megata-Mzinga.

g. Reasons for Severe Erosion in Africa

Severe erosion in sub-Saharan Africa is observed in subhumid and semiarid regions and in the eastern highlands. Gully erosion is a national catastrophe in southeastern Nigeria. Severity of erosion in all these regions is attributed to high population density, soils that are highly prone to erosion due to both harsh climate and intensive arable land use, uncontrolled grazing and excessive stocking rate, and soil profile characteristics that render them highly susceptible to undercutting and gully erosion. Erosion can, however, be controlled to tolerable levels by adopting appropriate land use and conservation-effective soil and crop management systems. Intensive land use is technically feasible with the technical knowledge that already exists.

Soil erosion is relatively less severe in the Congo Basin and in regions still covered by tropical rain-forest vegetation. Erosion can, however, become extremely severe when land is cleared by mechanized means involving heavy machinery. Postclearing soil and crop management are also important determinants of soil erosion. If the land is immediately planted to cover crops and to perennial (tree) crops, soil erosion is reduced to an extremely low level by the time tree canopy is fully developed or the ground is covered by a quick-growing legume (*Pueraria*, *Centrosema*, *Mucuna*, etc.). Once again, the technology to control erosion in the humid tropical regions exists.

2. Tropical Asia

Tropical Asia consists of two distinct regions: (1) South Asia and (2) Southeast Asia. They are the most densely populated and their soils the most intensively used in Asia. Consequently, some of the most spectacular erosion is observed there. Still, the region has made some remarkable achievements in agricultural production since the early 1960s.

a. Southeast Asia

Southeast Asia comprises the Philippines, Indonesia, Malaysia, Thailand, Burma, Vietnam, Laos, and Cambodia. Irrespective of slope, the soils are used intensively, and severe erosion is common on hillslopes cultivated for upland crops.

Soil erosion is a severe environmental problem in the Philippines. The mean annual rainfall in the Philippines ranges from 1900 to 4200 mm. Rains are characterized by high intensities (300 mm h^{-1}), thunderstorms and typhoons often causing daily rainfall amounts of 50 to 200 mm for a number of consecutive days. The National Economic Development Authority estimates that about 58% of the total land area of the Philippines are steeplands with slopes exceeding 11%.

Of the Philippines' 30 million hectares, 58% were estimated by the National Economic Development Authority (1983) as being susceptible to erosion. About 22% of existing and potential farmlands are susceptible to erosion, while about 30% of the estimated 17 million hectares of forestland suffers from some erosion. Cabrido (1981, 1985) estimated that 9 million of the 13 million hectares of alienable and disposable land throughout the country are already eroded. Cabrido (1985) identified 13 provinces in which more than half the total area is already severely eroded: Batanges (83%), Cebu (76%), Ilocos Sur (73%), La Union (70%), Batanes (68%), Bohol (6%), Masbate (66%), Abre (65%), Iloilo (63%), Cavite (60%), Rizal (56%), Capiz (55%), and Marinduque (51%). After analyzing sediment yield and erosion from 18 watersheds on Luzon Island, De Vera and Rewtarkulpaiboon (1981) estimated soil losses from 2 to 10 t ha^{-1} yr^{-1} and sediment yields from 1 to 22 t ha^{-1} yr^{-1}.

In Indonesia, Suwardjo et al. (1985) reported that 22 million hectares of arable land are severely eroded. Hardjowitjitro (1981) reported sediment yields of 120 t ha^{-1} yr^{-1} from the Cilutung River in Java. Deforestation and cultivation of steeplands have caused severe erosion in densely populated Java. Avalanches and landslides are common on steeply cultivated

Figure 6. Hills in the vicinity of Hyderabad, Central India, are severely eroded by land misuse.

hills (Kronfellner-Kraus, 1980; Legowo, 1981). Aitkin (1981) reported sediment yields from Java from 30 to 80 t ha^{-1} yr^{-1} and erosion rates of 60 to 120 t ha^{-1} yr^{-1}.

Soil erosion is equally severe in Camboida and Thailand. In Cambodia, Carbonnel (1964), after estimating total erosion from the Grend Lac watershed and evaluating sediment transport in 15 rivers, estimated mean erosion rate to be 0.3 t ha^{-1} yr^{-1}. Soil erosion since then has been drastically accelerated by expansion of agricultural and urban activities.

In Thailand, high and severe erosion rates are reported on 6.82 million hectares (13.27%) and 6.27 million hectares (12.19%) of land, respectively (Suebsiri, 1984). Suspended sediment load from some river catchments exceeds 4 t ha^{-1} yr^{-1} (Attaviroj, 1986). More severe erosion problems are in the northern highland region. Sabhasri (1978) reported that on steep slopes there, annual soil erosion of 1400 to 1800 t ha^{-1} yr^{-1} is common. Landslides are a severe problem on steep slopes. Jantawat (1985), also working in Thailand, reported sediment transport rates of as high as 3874 t km^2 yr^{-1} in the Lam Dome Noi River watershed, a tributary of the Mun-Chi River. The five regions of the Mun-Chi River watershed lose soil at the following rates: 0.1 to 20 t ha^{-1} yr^{-1} in the northern region, 0.1 to 40 t ha^{-1} yr^{-1} in the northeast, 0.2 to 6 t ha^{-1} yr^{-1} in the central plain, 0.3 to 4 t ha^{-1} yr^{-1} in the east, and 0.3 to 18 t ha^{-1} yr^{-1} in the south. Severe erosion in the northeastern region is attributed to extensive deforestation. Malaysia is rapidly expanding its upland agriculture. The projected growth rate for agriculture during the five-year period from 1986 to 1990 is 2.6%. The area under oil palm is expected to increase to 1.8 million hectares by 1990, and the area under cocoa will increase from 242,000 ha in 1984 to 343,000 ha in 1990. More land is also being cleared for food crops. Most of the agriculturally suitable land, however, has already been cleared (Yeop et al., 1982). Wong (1974) estimated that agriculturally unsuitable land in Peninsular Malaysia occupies 6.85 million hectares out of a total of 13.04 million hectares, and 79.4% of the unsuitable land comprises stony and steeplands. A similar situation exists in Sabah and Saravak. The expansion of agricultural land in Malaysia is, therefore, taking place on marginal soils and steeplands that are highly vulnerable to erosion. Mass movement is a severe problem on steep terrians. Furthermore, deforestation of steeplands coupled with high mean annual but seasonal rainfall of 3000 to 4000 mm causes severe soil erosion.

b. South Asia

The tropical regions of South Asia comprise central and southern India, Sri Lanka, and parts of Burma. In India, the states of Gujrat, Maharashtra, Andhra Pradesh, Tamil Nadu, and Kerala suffer from severe erosion. Some of the excessively grazed hills have been severely degraded (Figure 6). However, the sediment load of southern Indian rivers is not so high as that of the northern Indian rivers. Nevertheless, Narmada, Tapti, Krishna, Mahanadi, and Godavari have high sediment yields. The landscape of the central highlands of Sri Lanka has been subjected to severe erosion since the introduction of tea and coffee plantations in the early nineteenth century. As early as 1947, Burns estimated that as much as 25 million tons of soil was being washed off the surface of Sri Lanka each year into the sea. The highlands of Burma are also subjected to severe erosion; the watershed of the Irrawaddy River has a sediment yield of about 9 t ha^{-1} yr^{-1} (Holeman, 1968).

c. Reasons for Severe Erosion

Severe erosion in tropical Asia is apparently due to anthropogenic factors. Cultivation of steep slopes in the Philippines and Java, and deforestation in Sumatra and Malaysia, have been responsible for increased sediment load from these regions. Using steep slopes for production of food crop annuals by shifting cultivators in the highlands of Thailand is another reason for severe soil erosion in that region.

Accelerated erosion in Southeast Asia is observed primarily from lands that have not been terraced. Terracing hill slopes for rice cultivation has been a cultural tradition in this region, and soil erosion is not a severe problem on properly terraced and well-maintained rice paddies. Erosion becomes severe when hilly and undulating terrains are used for production of upland crops (corn, beans, cassava, etc.) by methods of soil and crop management that are incompatible with the ecological environment of the region.

Similar to the situation in tropical Africa, erosion can be drastically curtailed if the hill slopes are used for appropriate crops and farming/cropping systems. Terracing for rice paddies, conservation tillage for food crops with agroforestry, and use of cover crops under plantations are some of the proven technologies to reduce soil erosion hazard.

3. Tropical America and the Caribbean

Erosion rates are low in the forested regions of tropical America, particularly from the undisturbed parts of the Amazon Basin.

In forest catchments, observed rates of erosion measured in French Guyana were as low as 0.4 t ha^{-1} yr^{-1} (ORSTOM, 1975). Selby (1975) reported the mean erosion rate in the Amazon Basin to be 1 t ha^{-1} yr^{-1}. A survey by UNESCO/WMO (1974) reported sediment yields of 0.7 to 0.9 t ha^{-1} yr^{-1} for the entire Amazon Basin. Severe erosion, however, occurs with deforestation and in regions where agricultural activities have been extended to steep lands and mountainous terrain. Most soils of the Amazon region are well drained. Large-scale destruction of their natural vegetation may turn them into barren laterites or white sands on which substantial regrowth is very difficult (Sombroek, 1979). The severity of the soil erosion problem in these regions has long been recognized (Lal, 1977b).

Soil erosion maps published by FAO in 1954 indicated that as much as 10%–25% of the land in northeastern Brazil and in most of Central America was already subject to severe erosion. The highlands of Columbia, Ecuador, and Venezuela have severe erosion problems.

Ecuador is a small country (0.27 million km^2) with mountainous terrain and a severe erosion hazard (Portch and Hicks, 1980). Severe erosion problems also have been reported from Cuba (Sague Diaz et al., 1979). In Peru, erosion is estimated to affect between 50% and 60% of the whole country's surface. The entire Andean region is severely affected by erosion. The total land area in Andean Peru has soil losses ranging from 20 to 70 t ha^{-1} yr^{-1}. Low and Goh (1972) estimated the affected areas to be 178,000 km^2 or 14% of the total. Kaufmann (1981) observed that large mountainous areas of Columbia are already eroded or in danger of erosion. Erosion rates of 71.4 t ha^{-1} yr^{-1} were reported. The introduced large-scale mechanized agriculture is responsible for severe erosion in Brazil, especially in Parana and Rio Grande du Sul. Politano et al. (1980), after conducting an aerial photographic survey in Sao Paulo state in Brazil, identified seven currently prevalent erosion classes. The severe erosion of Santa Catorine soils is related to deforestation, fire, lack of suitable pasture species, and low soil fertility (Suckling, 1981).

Soil erosion is reportedly severe in the Caribbean countries (Ahmad, 1977). In eastern Barbados, south of the Scotland district, the Joe's River

Basin suffers from severe erosion, including channel incision, gullying, spawning of erosion cells, stream capture, and extensive mass movements (Carson and Tam, 1977; Tam, 1980). The high erosion potential of soils in Tobago has been reported by Ahmad and Breckner (1974). The loss of arable land through intermittent channel gullying and land slippage is serious in these regions. Severe erosion is caused by a combination of susceptible soils and parent materials, high and intense rains, erosion-prone land use, etc.

Soil erosion is also reportedly severe in Jamaica. About two-thirds of the country's bedrock is erodible limestone on mountainous terrain, 25% of the country's landscape has slopes exceeding 30°, and 52% has slopes exceeding 20°. Of the 33 principal watersheds, 18 contain severely disturbed areas, with at least 5 requiring priority control measures (Blustain, 1982).

a. Ecological Factors Affecting Soil Erosion in Tropical America

The present rates of soil erosion reported from tropical America are among the lowest in the world. These low rates are expected due to the fact that large areas of the Amazon Basin are still protected by the tropical rainforest. The ecological balance is, however, being rapidly disturbed due to deforestation for mining, infrastructure and road development, and agricultural production. Soil erosion is as much as or more severe from nonagricultural uses than from agricultural land use. Abandoned mines, faulty road construction, and poorly installed drain outlets from roads and urban centers contribute significantly toward sediment transport from these regions. Conversion of tropical rain forest for agriculture, however, can be done in a way that does not expose the fragile soils to harsh environments.

4. Tropical Australia

In Australia, tropical regions comprise the Northern Territory and parts of Queensland. Sheet and rill erosion are problems wherever the natural vegetation has been replaced by intensive land use. Croplands suffer from severe erosion in the Northern Territory. Dumsday (1973) observed that in Queensland about 47,000 km^2 are susceptible to gully erosion, and another 132,000 km^2 to sheet erosion. In northeast Australia, where annual rainfall averages 4175 mm, high rates of surface runoff have been recorded even from forested catchments (Gilmour and Bonell, 1977; Bonell and Gilmour 1978; Bonell et al., 1979).

Sediment yields from basins in the headwaters of the Fly River in Papua New Guinea have been reported by Pickup et al. (1981). Highest sediment sources appear to be landslides and slope wash, even beneath the tropical rain forest. Erosion rates are high in the mountainous areas, between 5 and 57 t ha^{-1} yr^{-1}. Erosion rates are lower downstream, 4 to 8 t ha^{-1} yr^{-1}.

a. Reasons for Severe Erosion in Tropical Australia

Severe erosion in tropical Australia is reported from Queensland and Northern Territory. High erosion rates are reported from sugarcane plantations established on slowly permeable vertisols. Erosion in canefields is by compaction due to heavy harvesting traffic, and by burning of trash that exposes these soils to high-intensity tropical rains.

Mulch farming and conservation tillage are apparently very effective in erosion control, even in vertisols. Cultural practices of harvesting should be developed that facilitate the use of trash and residue left on the soil surface as mulch.

5. Comparative Erosion Hazard Among Tropical Regions

Among the four tropical regions discussed, current erosion rates are highest in Southeast Asia and lower in the Amazon and Congo Basins. Except for the high-rainfall regimes of northern Australia and Papua New Guinea, soil erosion in tropical rain-forest regions is serious only when the forest cover is removed. Nevertheless, in tropical primary forest, erosion can be more than in forests of the temperature regions. There are three commonly cited reasons for this difference: (1) The ground flora in tropical forest is less developed because it receives less radiation; (2) the humus layer is thinner; and (3) rains are more frequent, more intense, and possess higher energy loads. High erosion under primary rain forest also results from the high energy of falling raindrops.

Soil erosion is generally worse in semiarid regions with wet-dry climates than in the humid regions of the tropics. Soil erosion and sediment yields are especially high in regions with mountainous terrain and where steep slopes are cleared for arable land use under unsuitable farming practices or are excessively grazed with excessive stocking rates.

B. Soil Erosion in the Subtropics

In the context of this volume, the subtropics are those regions lying approximately within the latitudes 23° north and 35° south of the equator. The most severe erosion problem exists in these middle latitudes. They are characterized by seasonal rainfall, undulating and steep terrains, soils that are highly susceptible to erosion, high population densities, and a long history of intensive land use. They also have been the cradle of civilizations and have witnessed the development of settled agriculture.

The subtropical regions of Asia are East Asia, China, and the Himalayan-Tibetan area.

1. East Asia

Japan, and both North and South Korea, suffer from spectacular erosion in the form of landslides and debris avalanches. The geology and landforms in

Japan favor rapid erosion and degradation. Destruction of Buddhist temples at Mt. Hiari by fire, and subsequent deforestation to reconstruct them during the past 100 to 500 years have been major causes of accelerated soil erosion in Japan (Lowdermilk, 1927). Deforestation of hill slopes has triggered the processes responsible for serious erosion (Woo, 1982). Rates of erosion from Japanese farmlands are estimated to be 0.7 to 7 t ha^{-1} yr^{-1} for crops, 7 to 21 t ha^{-1} yr^{-1} for orchards without mulching, and 21 to 28 t ha^{-1} yr^{-1} for bare lands (Kadomura and Yamamoto, 1978). The mean annual sediment yield from Japan is estimated to be 12 t ha^{-1} yr^{-1} (Kadomura, 1980). Especially high rates of sediment yield are reported in the central highlands and in the outer zone of southern Japan. Tanaka (1976) reported rates of erosion in the Tanzawa Mountains as high as 14 t ha^{-1} yr^{-1}. Rates of soil erosion hazard in Japan are estimated to be 0.14 to 1.4 t ha^{-1} yr^{-1} from grasslands and forestlands, 0.25 to 14 t ha^{-1} yr^{-1} for farmlands, 14 to 140 t ha^{-1} yr^{-1} from bare lands, and 140 to 1400 t ha^{-1} yr^{-1} from devastated and disturbed lands.

2. China

Soils in the loess plateau of the Yellow River Valley in northwestern China are perhaps the most severely eroded in the world (Lee, 1984). Deposition of sediments has caused the river to be suspended over the delta reach (Robinson, 1979). The arid loess region is characterized by sparse vegetation, frequent and intense rains in summer, and land form and slopes that are highly vulnerable to erosion by water (Zunghu, 1981). The Yellow River is about 0.5% silt-laden and produces the largest sediment load of any river in the world—1980 million tons averaging 28 t ha^{-1} of average annual suspended load. The gross upstream erosion is three to four times higher. Sediment-yield maps of the upper and middle reaches of the Yellow River Basin of China are reported by Gong and Xiong (1980). It is estimated that the average annual sediment load in Shansi Province exceeds 250 t ha^{-1}, with an average annual soil loss of 100 t ha^{-1} for the area (Robinson, 1979; Gong and Jiang, 1979). The maximum sediment concentration in the middle reaches of the river approaches 700 kg m^{-3}, that is, about 50% by weight.

Robinson (1979) estimated that 31% of China's 1973 million hectares that were once productive have now been abandoned because of excessive erosion. Shih-Yang and Te-Chi (1977) and Bowen and Shaozu (1982) reported that the loess area susceptible to erosion is as large as 430,000 km^2, of which 236,000 km^2 is severely gullied. In this region the river tributaries Huanguchan, Kenya, and Wuding have an annual erosion rate of 140 to 280 t ha^{-1}. Wuding River, one of the large tributaries on the middle reaches of the Huang He, contributes nearly one-sixth of the sediment load and one-fourth of the coarse materials from a drainage area of only 1/23 of the total catchment of 688,000 km^2 about Sanmenxia (Dequi et al., 1981).

The soils of loess plateau are also prone to severe gully erosion. Eliassen (1936) reported from his visit, "I have watched a gully 3 m deep and 6 m wide cut back more than 10 m and deepen from 3 to 4 m during a single rainstorm." The gully density in this region is 3 to 7 km km^{-2} with a length of the main gully ranging from 50 to 100 m. In northern Shaanxi province, the gullied area comprises 26% to 54% of small watersheds. The gully itself erodes 1.10 to 1.76 times more than ungullied land (Shih-Yang and Te-Chi, 1977; Hao et al., 1977).

China's loess valley is not its only severely eroded region. Soil erosion, both by wind and water, is serious in many other regions. Hilly and mountainous soils constitute about 65% of the 9.6 million km^2 of China's total land area (Howard, 1981). It is estimated that a total of 1.3 million km^2 of land currently is subject to wind and water erosion. Since 1950, conservation measures have been applied to only 400,000 km^2; meanwhile, other lands have become severely eroded (Barrow et al., 1982). Serious soil erosion is also observed in southern Yunnan, due to recent deforestation. Red soils in southern China are also highly susceptible to erosion.

3. The Himalayan-Tibetan Mountain Ecosystem

The Himalayan-Tibetan Mountain region is one of the most severely degraded in the world (Dent, 1984). Nepal, because of the mountainous terrain and a high population density, is highly vulnerable to erosion, especially in the mountainous regions of the Sivalik Hills in the Himalayas (Dregne, 1982). The country's rivers now carry 336 million tons of soil to the main river system entering India (Brown, 1981). It is estimated that 63% of the Sivalik zone, 86% of the Middle Mountain zone, 48% of the transition zone, and 22% of the High Himalayas have been reduced to poor or fair watershed conditions. The bed levels of Terai rivers are said to be rising by 15 to 30 cm annually (Dent, 1984).

It is estimated that 175 million of 328 million hectares of arable land in India are degraded by soil erosion, and about 6,600 million tons of soil are displaced and carried to the oceans annually (Brown, 1984; Dhruva Narayana and Sastry, 1985). Of the 175 million hectares affected by erosion, sheet erosion is severe on about 72 million hectares of red soils (4–10 t ha^{-1} yr^{-1}) and 88 million hectares of black soils (11–43 t ha^{-1} yr^{-1}); gully erosion is severe on 4 million hectares (more than 33 t ha^{-1} yr^{-1}) and hillside erosion on 13 million hectares (more than 80 t ha^{-1} yr^{-1}) (Dhruva Narayana and Sastry, 1985). Erosion, landslides, and landslips are severe in the Sivaliks (Murthy and Shankara-Narayana, 1977). Extensive and deep gullies have ruined large area in Uttar Pradesh, Himachal, Madya Pradesh, Rajsthan, Gujrat, Maharashtra, Punjab, Bihar, Tamil Nadu, and West Bengal (Singh and Wali, 1962; Patnaik, 1975; Lal and Banerji, 1974; Lal, 1977a). In India the Ganges deposits 1.6 billion tons of soil into the Bay of Bengal each year (Brown, 1984). Curry and Moore (1971) observed that the Bay of Bengal receives sediments at the rate of 10 t ha^{-1} yr^{-1}.

Abbas and Subramanian (1984) estimated sediment transport from the Ganges River Basin and reported that the total annual load at Calcutta, the mouth of the river, is 411×10^6 t (328×10^6 t sediment load and 83×10^6 t chemical load). This rate of material transport gives an erosion rate of 6 t ha^{-1} yr^{-1}, one of the highest in the world. Reservoir sedimentation is serious in all ecologies in India (Bowonder et al., 1985). Narayana and Babu (1983) have estimated that, on average, soil erosion from Indian land is 16.4 t ha^{-1} yr^{-1}. Only 29% of the total eroded soil is lost permanently to the sea. A majority of the remainder is deposited in reservoirs. Siltation of reservoirs in India is about 200% more than the design flow (Bali and Karale, 1977; Dent, 1984).

Soil erosion and degradation are equally severe in neighboring Pakistan (Brown, 1981). Wind and water erosion are reportedly destroying an estimated 12,000 to 30,000 ha of land each year in Punjab alone (Dregne, 1982). The map of sediment yield from northern Pakistan compiled by Ahmad (1960) indicated high erosion rates from the Ravi Basin, and in hills with heavy grazing pressure. Severe erosion occurs from land that has lost its forest cover (Mashur and Hanif, 1972).

4. West Asia and North Africa

The arid and semiarid lands of the world cover about 4.5 billion hectares or 33% of the total land surface, a major part of which lies in West Asia and North Africa. Arid regions of West Asia, the Middle East, and North Africa are characterized by rainfall of 100 to 400 mm per year, which is concentrated during two or three months, so flash floods from denuded hills cause severe sheet and gully erosion. Riquier (1982) reported that soil erosion rates by water in this region range from 10 t ha^{-1} yr^{-1} to more than 200 t ha^{-1} yr^{-1}. The FAO (1979) observed that in Africa north of the equator, 11.6% of the total land area is affected by water erosion. In the Near East, in comparison, 17.1% of the total land area is affected by water erosion. High erosion rates are especially prevalent in the coastal regions of northwest Africa, e.g., Tunisia, Morocco, Libya, and Algeria. Le Houerou (1976) estimated mean annual erosion rates of 7 to 21 t ha^{-1} yr^{-1} in the arid zone of North Africa. Walling (1984) estimated that the mean annual sediment load for Tunisia, Morocco, and Algiers is 10 to 50 t ha^{-1} yr^{-1}. The average annual suspended sediment load in the main Nile River is estimated to be 134 million tons (Shahash, 1977). On the basis of sediment load data, the rate of annual soil erosion in the Blue Nile and Atbara River Basins is about 1.5 to 3 t ha^{-1} yr^{-1}. However, total sediments originating from arable lands may be 10 times more. In fact, advancing desert over the fertile lands of the Nile Valley and Delta is reported to be a severe threat to the productivity of Egypt's agricultural land. Zachar (1982) reported that the land surface severely affected by erosion includes 15 million hectares in Iran and 4.0 million hectares in Iraq. About 35% of the land in Israel is severely eroded.

5. Southern Europe

Regions surrounding the Mediterranean are subject to high rates of erosion. The erosion map of Cyprus (Janssen, 1982) shows that the highest erosion rates are associated with the flysch rock types, vine cultivation on steep slopes, and high rainfall. In Greece, devastated land occupies at least 2.3 million hectares; 24.5% of its currently used land needs immediate protection (Zachar, 1982). It is estimated that in Spain some 19 million hectares are destroyed by erosion, and an additional 24.4 million hectares of arable land and 6.5 million hectares of forestland also suffer from erosion (Giordano, 1956). Studies conducted by the Agricultural University of Wageningen at Hornos and Salarmanca in the hilly regions of Spain showed soil erosion rates as high as 100 t ha^{-1} yr^{-1} (Anonymous, 1974). About 1600 million hectares of land are presumably degraded in Portugal. The mediterranean region of France also has some severely eroded lands (Zachar, 1982).

6. Australia

Australia, with a land area of 7.6 million km^2 or 5.5% of the earth's surface, is the driest continent; more than half the country receives less than 300 mm of rainfall a year. It is estimated that some 58 million hectares or about 7.5% of the continent is agriculturally productive and also experiencing severe soil erosion. Although most of the country is arid, localized individual rainstorms can be intense. Intensive cultivation in response to the growing world grain demand has resulted in erosion problems on at least 50% of the agricultural land in Australia (Brown, 1981).

Dumsday (1973) observed that in nonarid Australia the land area affected by moderate, severe, extensive gully erosion is 80,000 km^2 in New South Wales and 178,000 km^2 in arid Australia. In comparison, sheet erosion occurs on about 95,000 km^2 in New South Wales and 284,000 km^2 in Australia. The percentages of land area affected by both wind and water erosion in nonarid Australia were estimated to be 42% in New South Wales, 32% in Queensland, and 30% in arid Australia.

Large areas of New South Wales have been affected by soil erosion. Kalesk's survey (1945) stated that some 18 million hectares of land in the eastern and central parts of the state were affected by erosion. Dyson (1970) mapped areas in New South Wales severely affected by sheet, gully, or stream-basin erosion. Severe gully erosion in New South Wales is reported in the Bango Creek area. Gillespie (1981) reported that severe gully erosion has developed since the mid-1930s. During the past half-century more than 16,000 m^3 of soil has been eroded from the 800-m-long gully. The Murray Valley catchment area of the Keepit Dam in NSW is one of the most severely eroded regions (Higginson, 1974; Emery, 1975). Even range lands and national parks are not safe from this menace (Manson, 1980). The erosion hazard in national parks also has been increased by recurring bushfires. Good (1973) observed that areas of high potential soil erosion occurred where the tree canopy had been scorched, where ground flora destroyed, and where fire trails and access tracks had been constructed during fire fighting. Severe erosion is also reported in western Victoria (Marker, 1976), western Australia (McGhie, 1980), and in Tasmania (Colclough, 1978).

Tunnel erosion is severe in southeastern Australia, where it develops into gully erosion. The alpine and subalpine grasslands of southeastern Australia have severe erosion (Downes, 1959). Sheep grazing and logging of *Eucalyptus* forest have resulted in tunnel erosion and salting (Leitch, 1982).

Severe soil erosion observed in New Zealand is the result of deforestation. New Zealand has soils on steep slopes derived from pumice deposits in the Kaimanawa, Kaweka, Ruanine, and Wakarare Ranges at the heads of the Rangitikei, Mahaka, Tutaekuri, Nagaruroro, and Waipawa Rivers. Soil erosion is severe in all these localities, probably most severe in the Kaimanawa Range, where soils have been exposed to bedrock by severe sheet and rill erosion (Grange and Gibbs, 1950). Severe erosion in these ranges is attributed to intense rainstorms, unstable geologic setting, introduction of herbivorous animals, and changes in land use that cause channel degradation in lower catchment areas.

7. Midlatitude Regions in America

In South America, mountainous regions and intensively grazed lands of Chile and Argentina have suffered severe erosion (Bonfils et al., 1960; Peralta, 1980a, b). In Uruguay, erosion is estimated to affect about 80% of the land area, with 30% severely damaged (Uruguay, 1982). Central and North America, Mexico, and the southeastern United States have areas with severe erosion. By 1977, high erosion rates were reported from many areas in the southeastern United States (USDA, 1978).

The literature presented indicates that the most severe soil erosion is from regions with mountainous terrain and where land is intensively used for agriculture, orchards, and grazing. The most severely eroded soils of the world lie in the subtropical regions of middle latitudes.

a. Ecological Factors Responsible for Severe Erosion in the Subtropics

The so-called middle latitudes experience some of the highest erosion rates reported in the world. Notable among these are the regions involving the loess valley of China, the Himalyan-Tibetan ecosystem, the Mediterranean Basin, and the northwestern region of Africa. Severity of erosion in these regions is due to a multitude of interacting factors. First, many of these regions such as the Himalyan-Tibetan ecosystem, are geologically unstable. Second, the soils and the parent material are extremely vulnerable to the shearing force of running water, such as the loess soils of China and the lower Sivalik Hills of northern India. Third, the world's highest population density occurs in these regions. Consequently, even shallow, steep, and marginal lands are being used intensively. Fourth, accelerated erosion and natural resource degradation are also related to widespread poverty. Resouce-poor farmers cannot afford to invest in restorative measures. National economy and gross national product also play a major role in preserving the resource base. Finally, the monsoonal climate brings torrential rains at a time when the soils are devoid of any protective ground cover.

This is also the ecological region that will always suffer from accelerated soil erosion because all the principal factors of erosion are active. The main factor is the high population density. In addition to the production of food and fiber, the demand for fuel wood has caused denudation of steep hills. Furthermore, the stress on fragile resources by ever-increasing human population is highly accentuated by animal population.

This is the region where an aggressive and forward-looking resource management policy is urgently needed. Resource management policies are needed at both national and international levels. The general public must be made aware of the ethical problems of resource use and preserving the heritage for the generations to come. We must realize that depriving future generations of their rights is immoral and unethical. Furthermore, resource management policies must cut across political and ethnic boundaries. Watersheds and large river basins usually cover two or three countries. There must be political goodwill among the nations concerned to solve this severe problem of land degradation.

C. Erosion in Temperate Regions

1. Europe

In spite of its intensive agriculture, Europe's erosion rates are less than in Africa or Asia. Total wasteland area in Europe is about 8800 million hectares. In early attempts to estimate global soil erosion, Fournier (1959) reported an average yearly global soil loss of 84 t km^{-2} compared with 715 t km^{-2} for Africa. However, he reported very high erosion rates for orchards and croplands, and from steeplands. For example, erosion from vineyards in northeastern Italy has been reported from 0.2 to 47 t ha^{-1} yr^{-1} (Tropeans, 1984). The data by UNESCO & IAHS (1974) indicated vast areas of Italy with annual erosion rates exceeding 0.8 mm. The thin topsoil has long been eroded from the rock slopes of mountainous terrain in Italy, Austria, Switzerland, and France. The Swiss Alps are estimated to lose some 6 to 14 t ha^{-1} of sediments annually (Clark and Jager, 1969). Erosion rates of 24 to 30 t ha^{-1} yr^{-1} are reported from the Lamone and Savior regions of Italy (Fournier, 1959). In general, about three-fourths of Italian land surface is damaged by erosion. The highlands of Austria and Switzerland, where torrent erosion, debris flow, and avalanches are common, suffer from intense erosion. The Alps and Pyrenees regions of France have severe erosion hazards (Zachar, 1982). Soil erosion is less severe in the United Kingdom and in the Scandinavian countries.

Accelerated erosion is a serious problem in croplands of the USSR, spurred by the introduction of heavy farm machinery and vehicular traffic. About 27% of the USSR's arable land is severely or very severely eroded (Dregne, 1982). More than 55% of the arable land and 79% of the pastures are located on slopes steeper than 1° (Tregubov, 1981). Rivers draining the European side of the USSR annually carry about 535 million tons of sediments, including 10 million tons of K, 264,000 tons of P, and 1.2 million tons of N (Zachar, 1982). The mean average erosion from European USSR is 0.4 t ha^{-1} yr^{-1}. The most intense localized erosion occurs in the plains and mountainous regions of the European part of the USSR (Zachar, 1982). Rates of anthropogenic erosion are usually highest in dry regions and low in humid zones. Gully erosion is severe in regions of extensive loess deposits. Approximate areas affected by different erosion rates from arable lands in European USSR estimated by Zaslavskiy (1977) show that severely eroded lands occupy 11 million km^2 in Asian USSR and 4 million km^2 in European USSR. The mountainous region of Central Asia is severely affected; 36% to 97% of the total land surface is damaged by erosion. Zachar (1982) reported that areas affected by severe erosion are 35.7% of Kirghiz SSR, 87.9% of Uzekistan, 71.2% of Tadzhikistan, and 97.2% of Turkmenistan. The Amu Darya River annually delivers about 97 million tons of sediments to the Aral Sea. Intense wind erosion also is observed in Kazakhstan. Considering relief condition, Zaslavskiy (1977) estimated that erosion from meltwater runoff affects 8.5 million km^2, erosion from rainwater runoff 4.9 million km^2, and erosion from rainfall alone 1.6 million km^2. Wind erosion is a serious hazard in the Ukraine (Zachar, 1982), which has the largest area of soil damaged by erosion.

In a survey of the erosion hazard in Europe, Zachar (1982) reported that erosion-damaged areas are about 7.5 million hectares (30.8% of total land surface) in Romania, 4 million hectares by water erosion and 8 million hectares (total of 20%) by wind erosion in Poland, 2.9 million hectares by water and 1.6 million hectares by wind (total of 36%) in Czechoslovakia, 2.3 million hectares by water and 1.5 million hectares by wind (total of 40%) in Hungary, and about 73% of cropland is affected in Bulgaria.

2. North America

Soil erosion maps of Canada, on the basis of suspended sediment load, were prepared by Stichling (1973). The Cordillera of Canada has erosion rates exceeding 3 t ha^{-1} yr^{-1}. The Oldeman River Basin in Alberta also

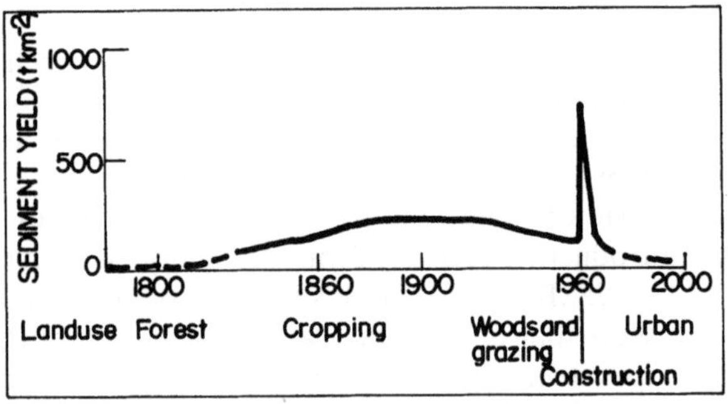

Figure 7. Effects of land use on soil erosion in North America (Wolman, 1967).

suffers from stream instability, undercut, massive landslides, gullying, and bank erosion due to logging, and clear cutting, and mining (Northwest Hydraulic Consultants, Ltd., 1980).

In spite of modern agriculture and high production, croplands in the United States have recently been subject to severe erosion (Figure 7). In fact, accelerated erosion and high yields go hand in hand. The total land and water areas of the United States are 0.91 billion hectares, of which croplands are estimated at 139.3 million hectares. During the last 200 years about one-third of the cropland has been subject to erosion (Pimental et al., 1976). The Second RCA Appraisal revealed that in 1982, sheet and rill erosion moved more than 1.8 billion tons of soil on nonfederal cropland (USDA, 1987). Similarly, sheet and rill erosion was estimated at an annual rate of 582 million tons from rangeland, 370 million tons from forestland, 180 million tons from pastureland, and 461 million tons from minor land uses. In all, sheet and rill erosion moved more than 3.4 billion tons of soil on nonfederal rural land in 1982 (USDA, 1987). The Mississippi carries 331 million tons of soil each year into the Gulf of Mexico (Brown, 1984). It is estimated that up to 50% to 60% of the land resources of the United States require amelioration and restoration of productivity (Kovda, 1977; Larson, 1981). A nationwide survey conducted in 1977 indicated that erosion rates exceeded 12.4 t ha^{-1} yr^{-1} on 33% of the corn land, 44% of the soybean land, 34% of the cotton land, and 39% of the sorghum land. Erosion exceeded the soil loss tolerance (T value) on more than 45.3 million cropland hectares (Larson, 1981). The area eroding at rates greater than tolerable level (T) includes 45.7 million hectares of land in the "e" subclass of the USDA's land capability classification system, and 22.4 million hectares of land in other subclasses of class I. For the nation as a whole, average sheet and rill erosion rates on cropland exceed T value on land in capability classes IV, VI, and VII (USDA, 1987). This soil loss excluded wind and gully erosion. An average of 15 t ha^{-1} yr^{-1} are eroded in Mississippi, 10 in western Illinois, and 10 in western Iowa (De Bivort, 1975). It is estimated that 2000 to 4000 ha of potential cropland are lost or removed from production annually in Iowa as a result of gully erosion. The National Resource Inventory (NRI) taken in 1977 showed the alarming magnitude of croplands plagued by different types of erosion (Crosson, 1985). Lee (1984), who also updated land use patterns and soil loss information on the basis of data for 1982, estimated the national average for sheet and rill erosion at 11.8 t ha^{-1} yr^{-1} on cultivated cropland, 3.5 t ha^{-1} yr^{-1} on pastureland, 5.7 t ha^{-1} yr^{-1} on grazed forestland, and 1.7 t ha^{-1} yr^{-1} on ungrazed forestland. In comparison, wind erosion averaged 8.2 t ha^{-1} yr^{-1} on cropland and 3.7 t ha^{-1} yr^{-1} on rangeland. According to Lee, about 56% of all U.S. cropland is eroding at levels less than or equal to T, and 44% at levels exceeding 2T. Sediment yield in catchments of less than 160 km^2 was compiled by Geiger (1957). Soils of the Palouse region, developed on loess, are some of the most vulnerable soils in the United States (Frazier et al., 1983). In spite of these high rates, percent loss in productivity in 100 years due to sheet and rill erosion is estimated at 1.46% (USDA, 1987).

Another realm of soil erosion is the pollution of natural waters and environments. Runoff from cattle feedlots is a major source of nonpoint pollution. Domestic animals in the United States produce more than a billion tons of excrement annually, and more than 400 million tons of urine (De Bivort, 1975).

a. Relative Erosion Hazard in Europe Versus North America

Soil erosion is comparatively less severe in continental Europe compared with other areas in temperate regions. An important reason for this is the gentle climate. Most rains are of relatively low intensity and low drop size. Furthermore, soils have high levels of organic matter content and possess favorable soil structure. Comparatively sound economy and favorable resource availability provide socioeconomic conditions that encourage farmers to invest in resource management.

In general, soil erosion is more severe in North America than in Europe. There are many reasons for this contrast. First, the climate is different. High-intensity rains are more common in North America than in Europe. Climatic extremes, hot summers and cold winters, increase soils' susceptibility to erosion by water. Use of heavy machinery and harvesting traffic, monoculture instead of ley farming, and continuous cropping have caused severe erosion and pollution of natural waters. The export-driven economy is another factor that cannot be ignored.

That soil erosion is a severe problem of land degradation in North America remains debatable. This is the region with the most advanced technology, the most resource-conscious farmers, and where the general public is aware of the problems of environmental degradation. There are established policies and regulations, and guidelines on do's and don'ts. Yet soil erosion in the region occurs more now than ever before. The magnitude of erosion can be reduced, but it may have both national and international repercussions.

IV. Wind Erosion

Wind erosion is a serious hazard in arid and semiarid lands. Almost 36% of the earth's surface, i.e., 50 million km^2, is subject to arid climates (Hagedorn et al., 1977). These regions are characterized according to Thornwaite's moisture index, which is

$$J_m = \frac{100s - 60d}{n}$$

where s = moisture surplus (precipitation in excess of evapotranspiration), d = moisture deficit (evapotranspiration in excess of precipitation), and n = moisture need. Based on this index, Meigs (1953) classified arid climates into three groups:

1. Semiarid regions, where J_m equals -20 to -40
2. Arid regions, where J_m equals -40 to -57
3. Extremely arid regions, where J_m is <-57

Larson and Pierce (1983) reported that wind erosion threatens 30 million km^2 or approximately 20% of the land area. The United Nations desertification map of the world indicated that 43% of the nondesert area of Africa, 32% of Asia, and 19% of South America are at risk of desertification. A survey of different ecologies has shown that wind erosion and desertification have affected 16.6 million km^2 in arid regions, 17.1 million km^2 in semiarid regions, and 4.0 million km^2 in subhumid regions. The United Nations Environmental Program, based in Nairobi, Kenya, has estimated that globally 80% of 3700 million hectares of rangeland, 60% of 570 million hectares of cropland, and 30% of 131 million hectares of irrigated lands are affected by some degree of desertification (Glantz, 1977; Karrar, 1984).

A. West Asia, and North and West Africa

Wind erosion and desertification are severe in the deserts of India, China (Luk, 1983), and Africa (FAO, 1979). The FAO (1979) estimated that areas affected by wind erosion are as much as 22.4% of Africa's land area north of the equator and 35.5% in the Near East. Destruction of vegetation cover by overgrazing has resulted in large areas of unproductive, abandoned lands in Iran (Pearse, 1971). In fact, desertification is taking its toll in most African countries lying on the fringes of the Sahara, which is expanding westward into Senegal and southeastward into Sudan. Brown (1981) reported that measurements taken in the Sudan in 1958 and again in 1975 indicated that the Sahara had annexed some 90 to 100 km during the

Figure 8. Severe wind erosion occurs in arid and semi-arid regions of Niger.

17-year period. Brown and Eckholm (1974) estimated that the Sahara's movement southward had accelerated all along its 5600-mi southern fringe from Senegal in the west to northern Ethopia in the east. The desert's rate of expansion is about 50 km yr^{-1}. Wind erosion is also serious in Mali, Mauritania, Niger, and Nigeria (Le Houerou, 1976; Dregne, 1982) (Figure 8). The dramatic Sahelian drought was the cause and consequence of severe wind and water erosion in this zone. Approximately 600,000 t of grains were provided as emergency relief aid in 1973 and again in 1974 to the six West African nations (Brown and Eckholm, 1974). Wind erosion and desertification are also severe in the Kalahari region in southern Africa (Grove, 1969) and in Botswana (Van der Poel and Timberlake, 1980).

B. North America

In Europe, Brown (1984) estimated that each year some 0.5 million hectares of cropland are abandoned because of severe wind erosion. The dust bowl that North America experienced in the 1930s was a costly lesson indeed. The dust storm of May 11, 1934, deposited some 12 million tons of topsoil in Chicago, and the whole eastern coast of the United States was covered with a fog composed of some 350 million tons of rich topsoil swept away from the Great Plains (Eckholm, 1976). Wind erosion is responsible for some 850 million tons of soil loss in the western United States alone

(Biswas and Biswas, 1978). Severe wind erosion is reported in parts of Ontario, Canada, especially on sandy loams and loams developed in fluvio-glacial outwash and glacio-lacustrine sediments (Nickling and FitzSimons, 1985).

The Second RCA Appraisal indicated that, in 1982, wind erosion in the United States moved some 1.2 billion tons of soil from cropland, 609 million tons from rangeland, 8 million tons from pastureland, 1 million tons from forestland and 162 million tons from miscellaneous land use (USDA, 1987). In 1982 wind erosion moved 2 billion tons of soil on non-federal rural land (USDA, 1987). It is estimated that wind erosion in the United States results in the loss of 2.4 kg ha^{-1} yr^{-1} of N (2.1%) and 0.6 kg ha^{-1} yr^{-1} of P (4.7%). The percent loss in productivity in the United States in 100 years due to wind erosion is estimated at 0.42%.

C. South America

Wind erosion also plagues the western pampas of Argentina. It is estimated that about 30% of the semiarid pampas was severely damaged by wind erosion during the first half of the twentieth century (Dregne, 1982). In the Patagonia region of Latin America, 78 million hectares in Argentina and 24 million hectares in Chile have been converted to semidesert by overgrazing and accelerated wind erosion (Biswas and Biswas, 1978). Wind erosion in the region is accentuated by strong winds and overstocking of sheep.

D. Australia

Wind erosion is also severe (over 26,000 km^2) in the Murray Mallece in southern Australia, where plant cover has been removed in favor of cultivation (Dregne, 1982). Overgrazing in the eastern rangelands of Australia has also led to some wind erosion problems. Dumsday (1973) observed that in nonarid Australia, moderate wind erosion occurs on about 51,200 km^2 in New South Wales, 22,500 km^2 in Queensland, and 99,300 km^2 in nonarid Australia. In contrast, the area affected by severe wind erosion amounts to about 1700 km^2 in New South Wales 2000 km^2 in nonarid Australia.

E. Causes of Accelerated Wind Erosion

Similar to water erosion, accelerated wind erosion and desertification are also human-created problems. Wind erosion is caused by aridization. The latter does not necessarily mean a decrease in annual rainfall or precipitation. It is the rainfall effectiveness that is drastically curtailed, due to land misuse.

Arid regions are ecologically fragile. The delicate balance between climate-vegetation-soil is easily disturbed by human activity, e.g., removal of native vegetation cover, intensive agriculture, uncontrolled and excessive grazing, and large unprotected fields devoid of protective shelter belts.

Severe wind erosion occurs in West Asia, Australia, the African Sahel, and the Great Plains region of the United States. The common denominator in all these regions is arid climate and intensive agricultural activity. Once again, the technology exists to curtail the erosion hazard. The adaption and implementation of these technological innovations, however, are socioeconomic and political issues that are often hard to overcome.

V. Soil Erosion in the Tropics versus Temperate Regions

A highly relevant question often asked is whether soil erosion is more serious in the tropics than in the temperate regions. It is a difficult question to answer. The survey of literature presented in this chapter indicates that soil erosion is almost equally severe in any region where land is misused. The answer is further confounded by the fact that severity of erosion depends on both the absolute quantity of soil eroded and the depth and quality of soil remaining. The quantity of soil eroded per unit of time and per unit of area may not necessarily be more from the tropics than from temperate regions. However, the consequences of erosion are generally more drastic in the tropics than in the temperate regions. The drastic erosion that causes productivity declines in soils of the tropics is caused partly by harsh climate and partly by low fertility and poor-quality subsoil. Management and input used are also at a lower level in the tropics than in temperate farming areas. The low productivity of the exposed subsoil and low inputs are the reasons that erosion is considered more severe in soils in the tropics than in temperate-zone soils.

Severe soil erosion in the tropics is observed wherever land is misused or shortcuts are made in land development and subsequent management. For example:

Accelerated erosion has been severe in only those large-scale mechanized schemes in Africa and elsewhere in the tropics that were hastily designed and implemented with utter disregard for soils and climate, without proper planning, and without a careful appraisal of biophysical and socioeconomic factors.

Severe erosion and erosion-related degradation in Central and South America and the Caribbean are attributed to the so-called hamburgerization of tropical rainforests into pastures regardless of slope steepness, soil suitability, or appropriate stocking rates. Erosion is indeed serious in Parana and other regions of central Brazil, where deep oxisols that developed on long and undulating terrains have been intensively used without effective erosion-control measures.

Erosion is observed to be serious on structurally unstable and predominantly low-activity alfisols in West Africa, where agriculture was utterly neglected during the oil-boom era, and where, regardless of the land's suitability, efforts are now being made to produce food by so-called modern techniques.

Severe erosion on vertisols in central India, Australia, and eastern Africa is attributed to the lack of appropriate technology to overcome the problem of trafficability on wet-clay soils to provide much-needed protective ground cover during the monsoons.

Gully erosion is also severe and even disastrous on soils developed on coastal sediments in southeastern Nigeria. Here social factors have played havoc with the natural environment. The planning and implementation of some development projects leave much to be desired. Those concerned with constructing new roads have often overlooked providing appropriate runoff outlets on highways and civil structures. Mining companies often abandon mines without restoring the land. A majority of the rural population in tropical Africa walks 1 or 2 km many times a day on steep terrain to fetch water for household use. Their sunken footpaths often turn into a disastrous gully virtually overnight. Footpaths-turned-gullies have devoured houses, buildings, roads, and anything else in their paths.

The severity of erosion in the tropics, and anywhere else, can be traced to people's helplessness, shortsightedness, poor planning, and other avoidable causes and perfectly manageable factors.

Not all soils of the tropics are shallow, fragile, or structurally unstable. Some are as stable and resistant to erosion as anywhere in the world. However, harsh climatic conditions, overpopulation, excessive use of resources beyond their capability, and the soil-food-population imbalance have rendered tropical ecosystems extremely vulnerable to soil erosion and erosion-induced soil degradation.

VI. Soil Erosion and Food Production in the Tropics

For two decades ending in 1980, food production grossly lagged behind the population needs in many tropical countries. In sub-Saharan Africa, food staples grew only 1.7% a year while population increased 3.2% a year (Paulino, 1986). Similar deficits have existed in eastern and southern Africa. Population growth also has exceeded food production growth in North Africa and the Middle East. Food production has to be increased substantially in many regions.

Although the continuing food deficit in Africa cannot be attributed entirely to erosion and erosion-induced soil degradation, there is a disturbing degree of correspondence between the areas affected by severe soil erosion and those prone to gross food deficits. Bringing new land under production has some flexibility, especially in Central Africa, South America, and the outer islands in Indonesia. But tropical rain forests cannot be justifiably converted to new arable lands when existing lands are underutilized or improperly utilized, and no attempts are made to restore the productivity of degraded lands. Even slight intensification of agriculture on existing lands that are not properly managed could substantially increase soil erosion risks.

Africa, in particular, faces the greatest challenge of breaking the vicious cycle of erosion-induced soil degradation and the resultant decline in crop productivity. Intensification is necessary to increase food production, even though it increases risk of soil erosion. Soil erosion decreases crop productivity, and that necessitates more intensification and bringing steep, shallow, and marginal land under cultivation. It is a vicious cycle, indeed. However, agricultural land use is not the only factor responsible for severe erosion. Urbanization and construction sites also cause very high rates of sediment transport.

VII. Soil Erosion in Different Geographic Regions

It is generally argued that net erosion decreases as rainfall increases. This hypothesis implies that soil erosion, by both wind and water, is more intense in desert and semiarid regions than in subhumid and humid regions (Hudson, 1971). The reason put forward is that in the desert and Sahel regions, for instance, lack of protective cover causes high soil erosion by both wind and water runoff, the latter from flash floods caused by high-intensity, isolated rainstorms. As the rainfall increases, however, vegetation cover also increases and protects the soil surface. In a forest region, with dense vegetation cover, erosion rates are particularly low. This concept is illustrated in Figure 9 from Kirkby (1980).

Although it is true that natural vegetation cover depends on the amount of rainfall, vegetative cover is increasingly being denuded by the ever-increasing population pressure. Deforestation in the equatorial region is estimated at 11 million hectares per year (Postel, 1984). Some ecologists fear that tropical forest reserves will disappear by 2040 A.D. (Grainger, 1980), and the land will be converted to arable and urban uses. The pressure to bring new land under agriculture is particularly high in the semiarid and subhumid ecological regions, regions more suited for food grain production than humid or arid environments are. Although natural vegetation cover was important in the past, its role has progressively decreased through encroachment by people, crops, and cattle.

The erosion potential with altered land use is high indeed for the humid regions. Soil erosion potential by water increases with increasing rainfall (Figure 9). Erosion by water, represented by the solid line in Figure 9, may

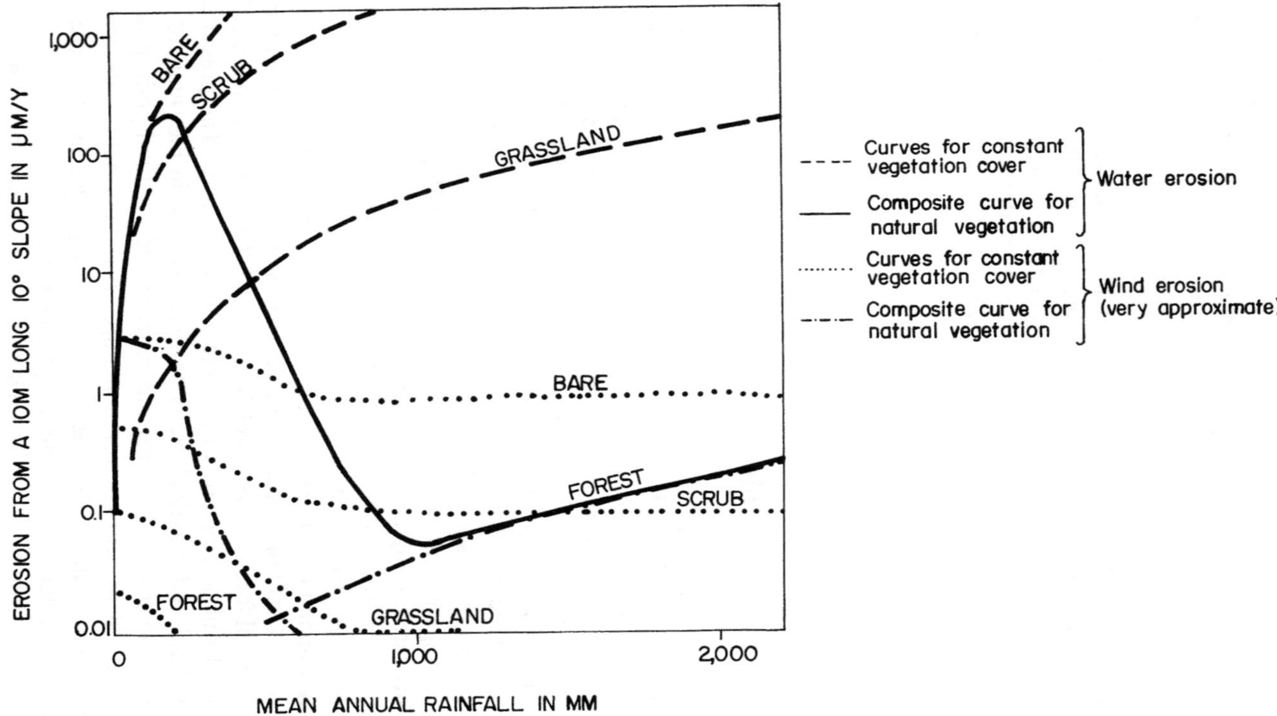

Figure 9. Soil erosion risks in relation to rainfall regime and vegetation. [M.J. Kirkby, The problem. pp. 1–12. In: *Soil erosion*, Kirkby and Morgan (eds.), copyright by John Wiley & Sons, Inc., 1980].

reach a plateau between 3000 and 5000 mm annual rainfall; however, high rainfall associated with typhoons and cyclones may further accentuate erosion by triggering mass movement along steep slopes (represented by the dotted line in Figure 9). Wind erosion decreases with increasing rainfall and is generally not severe in regions with annual rainfall of 1000 mm or more (Figure 9). In humid tropics, the severity of erosion and its consequences are particularly drastic for soils with shallow, effective rooting depths and where the chemical and physical condition of the subsoil are unfavorable—existing as large tracts of eroded and barren lands where lush green forest and savanna vegetation once prevailed.

Although wind erosion in the Sahara is more predominant than erosion by water, the Sahara also has erosion by water. The threshold amount or intensity of rainfall needed to cause erosion is lower in the Sahara and semiarid regions than in subhumid and humid regions, so erosion per unit of amount or per unit of energy of rain is more in arid and semiarid regions than in wetter climates.

The relationship between rainfall amount and erosion potential is not simple, but confounded by many interrelated factors. The tropics and subtropics have the world's highest annual rainfall and the most high-intensity rains. But the highest denudation rates by water erosion may not coincide with the highest rainfall amounts. And there are other important distinctions between temperate and tropical climates in erosion potential: First, snow-melt runoff is not a factor in the lowland tropics or subtropics. Second, the erosion potential per unit of rainfall energy or per unit of rainfall amount increases as mean annual temperature increases. That is why some ecologists argue that the global warming trends set in motion by deforestation and burning fossil fuels may increase erosion hazards.

VIII. Conclusions

Accelerated soil erosion is the response of an ecosystem to its misuse and is the most complete form of land degradation. It is complete because it affects soil properties and its life-support processes. Erosion depletes soil's organic matter and clay fractions; decreases soil's water and nutrients intensity and capacity factors; reduces effective rooting depth and plant available water reserves; exposes relatively infertile subsoil to the surface; and adversely affects plant growth and vigor. It is also the most selective process because it removes the clay, organic matter, and other colloidal fractions, leaving behind the coarse material and skeletal fractions. It also sets in motion other degradative processes, e.g., leaching and acidification, compaction and hardsetting, laterization, and biological degradation.

Erosion is a universal phenomenon. It has existed ever since the first particle of soil was developed. It is an inevitable happening, and at its natural rate, it is also a constructive process. It is accelerated erosion that is destructive. The accelerated process is related to both geophysical and anthropogenic factors. It is driven by human needs, greed, shortsightedness, poor planning, and cutting corners for quick economic returns.

Accelerated soil erosion is most severe in midlatitudes. It is not a coincidence that regions of high population density also experience the most severe rates of erosion. The erosion process is fueled by poverty and unsound economy. It is the desperate attempt to carve out a subsistence living that drives humanity to remove forest from steeplands, cultivate marginal lands for food crop production, and misuse fragile resources. Very high erosion rates in China, northern India, and in the Himalayan region are related directly to the high demands placed on these fragile ecosystems.

Accelerated erosion in the African Sahel, southeastern Nigeria, and East African Highlands is also related to the intensive and excessive use of these resources beyond their carrying capacity. Although the population pressure in these regions is not as high as in some countries of Asia, the human- and cattle-carrying capacity of these lands is relatively low. Rising social aspirations in some cases and desperate attempts merely to survive in others have created climate-soil-population imbalances leading to severe sheet, rill, and gully erosion. Although the rate of erosion (mm yr^{-1} or t ha^{-1} yr^{-1}) is low in Africa, the erosion per unit of production is probably higher in the African Sahel and East African Highlands than anywhere else.

Erosion rates are also excessive in North America. However, the region has lower erosion rates per unit of agricultural production compared with Africa or Asia. The erosion rates in North America are higher per unit of production than in many countries in Western Europe. Excessive erosion rates observed in the recent past are attributed to intensive land use, monocropping without frequent use of soil-conserving cover crops, replacement of ley farming by continuous cropping, and excessive and often unnecessary use of heavy machinery.

Is the world running out of good, arable land? Can the eroded and degraded lands be restored? Can the world feed its projected stable population of about 10.5 billion? These are important and pertinent questions. Accelerated soil erosion is apparently a big factor in the equation pertaining to these issues. The technical know-how to control erosion and to restore eroded lands is available not only for industrial economies, but also for low-resource farmers of Africa, Asia, and tropical America. The application and adaption of these technical solutions depend on local, regional, and international priorities, policies, and political goodwill. The problem of soil erosion at the regional scale of large watersheds involving river basins involves international understanding and commitment to address this serious issue. There is no doubt that the world has the capacity to feed itself provided that we have the commitment and dedication to achieve it. The commitment must be backed by thorough knowledge of potential, constraints of the natural resources, and how to use them for sustained production.

References

Abbas, N., and N. Subramanian. 1984. Erosion and sediment transport in the Ganges River Basin (India). *J. Hydrol.* 45:173–182.

Adu, S.U. 1972. Eroded savanna soils of the Nvrongo-Bawku area, Northern Ghana. *Ghana J. Agric. Sci.* 5:3–12.

Ahmad, N. 1960. Soil erosion by the Indus and its tributaries. *Pakist. Geogr. Rev.* 15(2):5–17.

Ahmad, N. 1977. Erosion hazard and farming systems in the Caribbean countries. pp. 241–249. In: D.J. Greenland and R. Lal (eds.), *Soil conservation and management in the humid tropics.* Wiley, NY.

Ahmad, N., and E. Breckner. 1974. Soil erosion on three Tobago soils. *Trop. Agric.* 51:313–324.

Ahn, P.M. 1975a. Erosion hazards and farming systems in East Africa. pp. 165–176. In: D.J. Greenland and R. Lal (eds.), soil conservation and management in the humid Tropics. Wiley, NY.

Ahn, P.M. 1975b. Erosion hazard and farming systems in East Africa. pp. 34–46. Dept. of Soil Sci., Nairobi Univ., Kenya, Nairobi.

Aitkin, A.P. 1981. Aspects of erosion and sediment transport in Java, Indonesia. pp. 81–91. In: T. Tinsanchali and H. Eggers (eds.), *Problems of soil erosion and sedimentation.* AIT, Bangkok.

Akrasi, S.A., and N.B. Ayibotele. 1984. An appraisal of sediment transport measurement in Ghanaian rivers. pp. 301–312. IAHS Publ. 144.

Anonymous, 1974. Hornos project report. Erosion study project, Spain. Agricultural University, Wageningen.

Attaviroj, S. 1986. Soil erosion and degradation in northern Thai uplands. *Proc. Int. Conf. on the Economics of Dryland Degradation and Rehabilitation, 10–14 March 1986, Canberra Australia, Annex.*

Balek, J. 1977. *Hydrology and water resources in tropical Africa.* Elsevier, Amsterdam.

Bali, Y.P., and R.L. Karale. 1977. A sediment yield index as a criterion for choosing priority basins. pp. 180–188. IAHS Publ. No. 122.

Barber, R.G. 1983. The magnitude and sources of soil erosion in some humid and semi-arid parts of Kenya and the significance of soil tolerance values in soil conservation in Kenya. pp. 20–46. D.B. Thomas and W.M. Senga, (eds.), In: *Soil and water conservation in Kenya.* University of Kenya, Nairobi.

Barrow, H.I., R.J. McCracken, R.J. Miller, W.R. Oschwald, J.N. Shiflet, and R.R. Willis, 1982. Report on China: An American view of how that nation manages its soil resources. *J. Soil Water Conserv.* 37(6):315–318.

Biswas, M.R., and A.K. Biswas. 1978. Loss of productive soils. *Int. J. Environ. Studies* 12:189–197.

Blustain, H. 1982. Social issues in technology choice: Soil conservation in Jamaica. *J. Soil Water Conserv.* 37(6):323–325.

Bonell, M., and Gilmour, D.A. 1978. The development of overland flow in a tropical rainforest catchment, *J. Hydrol.* 39:365–382.

Bonell, M., D.A. Gilmour, and D.F. Sinclair. 1979. A statistical method for modelling the fate of rainfall in a tropical rainforest catchment. *J. Hydrol.* 42:251–267.

Bonfils, C.G., J.E. Calcagus, and P.H. Etchevehere. 1960. Duelos Y erosion en la region pampeaua semiarida. *Rev. Invest. Agric.* (Buenos Aires) 13:321–396.

Bowen, G., and H. Shaozu, 1982. Gully correction and utilization in gullied rolling loess regions on the middle reaches of the Yellow River. Yellow River Conservancy Committee, Beijing, China.

Bowonder, B., K.V. Ramana, and T. Hanumanthae Rao. 1985. Sedimentation of reservoirs in India. *Land Use Policy* 2(2):148–154.

Brown, L.R. 1981. Eroding the base of civilization. *J. Soil Water Conserv.* 36(5):255–260.

Brown, L.R. 1984a. The global loss of top soil. *J. Soil Water. Conserv.* 36(5):255–260.

Brown, L.R. 1984b. The global loss of top soil. *J. Soil Water. Conserv.* 39:162–165.

Brown, L.R., and E.P. Eckholm. 1974. *By bread alone.* Praeger, NY.

Burns, R.V. 1947. Soil erosion in Ceylon with particular reference to floods. *Trop. Agric.* 103:240.

Cabrido, C.A. 1981. National soil erosion control management project and action programs (NASECOMPAP). National Environmental Protection Council, Quezon City, the Philippines.

Cabrido, C.A. 1985. An assessment of national soil erosion central management programs in the Philippines. pp. 13–20. In: E.T. Craswell, J.V. Remeniyi, and L.G. Nallana (eds.), *Soil erosion management*, Proc. Workshop PCARRD, 3–5 December, 1984, Los Banos, Philippines. ACIAR, Canberra.

Carbonnel, J.P. 1964. Erosion in Cambodia. *C.R. Acad. Sci. Paris* 259:3315–3318.

Carson, M.A., and S.W. Tam. 1977. The land conservation conundrum of Eastern Barbados. *Ann. Assoc. Am. Geogr.* 67:185–203.

Chakela, Q.K. 1981. Soil erosion and reservoir sedimentation in Lesotho. UNGI Rapport 54, Scandinavian Institute of African Studies, Uppsala, Sweden.

Charreau, G. 1968. Rain and erosion. *African Soils.* 13:241–55.

Christiansson, C. 1981. *Soil erosion and sedimentation in semi-arid Tanzania.* Studies on Environmental Change and Ecological Imbalance, Scandinavian Institute of African Studies, Uppsala.

Clark, S.P., and E. Jager. 1969. Denudation rate in the Alps from geochronologic and heat flow data. *Am. J. Sci.* 267:1143–1160.

Colclough, J.D. 1978. Soil Conservation in Tasmania. *J. Soil Conserv. Serv. New South Wales* 34(2):63–72.

Crosson, P. 1985. Impact of erosion on land productivity and water quality in the United States. pp. 217–236. In: S.A. El-Swaify, A.C. Moldenhauer, and A. Lo eds.), Soil Conserv. Soc. Am, Ankeny, IA.

Curry, J.R., and D.G. Moore. 1971. Growth of the Bengal deep sea fan and denudation in the Himalaya. *Am. Geolog. Soc. Bull.* 82:563–572.

De Bivort, L.H. 975. World agricultural development strategy and the environment. *Agric. Envir.* 2:1–14.

De Vera, M.R., and L. Rewtakulpaiboon. 1981. Analysis of sediment yield and estimated soil loss of selected philippine watersheds. pp. 107–115. In: T. Tingsanchali and H. Eggers (eds.), *Problems of soil erosion and sedimentation.* AIT, Bangkok.

Dent, F.J. 1984. Land degradation: Present status, training and education needs in Asia and the Pacific, UNEP Investigations on Environmental Education and Training in Asia and the Pacific, FAO Regional Office, Bangkok.

Dequi, J., D. Leidi, and T. Jiesheng. 1981. Soil erosion and conservation in the Wuding River Valley. pp. 461–479. In: R.P.C. Morgan (ed.), *Soil Conservation*. Wiley, Chichester, UK.

Dhruva Narayana, V.V., and G. Sastry. 1985. Soil conservation in India. pp. 3–9. In: El-Swaify, S.A. Moldenhauer, and A. Lo (eds.), *Soil erosion and conservation*. S.A. Soil Conserv. Soc. Am., Ankeny, Ia.

Downes, R.G. 1959. The ecology and prevention of soil erosion. *Biogeogr. Ecol. Austr.* 8:472–486

Dregne, H.E. 1982. Historial perspective of accelerated erosion and effect on world civilization pp. 1–14. ASA Special Publ. 45.

Dudal, R. 1982. Land degradation in a world perspective. *J. Soil Water Conserv.* 37:245–249.

Dumsday, R.G. 1973. The economics of some soil conservation practices in the wheat belt of Northern New South Wales and Southern Queensland—A modelling approach. University of New England, Farm Management Bull. 19.

Dunne, T. 1979. Sediment yield and landuse in tropical catchments. *J. Hydrol.* 42:281–300.

Dunne, T., and G.S.O. Ongweny. 1976. A new estimate of sedimentation rates on the upper Tana River. *Kenyan Geogr.* 2:20–35.

Dyson, J.R. 1970. Report on an erosion survey of the Pindari catchment area. Part 2. Erosion, land systems. *J. Soil Conserv. Serv. New South Wales* 26:205–221.

Eckholm, E.P. 1976. *Losing ground.* Norton, NY.

Edwards, K.A. 1977. Cultural practices and changes in catchment hydrology: A review of hydrologial research techniques as aids to development planning in the humid tropics. pp. 33–48. In: R. Lal and D.J. Greenland (eds.), *Soil conservation and management in the humid tropics*. Wiley, Chichester, UK.

Edwards, K.A. 1979. Regional contracts in rates of soil erosion and their significance with respect to agricultural development in Kenya. pp. 441–454. In: R. Lal and D.J. Greenland (eds.), *Soil physical properties and crop production in the tropics.* Wiley, & Sons, Chichester, UK.

Edwards, K.A., and J.R. Blackie. 1981. Results of the East African catchment experiments. pp. 163–188. In: R. Lal and E.W. Russell (eds.), *Tropical agricultural hydrology*. Wiley, Chichester, UK.

Eliassen, S. 1936. Soil erosion and river regulation with special reference to Yellow River. *J. Assec. Chinese Am. Eng.* 17:22–38.

Emery, K.A. 1975. Survey of erosion and land use within the catchment area of Keepit Dam. Part II. Topography, land use and erosion. *J. Soil Conserv. Serv. New South Wales* 31:33–58.

FAO. 1978. Methodology for assessing soil degradation. FAO/UNEP Expert Consultation, FAO, Rome.

FAO. 1979. Report on the agro-ecological zones project, Vol. 2. Results from South West Asia. World Soils Resources Rept. 48/2, FAO, Rome.

Finn, D. 1983. Land use and abuse in the East African region. *Ambio* 12(6):296–301.

Fournier, F. 1959. Study on the relationship between the erosion of soil by water and atmospheric precipitation. *C.R. Acad. Agric. Pans* 45:418–419.

Fournier, F. 1962. Map of erosion danger in Africa soµth of the Sahara. EEC Commission for Technical Cooperation in Africa, Paris.

Fournier, F. 1966. Use of hydrological basins for the study of soil erosion. *Bull. Ass. Geogr. Fr.* 349: 49–52.

Fournier, F. 1967. Research on soil erosion and soil conservation in Africa. *African Soils* 12(1): 53–96.

Frazier, B.E., D.K. McCool, and C.F. Engle. 1983. *J. Soil Water Conserv.* 38(2): 70–74.

Geiger, A.F. 1957. *Sediment yields from small watersheds in the United States.* Wallingford, UK.

Gillespie, P.H. 1981. Development of gully erosion at the head of Bango Greek near Yass, Sydney, New South Wales, Australia.

Gilmour, D.A., and M. Bonell. 1977. Streamflow generation processes in a tropical rainforest catchment-prelimary assessment. pp. 178–179. In: Institute of Engineers of Australia Symposium on the Hydrology of Northern Australia, Brisbain, Queensland, 1977 Natl. Conf. Publ. No. 77/5.

Giordano, G. 1956. *The mediterranean region.* pp. 317–352. World Geog. Soc., NY.

Glantz, M.H. (ed.). 1977. *Desertification, environmental degradation in and around arid lands.* Westview Press, Boulder, CO.

Golubev, G.N. 1983. Economic activity, water resources and the environment: A challenge for hydrology. *Hydrolog. Sci. J.* 28(1): 57–75.

Gong, S., and D. Jiang. 1979. Soil erosion and its control in small watersheds of the loess plateau. *Scientica Sinica* 22(11): 1302–1313.

Gong, S., and G. Xiong. 1980. The origin and transport of sediment of the Yellow River. Reprints of International Symposium on River Sedimentation, Chinese Society of Hydraulic Engineering.

Good, R.B. 1973. A preliminary assessment of erosion following wild fires in Kosciusko National Park, N.S.W. *J. Soil Conserv. Serv. New South Wales* 29(4): 191–199.

Grainger, A. 1980. The state of the world's tropical forests. *Ecologist* 10: 6–54.

Grange, L.I., and H.S. Gibbs. 1950. Soil erosion in New Zealand. Part I. Southern Half North Island. Soil Bur. Bull. No. 1, Soil Bureau, DSIR, New Zealand.

Griffith, J.F. (ed.). 1972. *Climates of Africa.* Elsevier, Amsterdam.

Grove, A.T. 1969. Landforms and climatic changes in the Kalahari and Ngamiland—A prelimary study. *Geogr. J.* 135: 191–213.

Grove, A.T. 1972. The dissolved and solid load carried by some West African rivers: Senegal, Niger, Benue and Shari. *J. Hydrol.* 16: 277–300.

Hagedorn, D.H., D.K. Giebner, O. Weise, D., Busche and G. Gruneit. 1977. *Dune stabilization.* GTZ, Eschborn, Germany.

Hao, C., H. Mai-Ting, C. Shih-Chi, L. Chen-Wu, H. Heng-Pin, C. Nai-Sen, and L. Pin-Wun. 1977. Regulation of sediments in some medium-size and small reservoirs on the heavily silt-laden streams in China. *Scientia Sinica* 20(1): 89–105.

Hardjowitjitro, H. 1981. Soil erosion as a result of upland traditional cultivation in Java Island. pp. 173–179. In: T. Tingsanchal and H. Eggers (eds.), *Problems of soil erosion and sedimentation.* AIT, Bangkok, Thailand.

Higginson, F.R. 1974. Soil erosion and siltation within the Murray Valley. *J. Soil Conserv. Serv. New South Wales* 30: 135–144.

Holdgate, M.W., M. Kassas, and G.R. White. 1982. The world environment:

1972–1982. Tycool International, Dublin.
Holeman, J.N. 1968. The sediment yield of major rivers of the world, *Water Resources Res.* 4:737–739.
Hove, A. 1977. Some aspects of current sedimentation. Depositional environment and submarine geomorphology of Kenya's submerged continental margins. In: C.O. Okidi (ed.), *Management of coastal and offshore resources in Eastern Africa, Workshop, Nairobi, April 26–29, 1977.* Institute of Development Studies, University of Nairobi, Occasional Paper No. 28.
Howard, P.M. 1981. The water erosion problem in the United States. pp. 25–34. In: *Proc. Am. Soc. Agric. Eng. Conf. on Crop Production with Conservation in the 80's.* Chicago, December 1980.
Hudson, N.W. 1971. *Soil conservation.* Batsford, London.
Janssen, M.B. 1982. Land erosion by water in different climates. Dept. of Physical Geography, Uppsala Univ., Uppsala, Sweden, UNGI Rapport 57.
Jantawat, S. 1985. An overview of soil erosion and sedimentation in Thailand. pp. 10–14. In: S.A. El-Swaify, W.C. Moldenhauer, and A. Lo (eds.). *Soil erosion and conservation.* Soil Conserv. Soc. Am., Anakeny, IA.
Kadomura, H. 1980. Erosion by human activities in Japan. *Geo. J.* 4:133–144.
Kadomura, H., and H. Yamamoto. 1978. Man-induced erosion: The rate of erosion and an example from Okinawa Island, southern Japan. *J. Geogr.* 87:1–15.
Karrar, G. 1984. The UN plan of action to combat desertification, and the concomitant UNEP campaign. *Environ. Conserv.* 11:99–102.
Kaufmann, R. 1981. Erosion control and conservation of small holdings in the high mountain areas of the Eastern Cordillera of Colombia. *Zentrum fur regionale Entwicklungsforschung Giessen* 8:207–216.
Kirkby, M.J. 1980. The problem. pp. 1–12. In: M.J. Kirkby and R.P.C. Morgan (eds.), *Soil erosion.* Wiley, Chichester, UK.
Kovda, V.A. 1977. Biosphere, soil cover and their changes. Meldinger fra Norges Landbrukshogskele, 56(4), Moscow University, Moscow.
Kowal, J. 1972a. The hydrology of a small catchment basin at Samaru, Nigeria. III. Assessment of surface runoff under varied land management and vegetation cover. *Samaru Res. Bull.* 149:120–133.
Kowal, J. 1972b. The hydrology of a small catchment basin at Samaru, Nigeria. IV. Assessment of soil erosion under varied land management and vegetation cover. *Samaru Res. Bull.* 149:134–147.
Kronfellner-Kraus, G. 1980. Erosion and torrent problems in Indonesia. *Centralblatt fur das Gesamte Forestwesen* 3:129–150.
Kuru, A. 1978. The environmental impact of agriculture in Ethiopia. Dept. of Environmental conservation, University of Helsiuki Viikki, Helsiuki, Finland.
Lal, R. 1977a. A brief review of erosion research in the humid tropics of South East Asia. pp. 231–240. In: D.J. Greenland and R. Lal (eds.), *Soil conservation and Management in the humid tropics.* Wiley, Chichester, UK.
Lal, R. 1977b. Review of soil erosion research in Latin America. pp. 231–240. In: D.J. Greenland and R. Lal (eds.), *Soil conservation and management in the humid tropics.* Wiley, Chichester, UK.
Lal, R. 1976. Soil erosion problems on an Alfisol in Western Nigeria and their control. IITA Monograph 1, IITA, Ibadan, Nigeria.
Lal, R. 1987. Soil degradation in relation to climate. International symposium on "Climate and Food Security," 6–9 February 1987, New Delhi, India.

Lal, R. 1988. Soil erosion by wind and water: problems and prospects. pp. 1–6. In: R. Lal (ed.), *Soil erosion research methods.* Soil Water Conser. Soc. Amer., Ankeny, IA.

Lal, V.B., and S. Banerji. 1974. Man's impact on erosion in the rural environment: The Indian experience. pp. 46–52. IAHS publication No. 113.

Lamb, R., and S. Milas. 1983. Soil erosion, real cause of Ethiopian famine. *Environ. Conserv.* 10:157–159.

Larson, W.E. 1981. Protecting the soil resource base. *J. Soil Water Conserv.* 36:13–16.

Larson and Pierce. 1983. The threat of soil erosion to long-term crop productivity. Science 218:458–465.

Le Houerou, H.N. 1976. Can desertification be halted? pp. 1–15. In: *Conservation in arid and semi-arid areas.* FAO Conservation Guide 3 FAO, Rome.

Lee, H.T. 1984. Soil conservation in China's loess plateau. *J. Soil Water Conserv.* 39(5):306–307.

Legowo, I.D. 1981. Volcanic debris control applied in Indonesia. *J. Hydrol* 20(1):71–79.

Leitch, C.J. 1982. Sediment levels in tributaries of the east Kiewa river prior to logging alpine ash. pp. 72–78. In: E.M. O' Loughlin and L.J. Bren (eds.), *First national symposium on forest hydrology.* Institute of Engineers, Barton, Act, Australia.

Low, K.S., and K.C. Goh. 1972. The water balance of five catchments in Salangor, West Malaysia. *J. Trop. Geogr.* 35:60–66.

Lowdermilk, W.C. 1927. Erosion control in Japan, *J. Assoc. Chinese Am. Eng.* 8:3–13.

Luk, S.H. 1983. Recent trends of desertification in the Maowusu Desert, China. *Environ. Conserv.* 10(3):312–224.

Lundgren, B. (ed.). 1975. Landuse in Kenya and Tanzania, the physical background and present situation and an analysis of the need for its rational planning, Royal College of Forestry, Stockholm, Sweden.

Makholibe, S. 1984. Suspended sediment transport measurement in Lesotho. pp. 313–321. IAHS Publ. 144.

Manson, D.A. 1980. Erosion control in Angourie National Park. *J. Soil Conserv. Serv. New South Wales* 36:23–32.

Marker, M.E. 1976. Soil erosion 1955 to 1976. A review of the incidence of soil erosion in the Dundas Tableland area of Western Victoria, Australia. *Proc. Roy. Soc. Victoria* 88(12):15–22.

Mashur, A., and M. Hanif. 1972. A study of surface runoff and sediment release in a clir pine area. *Pakistan J. Forestry* 22:113–141.

McGhie, D.A. 1980. The contribution of the mallet hill surface to runoff and erosion in the Narrogin Region of Western Australia. *Aust. J. Soil Res.* 18:299–307.

Meigs, P. 1953. World distribution of arid and semi-arid homoclimates. Arid zone Program 1, pp. 203–210. UNESCO, Paris.

Moore, T.R. 1979. Land use and erosion in the Machakos Hills. *Assoc. Am. Geogr. Ann.* 69:419–431.

Murthy, R.S., and H.S. Shankaranarayana. 1977. Utility of soil maps for watershed treatment in the Siwaliks. *J. Indian Soc. Soil Sci.* 25:313–316.

Narayana, D.V., and R. Babu. 1983. Estimation of soil erosion in India. *J. Irrigation and Drainage Eng.* 109(4):419–434.

National Economic Development Authority. 1983. National environment manage-

ment action plan. NEDA, Quezon City, The Phillipines.
Nickling, W.G., and J.G. FitzSimons. 1985. Relationship of soil type and agricultural systems to wind erosion in Southern Ontario. pp. 34–54. In: El-Swaify, W.C. Moldenhauer, and A. Lo (eds.), *Soil erosion and conservation. S.A. Soil Conserv. Soc. Am., Ankeny, IA.*
Northwest Hydraulic Consultants, Ltd. 1980. Erosional features and processes in the upper Oldman River Basin. Alberta Environment Planning Division, Canada.
Nyambok, I.O., and G.S.O. Ongweny, 1979. Geology, hydrology, soil erosion and sedimentation. pp. 17–37. In: R.S. Odingo (ed.), *An African dam. Ecological survey of the Kamburu Igharu hydro-electric dam area, Kenya.* Ecological Bulletin (Stockholm) No. 29.
O'Keefe, P. 1983. The causes, consequences and remedies of soil erosion in Kenya. *Ambio* 12(6):302–305.
ORSTOM. 1975. Quoted in *Tropical forest ecosystem*, UNESCO/UNEP/FAO, p. 263.
Osuji, G.E. 1984. The gullies of Imo. *J. Soil Water Conserv.* 39:246–247.
Oyebande, L. 1981. Sediment transport and river basin management in Nigeria. pp. 201–205. In: R. Laland E.W. Russel (eds), *Tropical agricultural hydrology.* Wiley, Chichester, UK.
Patnaik, N. 1975. Soil erosion: A menace to the nation. *Indian Farming* 24:7129.
Paulino, L.A. 1986. Food in the Third World, past trends and projections to 2000. IFPRI Research Report 52, Washington, DC.
Pearse, C.K. 1971. Grazing in the Middle East: Past, present and future. *J. Range Management* 24:13–16.
Peralta, P.M. 1980a. Geomorphology, soils erosion and soil use in the Fataleufa area. Region X (Chile). Fac. For. Sci., Univ. Chile, Santiago.
Peralta, P.M. 1980b. General considerations for the use of soils, principally forest soils in the Alto Palena and Chaiten areas, Region X. Fac. For. Sci. Univ. Chile, Santiago.
Pickup, G., R.J. Higgins, and R.F. Warner. 1981. Erosion and sediment yield in Fly River drainage basins, Papua, New Guinea. IAHS Publ. 132:438–457.
Pimentel, D., E.C. Terhune, R. Dyson-Hudson, S. Rochereau, R. Samis, E.Q. Smith, D. Denman, D. Reifschneider, and M. Shepard. 1976. Land degradation: Effects on food and energy resources. *Science* 194:149–155.
Politano, W., P.C. Corsini, and J.L. Donzelli, 1980. Anthropogenic erosion in the municipality of Monte Alto, Sao Paulo, Brazil. Alto, S.P. *Cientifica* 8(1/2):15–25.
Portch, S., and J.L. Hicks 1980. A soil conservation program for Ecuador. *J. Soil Water Conserv.* 35:243–244.
Postel, S. 1984. Protecting forests. pp. 75–92. In: *State of the world 1984*, L.R. Brown et al. (eds.). Worldwatch Institute, Norton, NY.
Randrianarijaona, P. 1983. The erosion of Madagascar. *Ambio* 12(6):308–311.
Rapp, A. 1972. Conclusions from the DUSER soil erosion projects in Tanzania. *Geogr. Ann.* 54A(3–4):377–379.
Rapp. A. 1975. Soil erosion and sedimentation in Tanzania and Lesotho. *Ambio* 4:154–163.
Rapp, A. 1977a. Methods of soil erosion monitoring for improved watershed management. FAO Conservation Guide No. 1, FAO, Rome.
Rapp, A. 1977b. Soil erosion and reservoir sedimentation—Case studies in Tanza-

nia. FAO Soils Bulletin No. 33:123–132.
Rapp, A., L. Berry, and P. Temple. Erosion and sedimentation in Tanzania. *Geogr. Ann.* 54A(3–4)
Rind, D. 1983. Global climate in the 21st century. *Ambio* 13(3):148–151.
Riquier, J. 1982. A world assessment of soil degradation. *Nature and Resources* (UNESCO) 18(2):18–21.
Robinson, A.R. 1979. Erosion and sedimentation problems and solutions in the Peoples Republic of China. Summer Meeting ASAE/CSAE, 1979 ASAE Paper No. 79-2058.
Robinson, D.A. 1978. Soil erosion and soil conservation in Zambia: A geographical appraisal. Zambia Geogr. Assoc. Occasional Study No. 9.
Roose, E. 1973. Seventeen years of experimental measurement of erosion and runoff on a sandy flurallitic soil in lower Ivory Coast. ORSTOM, Abidjan, Cote D'Ivoire.
Roose, E.J., and J. Godefroy. 1977. Pedogenesis of a reworked ferrallitic soil on schist, under forest and under a fertilized banana plantation in the lower Ivory Coast. *Cah. ORSTOM Pedol.* 15:409–436.
Rooseboom, A. 1978. Sediment discharge in South African rivers. Sedimentafuoer in Sinder-Afrikaanse riviere. *Water SA* 4(1):14–17.
Rossi, G., and J.N. Salomon. 1979. An example of accelerated erosion in Madagascar's sakasaka. *Z. Geomorphol.* 23:271–280.
Sabhasri, S. 1978. Effects of forest fallow cultivation on forest production and soil. pp. 160–184. In P. Kunstodter, E.C. Chapman, and S. Sabhasli (eds.), *Farmers in the forest.* East West Center, Honolulu.
Sague Diaz, H., L. Hernandez Martinez, J.M. Ortega Valdes, and L. Castres Lastres. 1979. *Water balance and erosion in Siera del Rosario.* Inst. de Hidroeconomia, Ministerio de la construcciou, La Habana, Cuba.
Selby, M.J. 1975. *The surface of the earth*, Vol. 1. Cassell, London.
Shahash, M.S.E. 1977. Erosion and solid matter transport in inland waters with reference to the Nile Basin. pp. 278–283. In: *Erosion and solid matter transport in inland waters*, Proc. Paris Symp., July 1977, IAHS Publ. No. 122.
Shih-Yang, K., and C. Te-Chi. 1977. Soil erosion and its control in small gully watersheds in the rolling loess area on the middle reaches of the Yellow River. Yellow River Conservancy Committee, Beijing, China.
Singh, J.S., and M.K. Wali. 1962. The problem of soil erosion in some parts of Kashmir Himalahas. *Proc. Natl. Acad. Sci. India B32*: 118, 124, R. Zh. 1963.
Sombroek, W.G. 1979. Soils of the Amazon region and their ecological stability. pp. 14–27. In: Annual Report 1979, Int. Soil Museum, Wageningen.
Stichling, W. 1973. Sediment loads in Canadian rivers. *Proc. Ninth Hydrol. Symp., Edmonton, May 8–9, 973, Fluvial processes and sedimentation.*
Stocking, M. 1984. Rates of erosion and sediment yield in the African environment. *IAHS Publ.* No. 144:285–293.
Strakhov, N.M. 1967. *Principles of lithogenesis*, Vol. 1. Oliver and Boddy, Edinburgh.
Stromquist, L. 1981. Recent studies on soil erosion, sediment transport and reservoir sedimentation in semi-arid central Tanzania. pp. 189–200. In: *Tropical agricultural hydrology*, R. Lal and E.W. Russell (eds.). Wiley, Chichester, UK.
Suckling, F.E.T. 1981. The scourge of santa cartorina. *Soil and Water (New Zealand)* 17(1):23–27.

Suebsiri, B. 1984. Soil conservation in Thailand. Proc. Fifth ASEAN Soil Conference, 10–23 June 1984, Bangkok, Thailand: E2.1–E2.6.

Suwardjo, H., M. Sudjadi, and M.S. Ross. 1985. Potentials and constraints and development strategies in agricultural land development in Indonesia. Inaugural Workshop IBSRAM Land Clearing and Development Soil Management Network, Jakarta, 26 August–3 September 1985.

Talbot, M.R., and M.A.J. Williams. 1978. Erosion of fixed dunes in the Sahel, Central Niger. *Earth Surf. Process* 3:107–114.

Tam, S.W. 1980. Causes of environmental deterioration in Eastern Barbados since colonization. *Agric. Envir.* 5:285–308.

Tanaka, M. 1976. Rate of erosion in the Tanzawa Mountains, Central Japan. *Geogr. Ann.* 58A:155–165.

Thomas, D.B. R.G. Barber, and T.R. Moore. 1980. Terracing of cropland in low rainfall areas of Machakos District Kenya. *J. Agric. Eng. Res.* 25:57–63.

Thomas, D.B., K.A. Edwards, R.G. Barber, and G.G. Hogg. 1981. Runoff, erosion and conservation in a representative catchment in Machakos District, Kenya. pp. 395–417. In: R. Lal and E.W. Russel (eds.), *Tropical agricultural hydrology*. Wiley, Chichester, UK.

Tregubov, T. 1981. Effective erosion control in the USSR. pp. 451–459. In: R.P.C. Morgan (ed.), *Soil conservation.* Wiley, Chichester, UK.

Tropeans, D. 1984. Rate of soil erosion processes on vine yards in Central Piedmont (NW Italy). *Earth Surf. Processes and Landforms* 9(3):253–266.

UNESCO and IAHS. 1974. Gross sediment transport into the oceans. IAHS, Wallingford, UK.

UNESCO/WMO. 1974. Records of the Int. Conf. on the results of the Intern. Hydrological Decade and on the future Programmes in Hydrology, Sept. 1974, Vol. I. Final Report.

Uriyo, A.P. 1983. Progress in management of tropical soils at IITA. pp. 347–365. In: J.C. Holmes and W.M. Tahir (eds.), *More food from better technology.* FAO, Rome.

Uruguay's Save Our Soils Compaign. 1982. *World Water (UK)* 5(8):22–23, 25.

USDA. 1978. 1977 National Resource Inventories. U.S. Govt. Printing Office, Washington, DC.

USDA. 1987. The Second RCA Appraisal: Soil, Water, and related resources on non-federal land in the United States: Analysis of conditions and trends. Washington, DC.

Van der Poel, P., and J. Timberlake. 1980. Sand dune stabilization. Directorate of Land Utilization. Ministry of Agriculture Gabrorone, Botswana.

Van Lancker, J., and H.Y. Bruggeman. 1982. Rain erosion in the Kizingwe watershed. Facultes de sciences Agronomiques, Rijksuniversiteit Gent, Bujumbura, Burundi.

Virgo, K.J., and R.N. Munro. 1978. Soil and erosion feature of the Central Plateau region of Tigrai, Ethiopia. *Geoderma* 20:131–157.

Walling, D.E. 1984. The sediment yield of African rivers. Publication 144. *IAHS*, Wallingford, UK., 265–283.

Watson, H.K. 1984. Veld burning and sediment yield from small drainage basins. IAHS Publ. No. 144:323–333.

Whitmore, J.S. 1972. Land-use planning to improve the water yield of catchments. Presented at Thirteenth Ordinary Meeting of Southern African Regional Com-

mission for the Conservasion and Utilization of the Soil, 18th to 22nd Sept. 1972.
Wolman, M.G. 1967. A cycle of sedimentation and erosion in urban river channels. *Geogr. Ann.* 49:385–395.
Woo, B.M. 1982. Soil erosion control in South Korea. *J. Soil Water Conserv.* 37(3):149–150.
Zachar, D. 1982. *Soil erosion.* Developments in Soil Science, 10. Elsevier Scientific, Amsterdam.
Zaslavskiy, M.N. 1977. Methological problems in evaluating and mapping eodibile soils. *Soviet Soil Sci.* 9:381.
Zunghu, Z. 1981. Regional geological and morphological characteristics and development of soil erosion on the Loess Plateau, China. *Acta Geol. Sinica* 55(4):308–320.

Soil Wetness and Anaerobiosis

N.R. Fausey* and R. Lal†

I. Introduction

Soil wetness brings to mind thoughts about the soil water regime as a soil classification parameter and about artificial drainage systems to moderate the frequency and duration of occurrences of excessive soil wetness related to high water table or saturated soil conditions. When excessive wetness occurs, air is excluded from the soil, and the beneficial exchange of gases

*USDA, ARS, Soil Drainage Research Unit, The Ohio State University, Columbus, OH 43210, U.S.A.

†Department of Agronomy, The Ohio State University, Columbus, OH 43210, U.S.A.

between the soil the atmosphere is precluded. With this phenomenon comes a root zone environment devoid of oxygen, and biological activities within the soil are confined to those that can obtain energy for respiration without free oxygen being present. This process is known an *anaerobiosis*.

As important as it is to understand the role of the soil water regime in the soil-forming process, and the inherent physical, chemical, and biological characteristics of soil formed under various soil water regimes, these subjects have little relationship to the degradation of soil due to soil wetness and anaerobiosis. Soil degradation due to soil wetness and anaerobiosis requires a change of the soil water regime from its natural state to a wetter state—no matter how wet the natural state. Great progress in understanding the capabilities and limitations of soil for various uses has been made through research and observation of soils formed under different soil water regimes, and some information has been obtained about changes in soil properties and processes when wet soils are drained, i.e., the soil water regime is altered to a drier or a less frequently wet status. However, there is very little known about the changes, usually assumed to result in degradation of productive potential, in soils where the soil water regime has become wetter.

An effort will be made to draw from both direct observation of changes in soils that have become wetter and indirect assumption of what would happen based on the changes created by draining soils from a wet state to a drier state should that condition be reversed. It is clear that the physical, chemical, and biological properties and processes all will be degraded by the onset and continuation of excessive soil wetness.

II. Causes of Soil Wetness

A. Short-Term Wetness Caused by Excessive Rainfall or Flooding

In rainfed areas, the unusual climatic events that occur as once in 20-, 50-, 100-year events can cause severe surface alterations. Intense monsoons, typhoons, hurricanes, and thuderstorms posses sufficient energy to erode topsoil and totally change the surface of the soil that developed over a long time period. In contrast, an area may be completely covered by the deposition of eroded materials from other areas. While soil wetness in those cases does not provide the energy for rearrangement of the landscape, the loss of soil strength and aggregation associated with the soil wetness definitely contributes to this degradation of the soil in place prior to the extreme events.

B. Groundwater Table Rise Caused by Irrigation and Canal Seepage

A rise in the groundwater table caused by irrigation and canal seepage probably provides the most visible illustrations of soil degradation associated with soil wetness. Millions of hectares of arable land have been rendered nonproductive as a result of this condition. The overapplication and/or leakage of water in permeable soils leads to net downward movement and accumulation of water at the normal groundwater table elevation. Consequently, the groundwater table begins to rise and will continue to rise unitl a new hydrologic balance is achieved. This new steady-state level is often very near to the soil surface, giving rise to phreatic vegetation, loss of trafficability, and salt deposition in the upper portion of the profile. The salt buildup results from evaporative concentration of salts present in the irrigation water.

Canal leakage contributes more of the water to groundwater table rise than does overapplication of irrigation waters, but both are significant amounts. This problem has been documented in great detail for the situation in the Indus plain in Pakistan and in Punjab province of western India by the White House–Department of Interior Panel on Waterlogging and Salinity in West Pakistan (1964). The soil is degraded largely due to salinization rather than excessive wetness and anaerobiosis. Elevated water tables also contribute to more frequent and longer duration periods with excess soil water and anaerobiosis following rainfall occurrences.

C. Perched Shallow Water Tables Caused by Soil Compaction

Frequent cultivation and trafficking on the soil can result in soil compaction. As the cultivation becomes more frequent and the axle loads get larger, the depth and intensity of compaction both increase. Ultimately the depth of compaction exceeds the depth of tillage, and a permanently compacted layer develops at the base of the tillage zone (Figure 1). This layer impedes percolation of water deeper into the soil profile, and a saturated zone builds up in cultivated soil above the compacted layer. This condition has been documented in some detail by Fausey (1987) and by Fausey et al. (1986).

This saturated zone reoccurs during each subsequent rainfall event if the surface infiltration is not impeded and the rainfall amount exceeds the available water capacity of the cultivated layer. Trafficability and cultivation are delayed and the probability of further compaction is enhanced as the urgency of avoiding late planting encourages farmers to traffic on the soil while this zone is still at a moisture level that is optimum for compaction (Figures 2 and 3). The high moisture level above the compacted layer promotes slaking and dispersion of soil aggregates, limits rooting depth, and creates an anaerobic environment that affects nutrient availability.

D. Groundwater Table Rise Due to Land Surface Management (Extended Fallow, Deforestation, Mining, etc.)

Various examples exist where the hydrology, including groundwater level and flow direction, is altered by land surface management. These hydrolo-

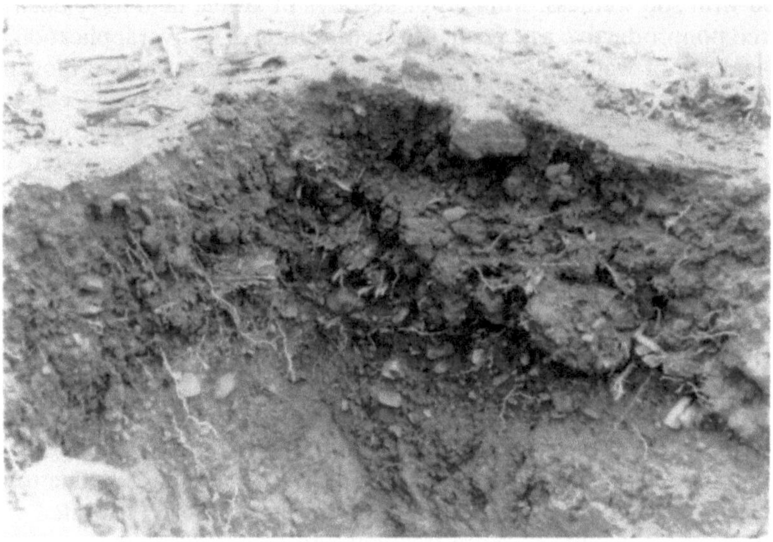

Figure 1. Formation of a permanently compacted layer at the base of the plow layer.

Figure 2. Traffic on a wet soil creates compaction in the wheel traffic zone.

Figure 3. Low soil infiltrability in the traffic zone creates transient inundation following a heavy rain.

Figure 4. Land surface management and land forming can cause excessive wetness in the near-surface horizon.

gic changes result, over time, in excessive wetness in the near-surface horizons of the soil (Figure 4). Ultimately the soil is degraded to the extent that its productive use is limited or eliminated.

Dryland farming typically involves alternating years of cropping and fallow. The fallow year or years are used to accumulate soil moisture by promoting infiltration while minimizing evaporative losses. The natural equilibrium hydrology is disrupted by this pattern. As this concept was extended from one year of fallow before each year of crop to two years of fallow before a crop, groundwater levels were found to increase to a level where hillside seepage developed. These seepage areas typically are very saline and nontrafficable. Doering and Sandoval (1976) have presented a good description of this condition.

Deforestation as a means of conversion of wooded areas to cultivated land has been practiced throughout all climatic zones. This conversion also disrupts the natural equilibrium hydrologic balance. Much attention has been given to the effects of this practice on soil erosion, but less attention has been given to the subsurface hydrologic impact. In rolling or hilly areas, this practice causes the occurrence of more seepage areas and increased stream flow. In flat areas, shallower water tables are the primary effect (Bettany et al., 1964). McGuiness and Harold (1971) have given a good description of this condition for sloping areas. The potential benefits resulting from water table lowering by afforestation have been demonstrated by the Tennessee Valley Authority (1962). Trafficability and rooting depth are both affected by these hydrologic changes.

Figure 5. Crop failure in a temporarily inundated field.

E. Impeded Surface Drainage Due to Construction of Highways, etc.

Improvements in infrastructure such as modern highways, flood-protection levees, and airport runways, to name a few, sometimes result in impeded surface drainage. One might be tempted to suspect inadequate design of the facilities as the cause, but frequently economic considerations dictate what is a cost-effective design. Highway design objectives involving uniform grade/slope and shortest distance typically disrupt the natural overland water flow patterns. Channels may be rerouted, culverts may be installed, and pumps may be required because the construction impedes the natural flow pathways. Waterlogging of the soil and inundation of the soil surface typically may occur in these areas where the outlets have been restricted by the construction. While these are temporary impoundments, the duration of excessive wetness is site-dependent, and soil degradation may result. Certainly trafficability by machines and animals will be reduced and anaerobic conditions may alter nutrient availability and rooting depth. Conacher and Dearden (1988) have documented some of these impacts within the rainforest at Gosford, Australia.

III. Effects of Soil Wetness

As has been suggested above, there are many diverse effects of soil wetness. One of these is anaerobiosis, a very significant effect from an agronomic or biological perspective, but of less significance from a physical perspective (Figures 5 and 6). In an attempt to identify and describe the effects

Figure 6. Agronomic impact of poor drainage in humid regions can be significant.

Figure 7. Wheel ruts created by sinkage of wheels on excessively wet soils.

of soil wetness, separation into physical, chemical, and biological effects will be used. These effects are so significant that they appear as separate chapters in this volume.

A. Physical Effects of Soil Wetness and Anaerobiosis

The major physical change that can be defined as soil degradation associated with soil wetness is the loss of strength in the soil. This is exhibited as the loss of unconfined compressive strength and is illustrated by sinkage of wheels and animal hooves (Figure 7). With sinkage comes loss of traction and incapacity to perform necessary operations. Unconfined compressive strength in cohesive soil is related to the consistency, which reflects the moisture content and at high moisture content is described by "very soft." This phenomenon is very straightforward, and anyone who has walked or driven on wet soils understands the nature of this degradation.

The long-term impact of this degradation depends on the management of the soil while it is too wet. If no energy is applied to the soil, no physical degradation will occur. But when raindrops, wheels, animal feet, or other energy-imparting instruments are applied to the soil, then detachment of particles and/or puddling may occur. This represent true degradation of the soil physical condition. Sharma and DeDatta (1985) documented that puddling decreased bulk density, percolation, and saturated hydraulic conductivity of the surface layer of dry compacted soil. Reid and Parkinson (1984) observed that cracking occurs in soils where the structure has been

Table 1. Comparison of physical properties of lacustrine clay soil in unnaturally wet (diked), near naturally wet (surface drained), and drier than natural (surface + subsurface drained) plots

Soil properties	Diked	Surface and subsurface drained	Surface drained only
Bulk density, g cm^{-3}			
(0–15 cm depth)	1.29	1.22	1.26
(15–30 cm depth)	1.36	1.32	1.33
1 bar moisture retention, %w/w			
(0–15 cm depth)	30.4	30.1	30.6
(15–30 cm depth)	29.2	29.6	29.6
Air-filled porosity, %v/v			
(0–15 cm depth)	12.5	17.0	14.5
(15–30 cm depth)	9.5	1.6	12.0
Unconfined compressive strength, kg cm^{-2}			
(0–15 cm depth)	2.5	1.6	2.3
(15–30 cm depth)	3.1	2.3	2.9
Surface penetration resistance, kg cm^{-2}	14.4	6.0	12.6
Hydraulic conductivty, cm h^{-1}	0.06	2.0	0.30

Source: Adapted from Hundal et al. (1976).

damaged by trampling, while the shrinkage (seasonal) on nontrampled areas was accomodated without cracking. This also is shown clearly in photographs presented by Hundal et al. (1976), showing crust cracking patterns under different drainage treatments. Undrained areas had the largest and thickest crust units and the widest cracks.

The undrained plots described by Hundal et al. (1976) were surrounded by a dike that prevented runoff. All precipitation was retained on the plots, and this created an unnaturally wet soil condition. Table 1 summarizes the comparison of results between the diked (undrained) plots and the undiked (surface drained) plots for six parameters that reflect the physical condition of the soil in these plots. Clearly, the excessive wetness in the diked plots resulted in an increase in bulk density, unconfined compressive strength and surface penetration resistance, and a decrease in air-filled porosity at 1 bar suction and saturated hydraulic conductivity.

Steinhardt and Trafford (1974) were able to measure differences in wheel sinkage, lateral heaving beside the ruts, penetration resistance, and wet density in replicated plots. From these measurements they showed that, on plowed clay soil, subsurface drainage reduced damage from tractor traffic, decreased wheel sinkage, and reduced compaction both below and 16 cm away from the track edge. As the soil matric suction increased in the range of 2 to 25 cm, each 10-cm increase was as effective in reducing

Table 2. Drainage effects on soil temperature in April 1984 at Columbus, Ohio

Distance from drain (m)	Mean daily maximum (°C)	Mean daily minimum (°C)	Summation of hourly readings for 30 days (°C)
0	13.4	6.0	5604
9	13.6	5.7	5540
18	13.7	5.7	5579
27	14.4	5.6	5715

rutting as if the wheel load were reduced by 670 kg. They concluded that for clay soils having temporary excess moisture, draining to lower the water table to 50 to 60 cm depth should be recommended to minimize structural damage.

Slaking is another manifestation of degradation by soil wetness. With prolonged hydration the water films surrounding mineral particles get larger and larger until the charges holding particles in loose arrangement are broken and separation of aggregates occurs. Francis and Cruse (1982) showed that aggregate stability is very sensitive to matric potential, especially at matric potential near zero. They concluded that management practices that result in slow internal drainage may have a significant negative impact on structural stability through their impact on increasing the matric potential. This has strong implications for detachment by raindrops. Abu-Sharar et al. (1986) reported that slaking may proceed without clay dispersion, but slaking releases dispersed clay concurrently with aggregate breakdown. These dispersed clays settle and aid in the formation of an impermeable or very slowly permeable zone at the base of the cultivated layer.

There has been a longstanding perception that wet soils are slow to warm up in the springtime. Recently Steenhuis and Walter (1986) questioned this concept and showed theoretically that the degree of wetness is not responsible for temperature differences in mineral soil. Our own observations of surface soil temperature near to and away from a subsurface drainpipe (Table 2) shows that in April the mean daily maximum soil temperature at 5 cm depth was highest at the greatest distance from the drain, as were the sum of hourly readings for the 30-day period. Consequently, the perception that excessive wetness will cause the soil to the warm up slowly in the spring is questionable. Scotter and Horne (1985) showed, by measurements and by a simulation technique based on laboratory-measured changes in volumetric heat capacity, thermal conductivity, and thermal diffusivity, that drainage does not have a measurable effect on soil temperature.

For every rule there must be an exception. Excessive wetness created by flooding has thus far been described as a soil-degrading agent. Flood fallowing, however, is an interesting soil management technique employed in the sugar industry in Guyana (Gumbs, 1982). It is a tecnhique that provides a very much improved and stable soil structure in a heavy soil, and is a very important reason for the success of continuous cultivation in the heavy clay soil of Guyana. In this technique, the soil is tilled and then flooded and left fallow for six to nine months. Final land preparation is done after the flood water has drained away.

B. Chemical and Biological Effects of Soil Wetness and Anaerobiosis

There are two major chemical effects that can be defined as soil degradation associated with soil wetness and anaerobiosis. One is the accumulation of salts at or near the surface in semiarid or arid regions under high-water-table conditions. The other is the change in solubility and chemical form of nutrients under anaerobic conditions. Soil salinization degrades the soil by rendering it unsuitable for crop production. Anaerobic conditions degrade the soil by rendering some nutrients unavailable, while others become available in potentially toxic quantities for crop production. Discussion of the salinity problem will be left to the chapter by Gupta and Abrol that focuses solely on that aspect of soil degradation.

It is not documented that the soil is degraded by the impact of soil wetness on biological properties and aspects of the soil. It is clear that under anaerobic conditions the organic matter content will increase over time. Soils formed under wet conditions have high organic matter content or frequently are organic soils. It is also clear that an anaerobic environment changes the relative balance of biological populations in the soil. Reducing conditions prevail, giving rise to denitrification and conversion of other nutrients to unavailable forms.

The oxidation-reduction (redox) potential decrease is the most dramatic and quantifiable short-term change that occurs in a soil as a result of waterlogging. Aerated soils characteristically have redox potentials in the range of +400 to +700 mv. Waterlogged soils may have redox potentials as low as −300 mv. Below +400 mv, soil is characterized as moderately reduced, and at −100 mv as highly reduced. As long as oxygen is available in the soil, other oxidized components of the soil are relatively safe from biological and chemical reduction because the gaseous oxygen serves as an electron acceptor, allowing chemical and biochemical reactions to proceed. Once oxygen is displaced and excluded from the soil by waterlogging, other oxidized components of the soil become the electron acceptors and are reduced. The next more easily reduced compound is nitrate, the plant-available form of nitrogen in the soil.

This reduction of nitrate is called denitrification, and it results in a very rapid escape of nitrogen from the soil as gaseous nitrogen. Because large quantities of nitrogen are needed for efficient crop production, this result can clearly be viewed as soil degradation because the supply of plant-available nitrogen is rapidly depleted from the soil by this process.

There is quite a lot of information available on the chemistry of submerged soils. Ponnamperuma (1972) provides a very complete review of the work on this subject up to the 1970s. His discussion on paddy soils probably has the most application to the subject of this chapter because these soils are managed for the wet cultivation of rice, where levees are built to impound water and soil puddling is done deliberately. During submergence, reduction occurs, and iron, manganese, silica, and phosphate become more soluble and diffuse to the surface or move by diffusion and mass flow to the subsoil. Wherever reduced iron and manganese reach an oxygenated surface, they precipitate with silica and phosphate.

Lal and Taylor (1969, 1970) reported the uptake of 5 macronutrients and 10 micronutrients by corn plants under high-water-table conditions compared to a well-drained condition in lysimeters. In the wet soils, uptake of Al, Fe, Mn, and Mo were increased while uptake of N, P, K, Zn, Cu, B, and Sn were decreased. Sharma and DaDatta (1985) reported that puddling increases Fe and Mn solubility and decreases the redox potential and the leaching losses of nitrate, ammonia, P, K, and Zn. Patrick and Henderson (1981) also confirmed that Fe and Mn are more soluble following depletion of oxygen from the soil. Unpublished data of Cooper and Fausey (1987) clearly show that this increased solubility of Mn can result in toxic uptake by soybeans under low soil pH.

IV. Reclaiming Wet Soils

Much controversy has surrounded the issue of converting wetlands to cropland. That issue is not the subject of this discussion; rather, when cropland is degraded by excessive wetness and anaerobiosis from whatever causes, reclamation procedures should be applied to minimize the adverse impacts of the wetness and to maintain a productive soil resource base.

Historically, as can be determined from evidence in ancient cultures (Adams et al., 1981), some system of surface drainage has been employed. These systems of raised areas and ditches allow for the rapid discharge of excess water from the land surface (or part of the surface), and for at least a portion of the soil surface to not be submerged for long period. An aerated zone can be maintained for extended periods with such a system.

Eventually, as a way to avoid taking land out of production, subsurface drainage systems were devised. Earliest instructions regarding design and construction are thought to be those by the Roman statesman, Cato (Weaver, 1964). Such systems could provide protection against excessive soil wetness by draining water from the soil profile as well as from the soil surface, thereby creating a deeper and more permanent aerated zone in the soil. Where adequate outlets were not present, pumps were devised as a way to provide an outlet and ensure less frequent and shorter duration periods of excessive soil wetness.

The importance of drainage to ameliorate excessive soil wetness has long been a subject of study (U.S. Department of Agriculture, 1987). The protection afforded by drainage against soil degradation as a result of excessive wetness varies from place to place according to the climate, the soil, and the type of farming to be carried out (Clark et al., 1988). Drainage is not a guarantee against soil degradation by wetness, but drainage can minimize periods of anaerobiosis, impove trafficability, aid in preventing salinization of the soil, and reduce soil erosion.

V. Conclusions

Productive soil is a vital resource for sustaining an acceptable standard of living for humanity. Strong efforts must be made to protect this resource from degradation by any cause. Excessive wetness degrades soil physically and chemically. Loss of soil strength encourages soil erosion, facilities soil compaction, and impedes trafficability. Chemical and biological reduction under anaerobic soil conditions diminishes the availability of many essential plants nutrients, especially nitrogen, and solubilizes some others to potentially toxic levels. Drainage is a management tool that can be applied to protect soil from degradation by excess wetness. Both surface and subsurface drainage provide some protection, and often both are required for the most effective control of excess soil water.

References

Abu-Sharar, T.M., F.T. Bingham, and J.D. Rhoades. 1986. Stability of soil aggregates as affected by electrolyte concentration and composition. *Soil Sci Soc. Amer. J.* 51:309–314.

Adams, R.E.W., W.E. Brown, Jr., and T.P. Culbert. 1981. Radar mapping, archeology, and ancient Maya land use. *Science* 213:1457–1463.

Bettany, E., A.V. Blackmore, and F.J. Hingston. 1964. Aspects of the hydrological cycle and related salinity in the Belka Valley, Western Australia. *Aust. J. Soil Res.* 2:187–210.

Clark, A.M., A.C. Armstrong, R.J. Parkinson, and I. Reid. 1988. Field drainage and land management, a comparison of four long term field trials. *Agric. Water Management* 14:113–124.

Conacher, P., and P. Dearden. 1988. Soil conservation integrated with road construction—A reality after 200 years. *Aust. J. Soil Water Conserv.* 1(1):17–23.

Cooper, R.L., and N.R. Fausey. 1987. Unpublished data.

Doering, E.J., and F.M. Sandoval. 1976. Hydrology of saline seeps in the Northern Great Plains. *Trans. ASAE* 19(5):856–861, 865.

Fausey, N.R. 1987. Impact of cultural practices on drainage of clay soils. pp. 288–

292. In: *Drainage design and management*. Proc. Fifth Natl. Drainage Symp., Amer. Soc. Argric. Eng., St. Joseph, MI.

Fausey, N.R., G.S. Taylor, and G.O. Schwab. 1986. Subsurface drainage studies in a fine textured soil with impaired permeability *Trans. ASAE* 29:1650–1653.

Francis, P.B., and R.M. Cruse. 1982. Soil water matric potential effects on aggregate stability. *Soil Sci. Soc. Amer. J.* 47:578–581.

Gumbs, F.A. 1982. Soil and water management features in Trinidad and Guyana. *Trop. Argric.* (Trinidad) 59(2):76–81.

Hundal, S.S., G.O. Schwab, and G.S. Taylor. 1976. Drainage system effects on physical properties of a lakebed clay soil. *Soil Sci. Soc. Amer. J.* 40:300–305.

Lal, R. and G.S. Taylor. 1969. Drainage and nutrient effects in a field lysimeter study: I. Corn yield and soil conditions. *Soil Sci. Soc. Amer. J.* 33:937–941.

Lal, R., and G.S. Taylor. 1970. Drainage and nutrient effects in a field lysimeter study: II. Mineral uptake by corn. *Soil Sci. Soc. Amer. J.* 34:345–248.

McGuiness, J.L., and L.L. Harrold. 1971. Reforestation influences on small watershed streamflow. *Water Resources Res.* 7(3):845–852.

Patrick, W.H., Jr., and R.E. Henderson. 1981. Reduction and reoxidation cycles of manganese and iron in flooded soil and in water solution. *Soil Sci. Soc. Amer. J.* 45:855–859.

Ponnamperuma, F.N. 1972. The chemistry of submerged soils. *Adv. Agron.* 24:29–96.

Reid, I., and R.J. Parkinson. 1984. The wetting and drying of a grazed and ungrazed soil. *J. Soil Sci.* 35:607–614.

Scotter, D.R., and D.J. Horne. 1985. The effect of mole drainage on soil temperatures under pastures. *J. Soil Sci.* 36:319–327.

Sharam, P.K., and S.K. DeDatta. 1985. Puddling influence on soil, rice development, and yield. *Soil Sci. Soc. Amer. J.* 49:1451–1457.

Steenhuis, T.S., and M.F. Walter. 1987. Will drainage increase spring soil temperatures in cool and humid climates? *Trans. ASAE* 29(6):1614–1645, 1649.

Steinhardt, R., and B.D. Trafford. 1974. Some effects of sub-surface drainage and ploughing on the structure and compactability of a clay soil. *J. Soil Sci.* 25:138–152.

Tennessee Valley Authority. 1962. Reforestation and erosion control influences upon the hydrology of the Pine Tree Branch Watershed, 1941–1960. TVA, Knoxville, TN.

U.S. Department of Agriculture, Economic Research Service. 1987. *Farm drainage in the United States: History status and prospects*, G.A. Pavelis (ed.). Misc. Publ. No. 1455.

Weaver, M.M. 1964. *History of tile drainage*. M.M. Weaver, Waterloo, NY.

White House–Department of Interior Panel on Waterlogging and Salinity in West Pakistan. 1964. Report of Land and Water Development in the Indus Plain.

Chemical Degradation of Soil

T.J. Logan*

I. Introduction

Land degradation had become a major global concern in recent years as a result of increasing demands on the land for food production and waste disposal. People are learning that the resiliency of soil is finite and that soil degradation is not easily reversed, if ever. The focus of land degradation in this century has been on soil erosion, as increasing areas of forest, grassland, and wetland have been cleared for crop production. Soil erosion represents the most complete form of land degradation—the remvoal of the

*Department of Agronomy, The Ohio State University, Columbus, OH 43210, U.S.A.

soil resource itself—and eroded sediment deposited on adjacent lands and drainageways can lead to further degradation. Soil erosion remains the focus of conservation efforts in the developed world, and much of the resources of agencies such as the U.S. Soil Conservation Service are devoted to reducing soil loss to "tolerable" limits (Follett and Stewart, 1985). Other developed countries have similar programs. Soil erosion from land clearing and improper management in the developing world—and in particular those of the subtropics and tropics—has reached massive proportions and is the subject of worldwide attention.

Of no less importance, but often overlooked, is the impact of chemical degradation on soil. Chemical degradation is defined here as the accumulated negative impact of chemicals and chemical processes on those properties that regulate the life processes in the soil. The soil here is viewed as a living organism, and as a living organism it has a complex system of self-regulation. A "healthy" soil has important chemical and biological attributes including nutrient supply, acid and base buffer capacity, organic matter decomposition, pathogen destruction, toxic metal inactivation, and toxic organic inactivation and degradation. These attributes are well expressed in a "healthy" soil, but their capacities are finite and can be overwhelmed by mismanagement.

The extent to which these processes are affected defines the extent of chemical degradation. As with soil loss, it is important to determine how much chemical degradation is tolerable. The difference, however, is that unlike soil loss, which is essentially irreversible in the short term, chemical degradation can in some cases be reversed; excess acidity can be easily neutralized, but accumulation of copper, lead, or cadmium to toxic levels is irreparable.

Natural processes can result in chemical degradation of soil. For example, fire can destroy built-up reserves of soil organic matter, changes in stream hydrology can result in restricted drainage and anaerobic soil conditions, and volcanic ash high in soluble salts can be deposited on fertile soil. The emphasis of this chapter, however, will be on anthropogenic impacts, particularly those of modern industrial people. Human-produced effects include depletion of the chemical stores of the soil through mining, nutrient depletion, and excessive leaching. They also include chemical contamination of soil from excessive or improper disposal of wastes. The scope of this chapter will be global and will include problems more common to the developed world, such as waste disposal and acid deposition, as well as those of the developing countries, including nutrient depletion and excess acidity. This chapter will not be an exhaustive review of the world literature—a task that would fill several volumes—but will emphasize the more pressing, contemporary problems. It will draw on both primary and secondary sources and will identify, where appropriate, important summaries of individual topics.

II. Chemical Processes in Soil

A. Chemical Weathering

The weathering of minerals in soil is a natural and relentless process, the net result of which is the gradual loss of base-forming cations (Na, K, Ca, Mg) and an accumulation of insoluble compounds of Si, Al, and Fe. This process is retarded by the accumulation of organic matter and the formation of metastable clay minerals, such as kaolinite, illite, smectites, or vermiculites, with net cation-exchange capacity (CEC). Organic matter serves to retain base and metal cations by cation exchange and by complexation. Weathering also results in a gradual shift of the equilibrium soil system to acidic conditions as base-forming cations are released by acid neutralization of primary minerals and are replaced by Al and H on exchange sites. With few exceptions, most anthropogenic processes that affect soil are acid-producing, and it is necessary that soils have acid-buffering capacities. Young, unweathered soils, or those formed from basic rocks, have greater capacity to buffer against acid input than older, more highly weathered soils. This is of particular importance—as will be discussed later—when assessing the impact of acid deposition on soils of the United States, Canada, and Northern Europe. Acid buffer capacity is also of importance, however, in soils of the humid tropics, where equilibrium pH is low (often <4.5) and strongly buffered by Al, creating conditions that are unfavorable for optimum crop production.

B. Organic Matter Accumulation and Loss

The importance of organic matter in soils has been the subject of international debate for many years (Stevenson 1982; Chen and Avnimelech, 1986). The role of organic matter in stabilizing soil aggregates and reducing soil erosion has been clearly demonstrated (Hayes, 1986). The importance of organic matter in soil chemical processes has also been well established, although the chemistry of soil organic matter and the interaction of organic matter with soil minerals and other soil chemical constituents is still not well understood. In general, however, soil scientists agree that increasing organic matter content improves the "health" of the soil.

The rate of accumulation or loss of organic matter from soil has been the subject of much research. Recent reviews by Stevenson (1982, 1986) detail the large literature on this subject. A common theme in this work is the concept of a steady-state equilibrium value for soil organic matter content. The collected data suggest that as organic matter accumulates in soil through additions of residues or manures, a steady-state equilibrium organic C value should be approached asymptotically. The addition of farmyard manure (FYM) at an annual rate of 35 mtons ha^{-1} to the long-term

Rothamsted plots (Jenkinson et al., 1987) has approximately trebled the organic C content of the surface soil in 140 years. However, equilibrium does not appear to have been reached. The plots receiving chemical fertilizer increased organic C levels compared to the unfertilized plots, and the level of organic C appears to have stabilized with fertilizer addition. The lack of equilibrium on the FYM plots may be a result of the large amount of carbon applied to the soil relative to increased biomass production with fertilizer alone and to the relatively cool temperatures of this part of England. Jenkinson et al. (1987) compared experiments on decomposition of ^{14}C-labeled ryegrass at Rothamsted and Ibadan, Nigeria. Decomposition was much more rapid at Ibadan, which has a mean annual temperature of 26°C compared to 9.3°C at Rothamsted.

The rate of decomposition of organic matter added to soil has also been studied extensively (Stevenson, 1986). Many of these studies suggest that 20% to 70% of added organic C is retained in soil following micobial decomposition, the remainder being evolved primarily as CO_2. Stevenson (1986) suggests an average retention of about 30%. Stevenson cites the work of Clark and Paul (1970), who partition soil organic C into three pools: decomposing plant residues and associated biomass, which turn over at least once every few years; microbial metabolites and cell-wall constituents stabilized in soil, which have a half-life of 5 to 25 years; and resistant soil organic matter fractions with half-lives that range from 250 to 2500 years. Goh (1980), citing the work of Greenland and Nye (1959), reported that organic matter in tropical forests decomposes more rapidly than savannah organic matter. However, under fallow conditions, organic matter accumulates more rapidly in the forest than in a savannah system. Goh estimated that it would take 7 to 16 years of fallow to rebuild organic C lost by cultivation of tropical forest versus 27 to 67 years to restore carbon levels in savannah.

An important question becomes how much the organic matter content of a soil has to be reduced before it is considered to be chemically degraded. Are higher levels of organic matter required for some soils than for others? The answer depends on the particular chemical process considered.

1. Cation-Exchange Capacity

Soil organic matter has a high cation-exchange capacity. Stevenson (1982) notes that humic acids have total acidities of 485 to 870 $cmol_c\ kg^{-1}$ and fulvic acids up to 1400 $cmol_c\ kg^{-1}$. These values represent potential exchange capacities, since only a fraction of the acidic functional groups are dissociated at normal soil pH,, and many are blocked by strongly complexed polyvalent cations. Effective CEC (ECEC) values for soil organic matter are apparently around 100–300 $cmol_c\ kg^{-1}$. In a detailed study of the effects of organic matter content on CEC, Helling et al. (1964) found that between 19% and 45% of soil CEC was attributable to organic matter in a pH range of 2.6 to 8.0. The contribution of organic matter to CEC increased with increasing pH as a consequence of the variable charge nature of organic matter exchange capacity. This study was conducted on soils of Wisconsin that contain significant amounts of 2:1 clay minerals with permanent negative charge. The contribution of organic matter to CEC of ultisols and oxisols, which contain little or no constant-change minerals, would be expected to be far greater. This has been confirmed by the work of Hallsworth and Wilkinson (1958, as cited by Stevenson, 1982) who showed that the contribution of organic matter to CEC of bauxitic soils of Australia was far greater than for less weathered soils. Morais et al. (1976) studied the effect of surface and subsurface organic matter content of two Brazilian oxisols on ECEC and found that organic carbon decreased by 65% to 85% from the A to B horizons; ECEC decreased by 68% to 72%. Thomas and Hargrove (1984) also indicate that 2:1 minerals interlayered with hydrous sesquioxide have very little ECEC at pH near 4, and for these soils organic matter may be the only significant source of CEC.

Thomas and Hargrove (1984) and Juo and Kamprath (1979) showed that much of the strong acidity associated with functional groups on soil organic matter is complexed with Al. The remaining functional groups are much weaker, and there is a reduction in the total charge available for CEC. It is apparent that organic matter is the principal source of CEC in soils without significant quantities of permanent-charge clay minerals, and losses of organic matter can significantly reduce the soil's ability to retain base cations and to buffer acidity.

2. Nutrient Mineralization

Organic matter contains in varying quantities all of the essential nutrient elements. In natural systems, the cycling of these nutrients by organic matter immobilization-mineralization is the principal mechanism for nutrient supply. In forest systems, for example, a large percentage of the available pool of N, P, S, Ca, Mg, K, and trace elements are in the standing crop and in the soil biomass (Jordan, 1985). In subsistence cropping systems of the tropics and subtropics, organic matter is the major source of nutrients and is an important reservoir for nutrients that would otherwise leach from the soil (Nye and Greenland, 1964; Jordan, 1985). Likewise, organic farming systems build and maintain organic matter carefully to optimize nutrient supply to crops (Lockeretz et al., 1984). In high-input systems that rely heavily on chemical fertilizer, organic matter still plays a beneficial role through organic N mineralization, which can vary from <10 to >100 kg N ha^{-1} yr^{-1} (Stanford and Smith, 1972). It also increases the availability of P in soils of high P-retention capacity by retaining P in a slowly mineralizable pool (Stevenson, 1986) and by interference of organic acids in P adsorption by soil minerals (Traina et al., 1986). Organic matter may also aid in trace nutrient availability through the formation of stable complexes of trace nutrient metals with fulvic acids (Ellis and Knezek, 1972; Leeper, 1972).

3. Trace Metal Complexation

One of the important attributes of soil organic matter is its strong complexaton of polyvalent metal cations, particularly Fe, Al, and the transition element metals (Lindsay, 1979; Stevenson, 1986). This has great significance for the inactivation of toxic levels of these metals in soil resulting from strongly acidic conditions (e.g., toxic Al^{3+} in acidic oxisols, ultisols, and spodosols), atmospheric deposition of metal from metal ore smelters (Whitby et al., 1976), or from application of metal-containing wastes such as sewage sludge (Logan and Chaney, 1983). The role of organic matter in ameliorating soil degradation by toxic metals will be discussed in later sections.

4. Partitioning of Organics into Soil Organic Matter

Many of the pesticides used today and many of the toxic industrial organic compounds discharged into the environment are aromatic or have some aromatic character. Compounds with aromatic rings are hydrophobic and have much stronger affinities for organic matter than for water. The mechanism for this affinity is believed to be twofold: (1) strong disassociation of the organic molecule with water as a result of an unfavorable entropy change; and (2) strong binding of the organic molecule to organic matter through short-range van der Waals forces and hydrogen bonds (Chiou et al., 1986).

Considerable work has been done on predicting the binding of organic compounds to soil, and these studies all demonstrate the overriding effect of soil organic matter on binding as well as the influence of the hydrophobic character of the chemical itself (Rao et al., 1983). It has been shown (Rao et al., 1983) that the partition coefficient (K) of an organic compound between soil and water is proportional to the soil organic C content, and that the effect of soil organic matter can be accounted for by calculating K_{oc} (where $K_{oc} = K/\%$ organic C). K_{oc}, then, is a property of the compound and is independent of the soil. Karickhoff (1981) has developed several empirical equations for a wide range of organic compounds and soils that relate K_{oc} to the water solubility of the compound (given as mole fraction, X_{sol}), its melting point (MP in °C), and its tendency to partition between water and a nonpolar solvent, octanol (K_{ow}). These are as follows:

$$\log K_{oc} = -0.921 \log K_{sol} - 0.00953(\mathrm{MP} - 25) - 1.405 \quad (1)$$
$$\log K_{oc} = -0.594 \log K_{sol} - 0.197 \quad (2)$$
$$\log K_{oc} = -0.987 \log K_{ow} - 0.336 \quad (3)$$

These relationships suggest that the retention and inactivation of toxic organics in soil will be favored by high soil organic matter and that these effects will be greatest for the most hydrophobic compounds. While partitioning of these potentially toxic compounds into soil organic matter reduces the potential for their uptake by plants and for leaching, there is much debate today as to the impact of this inactivation on microbial degradation of the compound.

C. Rates of Change of Chemical Processes and Their Reversibility

An important distinction between chemical and physical degradation is the rate at which chemical processes occur and the degree to which they are reversible. In general, with the exception of long-term weathering, chemical reactions in soil occur in time frames from instantaneous to several decades, with many processes functioning in time periods of no more than a few weeks. The overall rate of a chemical process in soil is often determined by microbiological processes that also occur in time periods of days and weeks. Some individual chemical processes are considered below.

1. Soil Acidity Buffering

Soil acidity buffering is primarily by ion exchange, Al hydrolysis, and dissocation of H on soil minerals and organic matter. Ion-exchange reactions are essentially instantaneous, but the overall process is limited by diffusion of ions in water films. Nevertheless, the process is extremely rapid if the source of acid or base is in a chemically reactive form. Thomas and Hargrove (1984) suggest that Al hydrolysis may be the rate-limiting step in neutralization of soil acidity and that adsorption of intermediate hydrolysis products, e.g., $[Al(OH)_2{}^+]_n$, reduces the hydrolysis and increases neutralization time. Thomas and Hargrove cite the work of Moschler et al. (1962), who found that the pH of Virginia soils with large capacity to adsorb hydroxy-Al was still increasing five years after lime addition. Soil pH never rose much above 7.0, and appreciable titratable acidity remained.

Soil pH is relatively easy to reverse if the change in total acidity or basicity is small compared to the buffer capacity. These changes would correspond to the sum of exchangeable H and Al or readily decomposable minerals such as calcite. Larger additions of acid or base would result in major changes in the chemistry of the soil. Large acid additions, such as would be encountered by high, countinuous inputs of strong acid deposition (Whitby et al., 1976) or oxidation of pyrite (Kelley and Tuovinen, 1988), would result in accelerated mineral weathering and a change in the suite of metastable and stable minerals in the soil. Large, strong base additions, such as partially neutralized $Ca(OH)_2$ wastes, would cause some solubilization of organic matter as well as mineral dissolution. Other minerals such as calcite would precipitate if Ca levels were high enough, and this would reduce the strong base content of the soil but would significantly change the dominant soil chemical reactions. and activities of soil chemical elements. Overcash and Pal (1979) point out that there are far fewer examples of alkaline wastes applied to land than there are acidic waste materials. However, residual lime materials are relatively common, and some may be used as substitutes for agricultural limestone (Che et al., 1988).

2. Oxidation-Reduction Potential

Oxidation-reduction (redox) reactions in soil are driven primarily by microbiological demand for terminal electron acceptors. The rates of redox reactions are essentially those, then, of bacterial processes that occur in time frames of days to weeks. Chemical oxidation and reduction can occur in soil (e.g., Cr^{2+} oxidation by manganese oxides), and the reaction is rapid. These processes are less significant, however, than the myriad of microbially driven reactions. The degree to which the redox potential of a soil will be poised at a given redox couple (e.g., Fe^{3+}/Fe^{2+}) will depend on the availability of the electron acceptor and the carbon supply to the bacteria (either carbohydrate or CO_2). Reversal of redox reactions is relatively rapid unless conditions for bacterial growth, such as temperature, substrate, water, pH, or osmotic pressure, are not limiting.

3. Adsorption-Desorption

The adsorption and desorption of ions, ion pairs, solution complexes, and organic compounds is one of the fundamental mechanisms by which the concentrations of these substances in soil solution are regulated (Sposito, 1984). Adsorption-desorption includes electrostatic ion exchange, formation of surface complexes, ligand exchange, trace metal chelation, and hydrophobic bonding of organics. The mechanisms governing these processes have been the subject of extensive research beyond the scope of any one review. The following sources are recommended:

Ion exchange: Sposito (1984); Bolt (1982); Gast (1977).
Ligand exchange: Sposito (1984); Bowden et al. (1980); Parfitt (1980); Hingston (1981).
Surface complexation: Schindler (1981); James (1981).
Trace metal adsorption: Leckie and James (1974); Stumm and Morgan (1981).

Adsorption-desorption reactions are rapid, and near-equilibrium is often reached in minutes, hours, or days. Many authors have reported the existence of a slow reaction for adsorption of some chemical constituents such as phosphate (Barrow, 1980). This slow reaction, which can persist for several years, produces relatively small departures from equilibrium and has been attributed to experimental artifacts, diffusion of adsorbate into the adsorbing solid, and rearrangement of the adsorbate on the solid surface. It is likely that any adsorption process that involves the formation of short-range bonds, as is the case with ligand exchange, surface complexation, or hydrophobic bonding, will involve a slow reaction as the initially adsorbed chemical species undergoes further chemical reaction with the surface.

It is also likely that adsorption reactions that involve the formation of short-range chemical bonds with surface constituents will also display considerable hysteresis in the desorption reaction. Hysteresis, in which desorption departs considerably from adsorption, is commonly observed in soil systems with inorganic ligands (e.g., PO_4), metals (e.g., Cu^{2+}), and organics (e.g., paraquat, PCB). Barrow (1980) has argued that the hysteresis effect for phosphate is a consequence of not allowing the slow reaction to be completed before conducting the desorption experiment. For phosphate this can be several years. The significance of the slow reaction for soil degradation is that adsorption of toxic chemicals such as the trace metals, ligands, and organics may not be rapidly or easily reversed, and a drastic change in the chemical environment may be required to accelerate desorption.

4. Precipitation-Dissolution

Precipitation reactions in soil are particularly important for conditions of high ion concentrations, and these conditions are likely to be common when reactive wastes are disposed of on land. Precipitation is environmentally significant because, once solubility of a pollutant in soil is controlled by the dissolution of a precipitated solid phase, further contamination will result in increased concentration of the pollutant only if a new solid phase is formed or if the precipitating soil reactant is exhausted. This tends to placed an upper bound on solution activities of chemicals added to soil.

III. Examples of Chemical Soil Degradation

A. Nutrient Depletion

Unlike natural systems, in which biomass production is in equilibrium with the nutrient reserves of the soil and with biological nitrogen fixation, crop and animal production systems result in net removal of nutrients from the soil. Whereas nitrogen can be replaced on a sustained basis through N fixation, other plant and animal nutrients must be supplied from soil reserves. The pool of available nutrients when depleted must be replenished from the unavailable soil reserves through mineralization of organic matter, dissolution of mineral precipitates, and desorption of strongly adsorbed chemical species. Degradation of the soil occurs when the total nutrient reserves are inadequate for biomass production or when the rate at which nutrients are mobilized is less than biomass demand. Three examples of long-term nutrient depletion will be considered, two from temperate regions and one from the tropics.

1. The Rothamsted Plots

The Rothamsted experiment station was established in 1843 by John Bennet Laws and is the oldest such facility in the world (Lawes Agricultural Trust, 1977). Between 1843 and 1856 Lawes and Gilbert started nine long-term experiments; some of the treatments were changed in the first few years, and one experiment was abandoned in 1878. Several modifications to treatments have been made over the years, but many of the original treatments have been maintained. Figures 1 and 2 show the changes in organic carbon on the Broadbalk wheat plots from 1843 to the present for unmanured, manured, and fertilized plots (Jenkinson et al., 1987). The results, based on nitrogen content of the soil, show that organic C and organic N levels have not decreased markedly with cropping in this period. Long-term additions of NPK have increased C and N levels only slightly compared to the unmanured plots (Figures 1). Manure additions, however, continue to increase the organic matter content of the soil. Figure 2 shows that carbon content has decreased steadily since manure additions (1852–1871) were terminated. Yet 20 years of manure addition have maintained organic matter (and organic N) levels above those of the unmanured plot 125 years after application ceased. Extractable P decreased from 7 mg kg^{-1} in 1951 on the unmanured plots to 2 mg kg^{-1} in 1974. On plots receiving

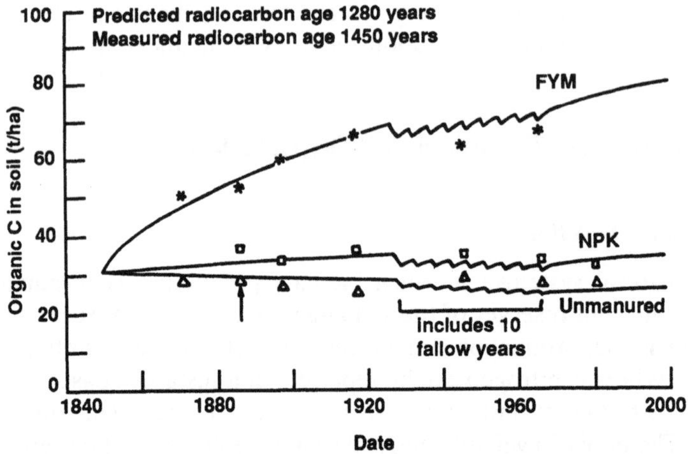

Figure 1. Organic carbon in the top 23 cm of continuous wheat plots at Rothamstad. Data was calculated from nitrogen content. Treatments include untreated, farmyard manure (FYM) (35 mtons ha^{-1} yr^{-1}) and NPK fertilizer (144 kg N, 35 kg P, and 90 kg K ha^{-1} yr^{-1}) (Jenkinson et al., 1987).

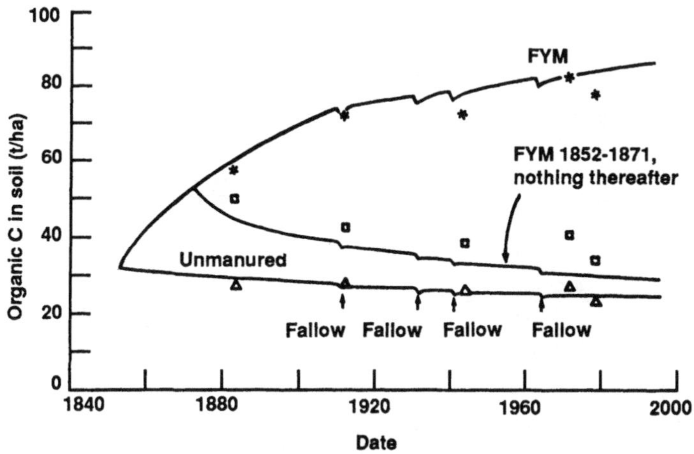

Figure 2. Organic carbon in the top 23 cm of continuous spring barley plots at Rothamsted. Farmyard manure (FYM) was applied at an annual rate of 35 mtons ha^{-1} (Jenkinson et al., 1987).

fertilizer from 1856 to 1901, P levels declined from 21 mg kg^{-1} in 1974. For the unmanured, fertilized, and FYM plots, exchangeable K levels declined from 74 to 69, 121, to 89, and 106 to 87 mg kg^{-1}, respectively. These results suggest that, over a period of several decades, P and K reserves have declined slowly; and where high fertility levels were previously established, these additions have continued to maintain levels of these available nutrients well above those of the unmanured plots.

2. The Morrow Plots

The Morrow plots were established at the University of Illinois at Champaign-Urbana in 1876 and are the longest continuously maintained plots in the United States. Much smaller (200 m^2 since 1904) than those at Rothamsted, the original treatments have been changed over the years. Several, however, have received no additions since the experiment started, and these serve to illustrate changes in nutrient and organic matter levels with continuous cropping versus those receiving various combinations of lime and fertilizer (Odell et al., 1982).

The Morrow plots are located on an aquic argiudoll, developed in loess over loam glacial till under prairie vegetation and somewhat poor natural drainage. These soils have high native fertility, and the results of more than 100 years of cropping without additions reveal that these reserves are still high. Table 1 shows differences in physical and chemical properties of the corn plot receiving no fertilizer or lime additions and that receiving periodic lime additions and annual applications of NPK (224 kg N ha^{-1}

Table 1. Physical and chemical properties of two of the Morrow plots 90 years after their installation

Horizon	Depth (cm)	Sand (%)	Silt (%)	Clay (%)	Bulk density (g cm^{-3})	Total porosity (%)	Available-moisture-holding capacity (cm cm^{-1})	Organic carbon (%)	pH	Base saturation (%)	Cation-exchange capacity ($cmol_c$ kg^{-1})	Exchangeable cations ($cmol_c$ kg^{-1}) Ca	Mg	K	Na
Subplot 3NC (continuous corn, no treatment)															
Ap	0–23	9.0	66.8	24.2	1.45	45.3	0.25	1.32	5.1	65	17.6	8.8	2.3	0.2	0.1
A12	23–30	8.4	64.8	26.8	1.34	49.4	0.24	1.49	5.3	71	18.0	9.7	2.7	0.2	0.1
B1	30–38	5.6	58.7	35.7	1.43	46.0	0.20	1.20	6.0	83	23.1	13.4	5.5	0.1	0.1
B21t	38–56	5.2	54.8	40.0	1.50	43.4	0.25	0.62	6.2	90	26.0	15.8	7.5	0.1	0.1
B22t	56–76	7.6	57.6	34.8	1.57	40.8	0.27	0.54	6.7	92	22.7	13.6	7.1	0.1	0.1
B3	76–124	12.7	60.9	26.4	1.53	42.3	0.27	0.24	7.5	100	16.8	11.2	5.8	0.1	0.1
IIC1	124–147	26.7	57.1	16.2	1.75	34.0	0.26	0.21	8.1	—	—	—	—	—	—
IIC2	147–183	26.4	55.6	18.0	1.87	29.4	0.26	0.13	8.3	—	—	—	—	—	—
Subplot 5NB (corn-oats-clover, LNPK 1955-present)															
Ap	0–25	6.4	67.1	26.5	1.39	47.5	0.25	1.94	5.8	72	21.5	10.4	4.5	0.4	0.1
A12	25–46	3.6	65.5	30.9	1.31	50.6	0.17	1.90	5.7	74	24.1	13.1	4.3	0.3	0.2
B1	46–58	2.8	62.9	34.3	1.40	47.2	0.18	1.24	5.7	79	25.8	14.3	5.7	0.4	0.1
B21t	58–74	1.8	58.0	40.2	1.61	39.2	0.27	0.58	6.0	88	28.9	16.2	8.6	0.5	0.2
B22t	74–94	2.7	63.9	33.4	1.64	38.1	0.30	0.28	6.6	93	25.7	15.2	8.3	0.5	0.1
B3	94–114	9.4	62.9	27.7	1.69	36.2	0.30	0.22	7.2	99	19.2	11.9	6.7	0.4	0.1
IIC1	114–175	15.3	71.1	13.6	1.57	40.8	0.31	—	8.1	—	—	—	—	—	—
IIC2	175–183	48.0	43.0	9.0	1.91	27.9	0.24	—	8.1	—	—	—	—	—	—

Source: Odel et al. (1982).

Table 2. Organic carbon and nitrogen contents of the Morrow continuous untreated corn plot and the corn-oats-clover plot (receiving manure, lime, and fertilizer) in 1944 and in 1972.[a]

Treatment	1944 Carbon ($mg\ kg^{-1}$)	1944 Nitrogen ($mg\ kg^{-1}$)	1972 Carbon ($mg\ kg^{-1}$)	1972 Nitrogen ($mg\ kg^{-1}$)
Sod (estern)	65,100	6,257	—	—
Sod (western)	55,100	5,473	—	—
Corn-oats-clover	53,700	5,096	44,600	4,130
Continuous Corn (untreated)	28,700	2,957	28,000	2,670

[a]C and N contents of the unplowed sod borders are shown for comparison.
Source: Odell et al. (1982).

yr^{-1}; 73 kg P ha^{-1} in 1955, 20 kg P ha^{-1} yr^{-1} from 1956–1966, and based on soil test after 1966; 93 kg K ha^{-1} in 1955, 28 kg K ha^{-1} yr^{-1} from 1956–1966, and based on soil test after 1966)(Odell et al., 1982). The results show for the surface 23 cm that the untreated plot had lower organic C, CEC, base saturation, and exchangeable cations than the plot receiving lime and fertilizer. The differences are not great, with the exception of organic C; and with the exception of nitrogen supply, nutrient reserves and pH in the untreated plot are not greatly limiting to crop growth. The 2.6-fold increase in corn yield with lime and fertilizer addition appears to be due primarily to a nitrogen response.

The Morrow plots were not sampled prior to initiation of the experiment in 1876. In 1944, however, the sod borders adjacent to the plots were sampled, and the organic carbon and nitrogen contents are compared for that year with those in the untreated continuous corn plot and the corn-oats-clover plot receiving lime, manure, and phosphorus (Table 2). Compared to the highest value for the sod border, organic C has declined by 18% for the corn-oats-clover plot, and by 56% for the untreated corn plot. The corresponding nitrogen reductions were 19% and 53%, respectively. By 1973, there had been a further reduction in C and N contents of the plots (Table 2), but as has been found in most studies of organic matter decomposition in cultivated soils, most of the loss of these nutrients occurred in the early years of the experiment.

The Morrow plots show—as does the Rothamsted experiment—that nutrient declines are inevitable when lands are cleared and cropped. The greatest losses are usually in those nutrients (carbon, nitrogen) that are bound in soil organic matter and are not readily retained by the mineral soil fraction. Phosphorus and the base cations are retained by mineral soil colloids, and decreases in these nutrients are less severe. Retention of base cations maintains base saturation of the exchange complex and prevents drastic decreases in soil pH.

The two experiments cited here also demonstrate that soil management, such as manure, lime, and fertilizer additions as well as rotations with legumes or small grains, can greatly reduce nutrient and organic matter losses following land clearing, although the new levels rarely approach those prior to land clearing. An exception was the FYM treament on the Rothamsted experiment (Figure 1), where carbon and nitrogen additions have apparently exceeded crop removal and mineralization rates and have resulted in a continuous increase in soil organic matter content.

The climatic conditions at Rothamsted and at Champaign-Urbana—a pronounced cold season with mean annual temperatures <10°C and rainfall of 700 to 800 mm—are conducive for biomass production and accumulation. These soils tend to be wet during winter and spring months, and this decreases the rate of soil organic matter decomposition following clearing (Jenkinson et al., 1987). Soil chemical degradation from nutrient depletion under these conditions can be expected to be less than those under hot and humid tropical conditions (see below).

3. Shifting Cultivation in the Humid Tropics

Very few long-term studies of nutrient cycling in disturbed tropical ecosystems have been reported. Of these, that of Greenland and Nye (1959) is notable. In their study of carbon and nitrogen changes in soils of temperate and tropical regions following clearing of forest or plowing of grasslands, they reported much higher carbon decomposition rates in tropical systems. Tropical forest carbon and nitrogen levels decreased rapidly in the first few years following clearing. This decline was greatest for bare fallow systems and crop rotations without legumes. Legumes in the rotation slowed the rate of net carbon decomposition, presumably because of the greater retention of carbon as the nitrogen input increased. Carbon declines with cultivation under tropical savannah, although much slower than under forest, resulted in lower levels after several years because of the much lower biomass production of tropical savannahs compared to tropical forests.

In addition to the loss of organic matter itself, nutrients mineralized from organic matter in tropical systems are subject to rapid leaching losses because of the greater rainfall and greater permeability of most tropical red soils. Lal (1985) reported on changes in soil chemical properties up to six years after clearing of secondary forest on an alfisol in Nigeria. Maize (*Zea mays*) was grown with or without tillage and fertilized annually with 100 kg N, 20 kg P, and 30 kg K ha^{-1}. The results are reported in Table 3. Six years after clearing, soil pH had declined from a precultivation value of 6.1 to 5.8 with no tillage and 4.4 with plowing. Organic C declined slightly in this period with plowing and actually increased for the no-till system. Total N

Table 3. Changes in soil chemical properties in the 0–10 cm layer with tillage after land clearing a secondary forest at Ibadan, Nigeria[a,b]

Property	Precultivation analysis, 1975		2 years		4 years		5 years		6 years	
	No tillage	Plowed	No tillage	Plowed	No tillage	Plowed	No tillage	Plowed	No tillage	Plowed
pH (1:1 in H_2O)	6.1 ± 0.3	6.1 ± 0.4	5.8 ± 0.6	5.8 ± 0.3	5.8 ± 0.14	5.8 ± 0.33	5.3 ± 0.15	4.7 ± 0.3	5.8 ± 0.3	4.4 ± 0.2
Organic carbon (%)	1.4 ± 0.4	1.7 ± 0.3	1.62 ± 0.52	1.92 ± 0.33	1.59 ± 0.18	1.37 ± 0.31	1.48 ± 0.14	1.35 ± 0.11	1.50 ± 0.37	1.23 ± 0.08
Total N (%)	0.34 ± 0.11	0.40 ± 0.08	0.18 ± 0.04	0.22 ± 0.04	0.29 ± 0.11	0.23 ± 0.12	0.191 ± 0.04	0.195 ± 0.02	0.195 ± 0.04	0.205 ± 0.04
Bray P (p.p.m.)	8 ± 4	8 ± 3	25.1 ± 5.0	11.5 ± 5.8	32 ± 28	11 ± 6	25 ± 7	43 ± 11	35 ± 8	56 ± 16
Exchangeable cations (mmol(+)kg^{-1})										
Ca^{2+}	60 ± 17	79 ± 18	60 ± 21	71 ± 18	61 ± 5	52 ± 13	24 ± 5	19 ± 3	30 ± 14	19 ± 4
Mg^{2+}	18 ± 6	20 ± 7	15 ± 7	19 ± 5	5.8 ± 0.7	5.5 ± 1.8	4.9 ± 1.6	3.3 ± 0.7	4.3 ± 2.3	1.3 ± 0.3
K^+	7 ± 2	8 ± 2	6.7 ± 2.8	9.5 ± 3.5	4.4 ± 0.4	3.8 ± 0.7	2.8 ± 0.6	1.7 ± 0.5	1.6 ± 0.5	1.8 ± 0.5
Na^+	—	—	0.9 ± 0.3	0.9 ± 0.5	1.1 ± 0.2	1.1 ± 0.2	0.4 ± 0.04	0.4 ± 0.1	2.4 ± 0.2	2.1 ± 0.2
Mn^{2+}	0.5 ± 0.2	0.5 ± 0.2	0.5 ± 0.3	0.3 ± 0.2	—	—	0.3 ± 0.2	1.2 ± 0.6	1.4 ± 0.5	3.0 ± 1.1
Total acidity	—	—	0.6 ± 0.4	0.5 ± 0.3	1.5 ± 0.6	—	0.5 ± 0.3	2.1 ± 1.7	3.6 ± 2.3	9.5 ± 4.5

[a]Each value is a mean of 12 composite samples consisting of 25 subsamples each.

[b]—Data not available.

Source: Lal (1985).

decreased by 49% for the plowed system and by 43% for the no-till system. The relatively small decline in C may be due to the fact that the secondary forest prior to clearing was only 15 years old. There were also major declines in exchangeable Ca^{2+} and Mg^{2+} six years after clearing, and the effect was more pronounced with plowing.

Organic matter in these soils is the major source of cation-exchange capacity, and organic matter decomposition further contributes to nutrient leaching losses. Organic matter can also reduce strong P adsorption by hydrous oxides of tropical soils by competing for adsorption sites (Sanchez, 1976), and soil humic acids bind trace nutrients that would otherwise be leached (Ellis and Knezek, 1972). Sanchez (1976) states that in tropical soils organic phosphorus may represent 60–80% of the total soil P and may be the main source of P to crops in zero-fertilizer crop-production systems. As discussed below, organic matter can reduce Al toxicity by complexing much of the Al^{3+} in soil solution.

Maintenance of organic matter appears to be more critical for tropical than for temperate region soils, as organic matter must be the primary source of available nutrients, provide exchange and retention capacity for soil nutrients, and neutralize toxic metals. Loss of soil organic matter, then, from long-term cultivation of tropical soils without residue management or green manuring, or from improper clearing of tropical lands (Lal et al., 1986), will result in highly degraded soil conditions for crop production.

B. Acidity and Toxic Aluminum

Excessive soil acidity manifested in low pH, low levels of exchangeable Ca, Mg, and K, and toxic levels of Al (and perhaps other metals) is detrimental to plant growth and to microbiological processes in the soil. Soil pH values much below 4 are extremely limiting to plant growth (Foy, 1984) and are unfavorable for the growth of many bacteria and actinomycetes (Alexander, 1977). In most cases, the problem of high acidity is a problem of high Al solubility. At soil pH much below 4.5, Al exists primarily as Al^{3+}. In this form it dominates the exchange complex, and Al^{+3} occupies a large part of the exchange complex in soils dominated by oxides and kaolinite. This results in low available reserves of K, Ca, and Mg and a high potential for leaching losses of these nutrients.

Evans and Kamprath (1970) and Cate and Sukhai (1964) have shown that Al in soil solution is usually <1 mg $liter^{-1}$ when Al saturation of the exchange complex is <60%. At saturations greater than this, soil solution Al increases rapidly. Total Al concentrations greater than 1 mg $liter^{-1}$ often result in toxicity to higher plants and reductions in crop yield. Plants exhibit a wide range of tolerance to Al, and native tropical species adapted to high-Al conditions will grow under conditions that would be fatal to corn, soybeans, or wheat. Pineapple, for example, native to northeastern

South America, is highly tolerant of Al, as are coffee, tea, rubber, and cassava (Sanchez, 1976). Advances have been made, however, in selection of Al-tolerant strains of crops such as wheat (Foy, 1984).

Adams and Lund (1966), and more recently Pavan et al. (1982), have demonstrated that plant toxicity to Al is a function of the Al^{3+} activity of the soil solution and not the total Al concentration. Al^{3+} activity is reduced by complexation with organic matter (McLean, 1976) and by ion-pair formation with ligands such as SO_4^{2-} (Pavan et al., 1982). Of these mechanisms, complexation by organic matter appears to be the most significant and adds support to the thesis of this chapter that chemical degradation of soil will depend to a large extent on management of soil organic matter.

Acidic conditions and high active Al concentrations are common to the highly weathered soils of the tropics but are also found in the spodosols of humid temperate regions. Sanchez (1976) has reported Al saturation data for tropical oxisols, ultisols, alfisols, and inceptisols. Oxisols generally have high Al saturations, as do ultisol subsoils. Tropical alfisols generally have low Al saturations but are poorly buffered, and base saturations can change readily with changes in management. The andepts tend to be buffered at pH above that of high Al solubility, and exchangeable acidity is often dominated by H^+ because of the high organic matter contents of these soils. McLean and co-workers (Bhumbla and McLean, 1965) reported high levels of exchangeable Al for acid spodosols of northeastern Ohio. Spodosols are to be found throughout the humid temperate regions of the world and are formed primarily under forest vegetation (McKeague et al., 1983).

Much attention has been placed in the last several decades in North America and Western Europe on the effects of acid deposition on terrestrial and aquatic ecosystems (Driscoll and Newton, 1985). Acid deposition, the so-called acid rain, is a complex mixture of inorganic and inorganic chemical compounds deposited in the form of the wetfall (precipitation) and dryfall (particulates and adsorbed gases). It is dominated in most cases by one or both of the strong acids H_2SO_4 and HNO_3, produced by hydrolysis of SO_2 and NO_x. The anthropogenic contribution to the total global S budget is 65 million metric tons S per year and represents slightly more than one-third of the total (Noggle et al., 1986). Natural sources include volcanic emissions and sea spray. Annual total S deposition in the United States ranges from <20 kg ha^{-1} yr^{-1} for primarily rural areas to as high as 30–170 kg ha^{-1} yr^{-1} near industrial sources (Olson and Rehm, 1986). Hutchinson and Whitby (1977), in their study of emissions from the large Coniston-Sudbury smelter in Ontario Province, Canada, showed that total SO_4 in rainfall during an approximately one-month period ranged from 5456 mg m^{-2} at a distance 0.8 km from the smelter to 131 mg m^{-2} at a point 105 km away. These enormous emissions were shown to have devastating effects on soils and vegetation in the affected area as a consequence of toxic levels of trace metals mobilized by the low pH of the soils as well as by the metals emitted by the smelter itself. This phenomenon will be discussed later.

Nitric acid is the other strong acid source in acid deposition. Nitrate in precipitation is a natural phenomenon (Tabatabai, 1985). Tabatabai (1985) summarized data on $NO_3{-}N$ in precipitation in the United States and found that annual loadings were generally <5–7 kg NO_3 –N ha^{-1} yr^{-1}.

Calculation of the total acidity added to soil as a result of nitric and sulfuric acids in precipitation suggests that the impact would be small compared to the acid-buffering capacity of most soils. It would amount to a lime requirement of <50–60 kg $CaCO_3$ ha^{-1} yr^{-1}, an amount approximately 1/100 of that required to neutralize the acidity of chemical fertilizer applied to a corn crop in the United States. There can also be positive nutrient response effects of these levels of *N* and *S* deposition. While the effects of acid deposition on forests and lakes of North America and Europe are extensive and significant, there appears to be little evidence to suggest that these levels of acid input will result in significant if any soil degradation.

C. Oxidation of Pyrite in Soils and Mine Spoils

Pyrite (FeS_2) is the most common reduced form of sulfur in rocks and soils. Sulfur in pyrite has an oxidation state of −1 and is found only in soils where reducing conditions are severe and prolonged (Lindsay, 1979). Pyrite is also found in a variety of ores and in particular in the high-sulfur coals of the eastern United States (Lundgren and Dean, 1979). Pyrite is highly stable under reducing conditions, and pyrite and other metal sulfides are extremely insoluble. When pyrite is exposed to oxidizing conditions such occur when pyritic soils are drained or when coal overburden spoil is removed and disposed of at the mine surface, pyrite oxidizes rapidly to sulfate with a large production of strong acid. The oxidation is a complex combination of bacterial and chemical reactions (Kelley and Tuovinen, 1988) and involves primarily the chemoautotrophic bacteria *Thiobacillus ferrooxidans*. The overall reaction has been given as:

$$2FeS_2 + 7.50_2 + H_2O = 2Fe^{3+} + 4SO_4{}^{2-} + 2H^+ \quad (4)$$

and involves both Fe^{2+} and $S_2{}^{2-}$ oxidation. Chemical oxidation of Fe^{2+} is apparently also an important process in pyrite oxidation and helps to catalyze bacterial oxidation.

The environmental significance of pyrite oxidation is the enormous production of strong acid, which can produce pH values <2.5, dissolves soil minerals and produces toxic concentrations of Fe, Al, Mn, and trace metals, and results in high salt levels. This environment is hostile to the growth of microorganisms, other soil fauna and flora, and higher plants. The results are acid sulfate soils that cannot be used for production of crops other than paddy rice, and the abandoned minelands of the eastern United

States, which often resemble moonscapes and produce highly toxic acid mine drainage (Brady et al., 1982).

The acid sulfate soils (also called cat clays or pegasse soils and classified as sulfaquepts) are located in the littoral zone of the eastern United States and South America, and in Southeast Asia and the Gambia in West Africa (Sanchez, 1976). They are generally used only for paddy rice production, where regulation of the water table can supress pyrite oxidation, or for acidophilic crops such as pineapple (as in Thailand) if sufficient toxic aluminum has been leached out of the rooting zone. In Guyana, brackish sea water has been used to reclaim acid sulfate soils by displacing exchangeable Al^{3+} prior to using them for paddy production. The trend to sodic conditions produced by use of sea water is less serious for paddy soils, as they are normally puddled to retain water.

Abandoned mine lands of the eastern United States are extensive and occupy millions of hectares. They are a product of unregulated surface mining prior to imposition of the federal mining law in 1977 that requires mine spoil to be buried after mining and the surface to be revegetated. Pyritic mine spoils are characterized as having pH <3–4, low levels of available P, K, Mg, and Ca, and high levels of Al, Fe, and Mn (Armiger et al., 1976). Under these conditions, vegetative cover is difficult to establish and erosion is a serious problem. Reclamation of these areas is expensive and involves the application of a layer of soil together with lime and fertilizer. Recently, researchers have successfully experimented with the use of amendments such as manure, compost, sewage sludge, papermill sludge, municipal solid waste, and flyash for reclamation of these materials (Sopper and Seaker, 1983). The key to reclamation is reduction of pyrite oxidation by burial of the spoil and establishment of an active biological surface zone. Reclaimed lands, although successfully vegetated, remain fragile, as the depth of rooting and nutrient supply is limited by the depth of reclamation, which is rarely more than about 25 to 30 cm. These lands should probably be kept permanently from agricultural production with the exception of occasional and well-managed livestock grazing.

D. Land Disposal of Wastes

Land has become the final solution to many of society's waste disposal problems. As the world comes to rely more and more on technology, the waste products of that technology accumulate. These products are increasingly diverse and many are potentially toxic to humans and to the ecosystem. Other mechanisms of waste disposal, such as incineration, ocean dumping, or deep burial, are expensive and have their own environmental problems. Soil has always been used for disposal of wastes, but unti recently with little regard for or knowledge of the assimilatory capacity of the soil. Disposal of organic wastes such as manures, food processing by-products, and domestic sewage have little permanent negative impact on soil. The organic material is degraded to CO_2 and humus, and some or all of the nutrients are retained. Excess nutrients that are not retained by the soil are lost by runoff or leaching, which, although an environmental problem, does not result in soil degradation per se. In general, the net effect of organic waste disposal on land is favorable. Organic matter and nutrient contents are increased; increased organic matter increases CEC, acid buffer capacity, and metal complexation; overall soil micro- and macrobiological processes are stimulated. There are, however, effects on soil from waste disposal that are negative. Many are a consequence of industrial constituents in the wastes, including toxic trace metals, pesticides, toxic synthetic organics, and radioactive elements. Other effects are nutrient imbalances caused by excessive loading of waste on land. Several of these will be discussed below.

1. Livestock Waste Disposal

Livestock wastes have traditionally been returned to land for grain or hay production in an essentially closed system. The net export of nutrients in meat, milk, or eggs was replaced by addition of fertilizer, nitrogen fixation, or by mining of soil nutrient reserves deeper in the soil profile. With today's large confined livestock operations, however, the nutrient cycle has been broken. Livestock feed in the form of grain, hay, or pelletized feed is shipped from the major grain-producing areas to areas where the livestock are confined. Cattle feedlots in the United States may handle as many as 100,000 animals, 100- to 500-cow dairy operation are becoming common, as are poultry operations with several million birds. Manure disposal problems arise when these operations attempt to dispose of the manure on limited nearby land. Not only are annual loadings in excess of crop nutrient requirements, resulting in potential for nitrate leaching, but repeated applications to the same soil can result in very high accumulations of nutrients, trace elements and sodium, and soluble salts.

Olsen and Barber (1977), in their review of the classic manure experiment at Rothamsted, indicated that 111 years of application of 3.14 metric tons ha^{-1} yr^{-1} of FYM had resulted in the precipitation of octacalcium phosphate, $Ca_4H(PO_4)_3 \cdot 2.5H_2O$, a metastable mineral with solubility significantly greater than hydroxyapatite or variscite, which are minerals found in soils under more normal conditions. Similar results were observed over the same period from an annual application of 33 kg P ha^{-1} of superphosphate. Precipitation of trace nutrients with these secondary minerals may explain findings of phosphorus-induced Zn deficiency found on calcareous or other high-pH soils receiving large applications of manure or P fertilizer (Olsen and Barber, 1977).

Another problem associated with large applications of manures is an imbalance of K^+, Ca^{2+}, and Mg^{2+} resulting from the high K content of manures relative to that of Ca and Mg. Stewart and Meek (1977) summarized Na, K, Ca, and Mg contents of livestock wastes. Beef manure had 0.65%, 1.54%, 1.46%, and 0.71% Na, K, Ca, and Mg, respectively; for dairy manure the corresponding values were 0.40%, 1.72%, 1.93%, and 0.86%; and poultry manure had values of 0.40%, 1.51%, 1.55%, and 0.35%. The K content was almost always higher than that of Ca or Mg, and Na content was significant although not high enough to cause a sodic problem (Stewart and Meek, 1977).

The primary problem associated with improper livestock waste disposal—and disposal of other organic wastes as well, although materials such as sewage sludge and effluent are more effectively regulated—is the off-site water pollution from manure nutrients, biological oxygen demand (BOD), and pathogens (Gilbertson et al., 1979). This is usually seen as surface-water contamination, but deep leaching of nitrate—with the potential to contaminate at least shallow groundwater—has been reported from large cattle and dairy feedlots (Mielke et al., 1974; Adriano et al., 1973). Runoff from confined animal feedlots is a source of concentrated nutrients, BOD, and coliform bacteria. These produce large numbers of local fish kills in the United States in small streams and ponds receiving manure runoff. Ammonia in manure is often the cause of fish kills, but the high BOD may greatly reduce the dissolved oxygen content of the stream. In areas of high livestock density, manure is one of the largest sources of phosphorus being discharged to lakes and contributing to cultural eutrophication (PLUARG, 1978; Miller et al., 1982).

Direct discharge of manure from feedlots and barnlots is the most significant source of water pollution from livestock activities, but runoff from agriculatural land can also be a problem at high rates of manure application and when manure is applied to frozen soil, applied in liquid form so as to exceed the infiltration capacity of the soil, or applied without incorporation to steep slopes. Walter et al. (1987) have shown, however, that surface-applied manure (assumed to be dry and with some bedding material such as straw) can reduce runoff and nutrient losses in initial runoff because it acts as a surface mulch. Nutrient losses from subsequent runoff events were higher from the manured plots, however. Walter et al. (1987) have argued that manure can be utilized on agricultural land in conjunction with conservation tillage without environmental impairment if loading rates are not excessive and if hydraulic loading is not exceeded. Since continuous no-till is rarely recommended and there is a current emphasis on crop rotations, it should be possible to integrate manure application with most tillage and cropping systems. Manure could be applied with disk or chisel plow incorporation before a grain crop and followed by no-till for several years. There would still be a net conservation benefit compared to a continuous-plow system, and the grain crop would benefit more from the manure nitrogen than from legumes in the rotation.

2. Disposal of Municipal Sewage Sludge

The last two decades have seen an enormous increase in the production and disposal of municipal sewage sludge in the United States, Canada and

Western Europe resulting from imposition of sewage treatment regulations (Page et al., 1987). With the parallel environmental concern over sludge disposal by incineration, ocean dumping, and landfilling, application of sludge to land has become an attractive disposal alternative for those communities with access to crop, forest, or disturbed land. Municipal sewage sludge contains nutrients (on average about 3%, 2.5%, and 0.3% N, P, and K, respectively), trace organics, and trace elements. Sludges contain thousands of synthetic organic compounds, and concentrations vary tremendously among sludges. Jacobs et al. (1987) reported trace organic analyses for 219 compounds in U.S. sludges. They found that 32% were below detection limits, while about 25% were detected in more than 50% of the sludges. Ninety percent of the compounds occurred at concentrations <10 mg kg^{-1}, 10% had median concentrations of 11 to 100 mg kg^{-1}, and only one compound, a phthalate ester used in the manufacture of polyester, had a concentration >100 mg kg^{-1}. Most organics appear to be strongly held by sludge organic matter, presumably through hydrophobic bonding, and there is little evidence for significant plant uptake of sludge-borne organics (Jacobs et al., 1987).

Sewage sludges contain varying concentrations of almost all chemical elements. Typical values are summarized in Table 4. Of those reported, only a few have been shown to be of concern to the soil-plant-animal system. Metals such as Zn, Cu, and Ni, which can be phytotoxic when applied

Table 4. Ranges and median concentrations of trace elements in dry digested municipal sewage sludges

	Reported range (mg/kg dry sludge)		
Element	Minimum	Maximum	Median
As	1.1	230	10
Cd	1	3,410	10
Co	11.3	2,490	30
Cu	84	17,000	800
Cr	10	99,000	500
F	80	33,500	260
Fe	1,000	154,000	17,000
Hg	0.6	56	6
Mn	32	9,870	260
Mo	0.1	214	4
Ni	2	5,300	80
Pb	13	26,000	500
Sn	2.6	329	14
Se	1.7	17.2	5
Zn	101	49,000	1,700

Source: Logan and Chaney (1983).

to soil in an inorganic form, have not been shown to exhibit phytotoxicity when applied with sludge even at very high sludge application rates (>500 metric tons ha^{-1}) (Logan and Chaney, 1983; Page et al., 1987). A number of other metals (including Cd, Pb, As, Se, Mo, and Zn) are potentially toxic to livestock and to humans, and metals such as Cd and As have been shown to interfere with some microbiological processes. The risk from a particular metal normally occurs only at very high concentrations in sludge (those greater than the 95th percentile of the ranges shown in Table 4) and at high sludge application. Two metals, however, are of sufficient concern to human health to pose some risk at lower concentrations and at lower sludge application rates.

Lead (Pb) is a metal that has been shown to cause brain damage in humans. Although automobile emissons and Pb-based paints are the primary sources of anthropogenic Pb, sludges can contain Pb at concentrations as high as 26,000 mg kg^{-1} (Table 4); the median concentration reported in Table 4 was 500 mg kg^{-1}. Lead is ingested primarily through hand-to-mouth activity of children, and a small proportion of the child population suffers from "pica," a condition in which the child ingests up to 100 to 250 g soil day^{-1} (Logan and Chaney, 1983). Recent studies suggest that soil Pb concentrations >100 to 200 mg kg^{-1} could potentially cause brain damage in pica children (R.L. Chaney, USDA-ARS, Beltsville, MD, personal communication).

The other metal in sludge of some risk to humans is cadmium (Cd). Cadmium is accumulated by humans over their lifetime, and at daily ingestion rates of 70 to 200 μg day^{-1} over a 30- to 50-year period has the potential to cause mild renal dysfunction (Ryan et al., 1982; Logan and Chaney, 1983; Chaney et al., 1987). The daily allowable dietary intake (ADI) is presently set at 71 μg day^{-1}, and it is estimated that the U.S. population may be receiving up to 60% of the ADI (Chaney et al., 1987). The value is higher for smokers, who ingest some Cd from tobacco. Cadmium application to land in sludge is presently regulated by the United States, Canada, Japan, and the Western European countries (Logan and Chaney, 1983). The regulations vary considerably, the Japanese and Europeans being the most restrictive and the United States the most lenient. At present, sludge-applied Cd cannot exceed 2.5–10 mg kg^{-1} above background values in soil (Logan and Chaney, 1983). A summary of research findings by North American scientists (Page et al., 1987) concluded that the risk to humans at these levels of soil Cd contamination was low.

E. Other Examples of Chemical Soil Degradation

The ways in which soil can be chemically degraded are as numerous as people's myriad activities. Many of these are the recent product of the industrial revolution, and we do not fully understand the long-term effects of these assaults on the soil system. A few are mentioned below. Soil salinity, a major example of chemical degradation, is the subject of another chapter.

1. Subsidence of Organic Soils

Organic soils (histosols) occur on all continents (Everett, 1983), and are most common between latitudes 70° and 90° in both the Southern and Northern hemispheres. Sixty-four percent of histosols occur in the Soviet Union. Histosols are formed in low-lying areas where biomass production is high and biological respiration is severly oxygen-limited by high water tables. Accumulated plant biomass is partially degraded to give a range of humic materials that range from fibric (little decomposed) to sapric (highly decomposed) (Everett, 1983). Histosols are drained and converted to agricultural production to take advantage of the large store of accumulated nutrients. The physical, biological, and chemical processes that resulted in histosol formation are reversed with drainage, and the organic matter begins to decompose rapidly. Everett (1983) reports data from studies in the United States and Europe and gives subsidence rates of 0.4 to 9.1 cm yr^{-1}. These compare with rates of peat accumulation of 0.02 to 0.08 cm yr^{-1} (one tropical marsh has a rate of peat accumulation of 2 cm yr^{-1}). The initial rapid subsidence is complete within 4 to 10 years of drainage installation. Contributing to subsidence is the loss of soluble organic matter, reported by Everett (1983) to be as high as 2 metric tons ha^{-1}; this estimate was based on a laboratory study and field values may vary. Of both agronomic and environmental significance is the large losses of nutrients in drainage waters with subsidence of these organic soils.

2. Soil Degradation from Metal Ore Smelters

A major source of chemical soil degradation in many parts of the world is deposition of acidity and trace metals from metal ore smelters. Here the soil downwind of the smelter is affected by both acid and high levels of potentially toxic trace metals. A classic example is the large nickel smelter near Sudbury, Ontario, Canada (Hutchinson and Whitby, 1977; Freedman and Hutchinson, 1981). Nickel and copper concentrations in soil attained levels as high as 9000 and 7000 mg kg^{-1}, respectively, compared to normal background levels of about 20 mg kg^{-1} for these two metals (Logan and Miller, 1983). As previously reported, pH in the vicinity of the smelter was as low as 3 or less. These conditions gave rise to highly phytotoxic metal levels in local native vegetation and produced large areas devoid of vegetative cover. Soil micro- and macrobiological processes were also affected.

Asami (1988) reviewed a number of causes of ecosystem degradation by mining operations around the world, with particular attention to those in Japan. Several of the Japanese cases were severe enough to cause some of the world's only examples of human environmental metal toxicity. Documentation of Cd toxicity in rice farmers using metal-contaminated irrigation water gave the world the best toxicological data on Cd-induced itai-itai bone disease and renal dysfunction. Dilute acid-extractable soil Cd levels ranged from 8 to 865 mg kg^{-1} in areas with itai-itai. This compares to average total Cd in soils of about 0.2 to 0.3 mg kg^{-1}. About 60% of the total Cd in these rice soils was acid-extractable.

3. Dredge Spoil Disposal

In this example of chemical soil degradation, the problem is one of contamination of new and existing soil. It is no accident that much of human activity is centered on major rivers, deltas, and embayments. These important water resources serve as water supplies, transportation avenues, and, unfortunately, convenient sewers. In fluvial environments such as deltas, river mouths, embayments, and coastal areas, sediments are continuously deposited. While some sediment is resuspended and ultimately discharged into the ocean, much of it accumulates. If accumulation is great enough to hamper navigation, the sediment may be dredged and some of it placed on land or pumped into diked areas to form new land. Pollutants discharged as sewage or industrial effluent react with deposited sediments. In the water environment, these pollutants are often in near equilibrium with the water column, and sediments serve generally to improve water quality by scavenging pollutants. The anaerobic environment of most sediments serves to immobilize metals through precipitation of insoluble metal sulfides (Bourg, 1988; Kersten, 1988) and binding of organic pollutants to sediment organic matter.

When dredged sediments are disposed of by diking or by spreading on land, oxidation occurs by complex chemical and biological reactions. Metal sulfides are oxidized to produce more soluble metal forms, and sulfide oxidation results in acid generation and enhanced metal availability. Mineralization of organic matter also results in liberation of complexed metals and sorbed organics. These reactions produce conditions that make vegetation of the dredged spoil difficult and also produce leachate that may be hazardous. Such conditions are exemplified by those found at the mouth of the Rhine River in The Netherlands, where the Port of Rotterdam is facing the difficult task of safely dredging and disposing of contaminated sediment. A recent symposium on the Rotterdam case has been published and offers important insights into the problems of dredged sediment disposal (Salamons and Forstner, 1988).

4. Radionuclide Soil Contamination

Since atmospheric nuclear weapons testing started in earnest in the 1950s, soil has become progressively contaminated with long-lived radionuclides. Of these, Cs-137 and Sr-90 are significant because of their long half-lives and their interaction with the plant-animal system. Cs-137 accumulates in surface soil, where it is strongly bound by adsorption to clay minerals and sesquioxides and by chelation with organic matter (Kabata-Pendias and Pendias, 1984). Because of its relative immobility in soil and the precise dates of atmospheric nuclear testing, Cs-137 is a useful tool for studying soil erosion (Ritchie et al., 1974).

Atmospheric nuclear weapons testing produced global low-level soil contamination, but excluding ground-zero test sites such as Bikini in the South Pacific, risk to the food chain and to the ecosystem has abated as a result of nuclide decay, erosion, and mixing of the soil surface and leaching to lower depths. Of far greater concern today is the disposal of high-level nuclear waste from the production of nuclear weapons and by nuclear power plants. The primary waste products are the transuranic radionuclides, isotopes produced by decay of uranium and plutonium. These nuclides include various isotopes of plutonium (Pu), americium (Am), curium (Cm), and neptunium (Np) (Kabata-Pendias and Pendias, 1984). Of these, Pu-239 and Am-241 are most important. The behavior of these isotopes, and of Pu-239 in particular, have been reviewed by Wildung and Garland (1980). The major form of Pu in waste is the oxide PUO_2, although $Pu(NO_3)_4$ can be formed in aerosols from power plant discharges. Like other heavy metals, the transuranic nuclides are retained in soil by adsorption, precipitation, and organic matter chelation. Wildung and Garland (1980) report that Pu can be taken up by plant roots and translocated to shoots, but Kabata-Pendias and Pendias (1984) found that little was known about the transfer of these radionuclides from contaminated soil to plants.

A major effort is being made in the United States to site and develop one or more controlled high-level nuclear waste disposal factilities, but public resistance has been great. The nuclear nations will have to develop such facilities—although at enormous cost—if nuclear power generation is to be a viable alternative for the future. Of greater concern to long-term and widespread soil contamination by radionuclides is the type of nuclear power plant accident that occurred at Chernobyl in the Soviet Union. The extent and degree of soil contamination by this single episode has yet to be discovered or fully revealed, but the implications of available information are that thousands of hectares in the vicinity of the plant are contaminated without repair for hundreds to thousands of years, while a larger area will have to be closely monitored indefinitely for ecological damage.

IV. Prevention and Restoration of Chemically Degraded Soils

A. Prevention of Chemical Degradation

Essential to prevention of soil chemical degradation is the knowledge that soils vary considerably in their ability to withstand chemical insult. The chemical buffering capacity of most soils is great but finite and can be overwhelmed. The diversity and activity of soil macro-and microfauna are also important elements of a chemically "healthy" soil. Prevention of chemical degradation requires that chemical impact not exceed the capacity of the soil to buffer chemical change. Soils with high buffer capacities can withstand more chemical insult than those with low buffer capacities.

As previously discussed, several soil chemical processes are important in chemical buffering. These include acid-base buffering, precipitation and dissolution, adsorption and desorption, and complexation.

1. Acid-Base Buffering

The soil's ability to buffer additions of acid or base is in the short term a function of its cation-exchange capacity and base saturation. This in turn is a function of the mineralogy and organic matter content of the soil as well as the soil's base status. In the long term, or where acid additions are massive, the content of residual base-containing minerals will determine the soil's ability to neutralize acidity. The weathering of these soil minerals will, however, drastically alter the chemical character of the soil. In the case of large additions of strong base, a major initial effect will be solubilization of both organic matter and soil minerals.

The long-term effects of large additions of acid or base will be determined to some extent by the degree of leaching associated with the chemical insult. If leaching is restricted, the acid or base addition will be localized in the surface soil and the impact there will be large. Ions released into the soil solution will recombine to form a new suite of minerals. If the soil is sufficiently permeable to permit rapid leachnig, the overall impact may be lessened as the acid or base addition will react with a larger soil volume. If off-site effects such as groundwater contamination are less of a concern than the impact on the soil itself, one means of remediating chemical contamination is by deep plowing. The contaminant is directly diluted by the larger volume of soil with which it is mixed, and there is a greater mass of soil with which to buffer the impact of the pollutant.

2. Precipitation-Dissolution

Soils with high contents of reactive cations are able to buffer large additions of ligands such as phosphate, arsenate, and selenite by precipitation, while metals can be precipitated with sulfide under reducing conditions (Bourg, 1988) and as coprecipitates with Fe, Al, Mn, Ca, and Mg compounds. The Fe and Al minerals precipate at low pH (<5) and those of Ca and Mg at pH >6 (Lindsay, 1979).

Precipitation is favored in high-activity clay soils, those with high contents of weatherable minerals, and usually by near-neutral to high pH.

3. Adsorption-Desorption

Metals and ligands are readily removed from solution by adsorption to clay minerals, oxides, and $CaCO_3$, and desorption is often much slower than adsorption (i.e., there is often a marked hysteresis in the adsorption-desorption curve). Adsorption is favored by the presence of amorphous oxides and free $CaCO_3$. Adsorption of metals is favored by pH >6, as the adsorption edges (the pH at which maximum adsorption occurs) of most metals are found to be above this value (Schindler, 1981; Bourg, 1988). Ligands are also strongly adsorbed to oxide surfaces, but adsorption generally decreases with increasing pH as the charge on both the ligand and the oxide becomes more negative as pH increases (Hingston, 1981).

4. Complexation

Polyvalent metals, including the so-called heavy metals, are strongly complexed with humic materials in soil. Organic matter content is often highly correlated with metal-binding capacity (Logan and Chaney, 1983), and binding increases with increasing pH as a result of dissociation of acidic functional groups on humic acid.

There are several examples of soil buffer capacity being used to prevent chemical degradation by limiting chemical loading. The U.S. Environmental Protection Agency uses CEC as a parameter to limit cumulative additions of cadmium (Cd) to soil from sewage sludge (Logan and Chaney, 1983). CEC is believed by many, however, to be a surrogate for Cd binding by organic matter and perhaps by oxides (Logan and Chaney, 1983; Sommers et al., 1987). McFee (1980) has screened soils in the United States for sensitivity to acid precipitation on the basis of $CaCO_3$ content and CEC. Klopatek et al. (1980) considered CEC and base saturation in their senstivity scheme. As previously discussed, partitioning of organic compounds in soil is highly correlated with soil organic matter.

While parameters such as CEC and organic matter content can be used to screen soils for their resistence to chemical insult, an equally important question is how soil chemistry principles can be used to remediate cases of chemical degradation.

B. Principles of Chemical Degradation Remediation

As previously suggested, chemical soil degradation is generally less irreversible than physical degradation. This is a result of chemical buffering in soil, the role of microbiological processes in soil chemical reactions, and the generally rapid rate of soil chemical reactions. Some general approaches can be used to reverse or ameliorate chemical degradation.

1. Modify Soil pH

Increases or decreases in soil pH have profound effects on the soil chemical system. If the desired chemical change is short-term (e.g., favoring pesticide degradation), pH control can be effective and readily attained. If long-term pH control is required (to reduce the bioavailability of the heavy metals, for example), the natural acid-buffering capacity of the soil must be considered. Sewage sludge application to agricultural soils in the United States requires that soil pH be maintained at 6.5 whenever food-chain crops are grown (Logan and Chaney, 1983).

2. Regulate Soil Redox Status

Oxidizing conditions can favor microbial degradation of some pollutants, while reducing conditions can favor precipitation of heavy metals by sulfide, produce denitrification of high levels of nitrate, or reduce acid production from pyrite oxidation. Paddy rice production on acid sulfate soils is a good example of redox control of soil chemical degradation, while pyrite-containing coal mine spoil is buried as soon as possible after mining to reduce oxidation.

3. Maintain or Increase Soil Organic Matter

Organic matter stimulates biological activity and inactivates metals and organic compounds. It also adds CEC, increases pH buffering, and immobilizes nutrients against excessive leaching. A "healthy" soil is one with an active organic matter pool, and soils should be managed to maintain existing organic matter levels through recycling of organic residues. Soil, or spoil materials such as mine tailings and dredged sediments with little or no organic matter content, should be amended with organic materials (sludge, manure, compost) and a permanent vegetative cover established.

4. Maintain Leaching Fraction

Some soil chemical degradation problems can be alleviated by leaching if they involve soluble pollutants that are held weakly by the soil or are readily displaced. Examples are total soluble salts (salinity), exchangeable Na^+ (sodicity), and boron toxicity. The fate of the leachate must also be considered in determining overall environmental impact. Irrigation tailwaters in California's Central Basin have been shown to be high in selenium (Se) and Cd. Migratory birds use the reservoirs that receive this drainage and have been poisoned by Se and Cd (Bureau, 1985).

5. Promote Volatilization

Volatile soil contaminants, such as excessive amount of NH_3, some pesticides and toxic synthetic organics, solvents, and radioactive gases, can be stripped from the surface soil by promotion of volatilization. Volatilization can be promoted by changes in pH (as in the case of NH_3), by drying the soil, and by deep plowing to expose more of the soil to the atmosphere. While this technique may be justified for localized soil contamination, atmospheric pollution will preclude large-scale use of this approach.

V. Conclusions

This review suggests that chemical processes at work in soil are as essential for maintenance of the life of the soil as are physical and microbiological processes. The review of the literature also demonstrates clearly the strong interdependence of chemical and biological soil processes. The soil is indeed a living organism, and as such is subject to starvation in the form of nutrient depletion and poisoning by a complex array of natural and synthetic toxic chemicals. The review also strongly demonstrates the critical role of soil organic matter in ameliorating soil chemical degradation. In almost every instance of chemical degradation reviewed here, increasing or maintaining the organic matter content of the soil appeared to be fundamental to alleviation of the problem. It is also clear, however, that an active biological system—stimulated by active organic matter in the soil and producing new stores of soil organic matter—is also essential for the maintenance of a chemically healthy soil.

Chemical processes in soil are dynamic, generally rapid, and usually reversible. Soil is rarely so far thermodynamically removed from chemical equilibrium that chemical degradation cannot be reversed. The challenge grows daily, however, as people rely more than ever on soil for the production of food and fiber, for disposal of wastes, and to buffer the system against environmental damage.

References

Adams, F., and Z.F. Lund. 1966. Effect of chemical activity of soil solution aluminum on cotton root penetration of acid subsoils. *Soil Sci.* 101:193–198.

Adriano, D.C., P.F. Pratt, and S.E. Bishop. 1973. Fate of inorganic forms of N and salt from land-disposed manures from dairies. *Proc. Int. Symp. on Livestock Wastes*. Am. Soc. Agric. Eng., pp. 243–246.

Alexander, M. 1977. *Introduction to microbiology*. Wiley, NY.

Armiger, W.H., N. Jones, Jr., and O.L. Bennett. 1976. Revegatations of land disturbed by strip mining of coal in Appalachia. *USDA Bull. ARS-NE-71*.

Asami, T. 1988. Soil pollution by metals from mining and smelting activities. pp. 143–169. In: W. Salomons and U. Forstner (eds.), *Chemistry and biology of solid waste. Dredged materials and mine tailings*. Springer-Verlag, NY.

Barrow, N.J. 1980. Evaluation and utilization of residual phosphorus in soils. pp. 333–359. In: F.E. Khasawneh, E.C. Sample, and E.J. Kamprath (eds.) *The role of phosphorus in agriculture*. Amer. Soc. Agron., Madison, WI.

Bhumbla, D.R., and E.O. Mclean, 1965. Aluminum in soils: VI. Changes in pH-dependent acidity, cation-exchange capacity, and extractable aluminum with additions of lime to acid surface soils. *Soil Sci. Soc. Amer. Proc.* 29:370–374.

Bolt, G.H. 1982. The ionic distribution in the diffuse double layer. pp. 1–26. In: G.H. Bolt (ed.), *Soil chemistry. B. Physico-chemical models*. Developments in Soil Science 5B, Elsevier, NY.

Bourg, A.C.M. 1988. Metals in aquatic and terrestrial systems: Sorption, speciation and mobilization. pp. 3–32. In: W. Salomons and U. Forstner (eds.), *Chemistry and biology of solid waste. Dredged material and mine tailings*. Springer-Verlag, NY.

Bowden, J.W., A.M. Posner, and J.P. Quirk, 1980. Adsorption and charging phenomena in variable charge soils. pp. 147–166. In: B.K.G. Theng (ed.), *Soils with variable charge*. New Zealand Soc. Soil Sci., New Zealand Soil Bureau, Lower Hutt, New Zealand.

Brady, K.S., J.M. Bigham, and T.J. Logan. 1982. The impact of colloidal iron oxide on water and sediment quality in streams receiving mine effluents: A case study of Black Fork Creek, Ohio. In: N.E. Smeck and P. Sutton (eds.), *Abandoned Mine Reclamation Symp*. 17:1–9. All-Ohio Chapter Soil Conserv. Soc. Amer., Columbus, OH.

Bureau, R.G. 1985. Environmental chemistry of selenium. *Calif. Agric*. 39 (July–August): 16–18.

Cate, R.B., Jr., and A.P. Sukhai. 1964. A study of aluminum in rice soils. *Soil Sci.* 98:85–93.

Chaney, R.L., R.J.F. Bruins, D.E. Baker, R.F. Korcak, J.E. Smith, and D. Cole. 1987. Transfer of sludge-applied trace elements to the food chain. pp. 67–99. In: A.L. Page, T.J. Logan, and J.A. Ryan (eds.), *Land application of sludge. Food chain implications*. Lewis Pubs., Chelsea, MI.

Che, M.D., T.J. Logan, S.J. Traina, and J.M. Bigham. 1988. Properties of water treatment lime sludges and their effectiveness as agricultural limestone substitutes. *J. Water Poll. Control Fed*. 60:674–680.

Chen, Y. and Y. Avnimelech (eds.). 1986. *The role of organic matter in modern agriculture. Developments in plant and soil sciences*. Martinus Nijhoff, Dordrecht, Netherlands.

Chiou, C.T., R.L. Malcolm, T.I. Brinton, and D.E. Kile. 1986. Water solubility enhancement of some organic pollutants and pesticides by dissolved humic and fulvic acids. *Envir. Sci. Technol*. 20:502–508.

Clark, F.E., and E.A. Paul. 1970. The microflora of grassland. *Adv. Agron*. 22:375.

Driscoll, C.T., and R.M. Newton. 1985. Chemical characteristics of Adirondack lakes. *Envir. Sci. Technol*. 19:1018–1024.

Ellis, B.G., and B.D. Knezek. 1972. Adsorption reactions of micronutrients in soils. pp. 59–78. In: J.J. Mortvedt, P.M. Giordano, and W.L. Lindsay (eds.), *Micronutrients in agriculture*. Soil Sci. Soc. Amer., Madison, WI.

Evans, C.E., and E.J. Kamprath. 1970. Lime response as related to percent aluminum saturation, soil solution aluminum and organic matter content. *Soil Sci. Soc. Amer. Proc*. 34:893–896.

Everett, K.R. 1983. Histosols. pp. 1–53. In: L.P. Wilding, N.E. Smeck, and G.F. Hall (eds.) *Pedogenesis and soil taxonomy. II. The soil orders*. Elsevier, NY.

Follett, R.F., and B.A. Stewart (eds.). 1985. *Soil erosion and productivity*. Amer. Soc. Agron., Madison, WI.

Foy, C.D. 1984. Physiological effects of hydrogen, aluminum, and manganese toxicities in acid soil. pp. 57–97. In: F. Adams (ed.), *Soil acidity and liming*, 2d ed. Amer. Soc. Agron., Madison, WI.

Freedman, B., and T.C. Hutchinson. 1981. Sources of metal and elemental contamination of of terrestrial environments. pp. 35–94. In: N.W. Lepp (ed.), *Effect of heavy metal pollution of plants. Vol. 2. Metals in the environment.* Applied Science Pubs., London.

Gast, R.G. 1977. Surface and colloid chemistry. pp. 27–73. In: J.B. Dixon and S.B. Weed (eds.), *Minerals in soil environments.* Soil Sci. Soc. Amer., Madison, WI.

Gilbertson, C.B., F.A. Norstadt, A.C. Mathers, R.F. Holt, A.P. Barnett, T.M. McCalla, C.A. Onstad, and R.A. Young. 1979. Animal waste utilization on cropland and pastureland. EPA-600/2-70-59.

Goh, K.M. 1980. Dynamics and stability of organic matter. pp. 373–393. In: B.K.G. Theng (ed.), *Soils with variable charge.* New Zealand Soc. Soil Sci., Soil Bureau, Lower Hutt, New Zealand.

Greenland, D.J., and P.H. Nye. 1959. Increases in carbon and nitrogen contents of tropical soils under natural fallows. *J. Soil Sci.* 10:284–299.

Hallsworth, E.G., and G.K. Wilkinson. 1958. The contribution of clay and organic matter to the cation exchange capacity of the soil. *J. Agr. Sci.* 51:1–3.

Hayes, M.H.B. 1986. Soil organic matter extraction, fractionation, structure and effects on soil structure. pp. 183–208. In: Y. Chen and Y. Avnimelech (eds.), *The role of organic matter in modern agriculture.* Martinus Nijhoff, Dordrecht, Netherlands.

Helling, C.S., G. Chesters, and R.B. Corey. 1964. Contribution of organic matter and clay to soil cation-exchange capacity as affected by the pH of the saturating solution. *Soil Sci. Soc. Amer. Proc.* 28:517.

Hingston, F.J. 1981. A review of anion adsorption. pp. 51–90. In: M.A. Anderson and A.J. Rubin (eds.), *Adsorption of inorganics at solid-liquid interfaces.* Ann Arbor Science, MI.

Hutchinson, T.C., and L.M. Whitby. 1977. The effects of acid rainfall and heavy metal particulates on a boreal forest ecosystem near the Sudbury smelting region of Canada. *Water, Air and Soil Pollut.* 7:421–438.

Jacobs, L.W., G.O. O'Connor, M.A. Overcash, M.J. Zabik, and P. Rygiewicz. 1987. Effects of trace organics in sewage sludges on soil-plant systems and assessing their risk to humans. pp. 101–143. In: A.L. Page, T.J. Logan and J.A. Ryan (eds.), *Land application of sludge. Food chain implications.* Lewis Pubs. Chelsea, MI.

James, R.O. 1981. Surface ionization and complexation at the colloid/aqueous electrolyte interface. pp. 219–261. In: M.A. Anderson and A.J. Rubin (eds.), *Adsorption of inorganics at solid-liquid interfaces.* Ann Arbor Science, MI.

Jenkinson, D.S., P.B.S. Hart, J.H. Rayner, and L.C. Parry. 1987. Modelling the turnover of organic matter in long-term experiments at Rothamsted. In: J.H. Cooley (ed.), *Soil organic matter dynamics and soil productivity*, 15:1–8. Int. Assoc. for Ecology, Univ. Georgia, Athens, GA.

Jordan, C.F. 1985. *Nutrient cycling in tropical forest ecosystems.* Wiley, NY.

Juo, A.S.R., and E.J. Kamprath. 1979. Copper chloride as an extractant for estimating the potentially reactive aluminum pool in acid soils. *Soil Sci. Soc. Amer. J.* 43:35–38.

Kabata-Pendias, A., and H. Pendias. 1984. *Trace elements in soils and plants.* CRC Press, Boca Raton, FL.

Karickhoff, S.W. 1981. Semi-empirical estimation of sorption of hydrophobic pollutants on natural sediments and soils. *Chemosphere* 10:833–846.

Kelley, B.C., and O.H. Tuovinen. 1988. Microbiological oxidations of minerals and mine tailings. pp. 33–53. In: W. Salomons and U. Forstner (eds.), *Chemistry and biology of solid waste. Dredged material and mine tailings*. Springer-Verlag, NY.

Kersten, M. 1988. Geochemistry of priority pollutants in anoxic sludges: Cadmium, arsenic, methyl mercury, and chlorinated organics. pp. 170–213. In: W. Salomans and U. Forstner (eds.), *Chemistry and biology of solid waste. Dredged material and mine tailings*. Springer-Verlag, NY.

Klopatek, J.M., W.F. Harris, and R.J. Olsen. 1980. A regional ecological assessment to atmospheric deposition. Effects on soil systems. In: D.S. Shriner, C.R. Richmond, and S.E. Lindberg (eds.), *Atmospheric sulfur deposition—Environmental impact and health effects*. Ann Arbor Sci. MI.

Lal, R. 1985. Mechanized tillage systems effects on properties of a tropical alfisol in watersheds cropped to maize. *Soil and Tillage Res*. 6:149–161.

Lal, R., P.A. Sanchez, and R.W. Cummings. Jr (eds.). 1986. *Land clearing and development in the tropics*. A.A. Balkema, Rotterdam, Netherlands.

Lawes Agricultural Trust. 1977, Rothamsted experimental station. Guide. Harpenden, Herts, UK.

Leckie, J.O., and R.O. James. 1974. Control mechanisms for trace metals in natural waters. pp. 1–76. In: A.J. Rubin (ed.), *Aqueous-evironmental chemistry of metals*. Ann Arbor Science, MI.

Leeper, G.W. 1972. Reactions of heavy metals with soils with special regard to their application in sewage sludges. Dept. of the Army, Corps of Engineers, Contract No. DACW 73-73-C-0026.

Lindsay, W.L. 1979. *Chemical equilibria in soils*. pp. 282–298. Wiley, NY.

Lockeretz, W., G. Shearer, D.H. Kohl, and R.W. Klepper. 1984. Comparison of organic and conventional farming in the Corn Belt. pp. 37–48. In: D.F. Bezdicek, J.F. Power, D.R. Keeney, and M.J. Wright (eds.), *Organic farming: Current technology and its role in a sustainable agriculture*. Am. Soc. Agron., Madison, WI.

Logan, T.J., and R.L. Chaney. 1983. Utilization of municipal wastewater and sludge on land—metals. pp. 235–326. In: A.L. Page, T.L. Gleason, J.E. Smith, Jr., I.K. Iskandar, and L.E. Sommers (eds.), *Proc. Workshop on Utiliz. Munic. Wastewater and Sludge on Land*. Univ. California, Riverside.

Logan, T.J., and R.H. Miller. 1983. Background levels of heavy metals in Ohio farm soils. Ohio Agric. Res. Dev. Center Res. Circular 275, Wooster, OH.

Lundgren, D.G., and W. Dean. 1979. Biogeochemistry of iron. In: P.A. Trudinger and D.J. Swaine (eds.), *Biogeochemical cycling of mineral-forming elements*. Elsevier, Amsterdam, Netherlands.

McFee, W.W. 1980. Sensitivity of soil regions to acid precipitation. EPA-600/3-80-013.

McKeague, J.A., F. DeConnick, and D.P. Franzmeier. 1983. Spodosols. pp. 217–252. In: L.P. Wilding, N.E. Smeck, and G.F. Hall (eds.), *Pedogenesis and soil taxonomy. II. The soil orders*, Elsevier, New York, NY.

McLean, E.O. 1976. Chemistry of soil aluminum. *Comm. Soil Sci. Plant Anal.* 7:619–636.

Mielke, L.N., N.P. Swanson, and T.M. McCalla. 1974. Soil profile conditions of cattle feedlots. *J. Envir. Qual.* 3:14–17.

Miller, M.H., J.B. Robinson, D.R. Coote, A.C. Spires, and D.W. Draper. 1982.

Agriculture and water quality in the Canadian Great Lakes Basin: III. Phosphorus. *J. Envir. Qual.* 11:487–497.

Morais, F.I., A.L. Page, and L.J. Lund. 1976. The effect of pH, salt concentration, and nature of electrolytes on the charge characteristics of Brazilian tropical soils. *Soil Sci. Soc. Amer. J.* 40:521–527.

Moschler, W.W., R.K. Stivers, D.L. Hallock, R.D. Sears, H.M. Camper, M.J. Rogers, G.D. Jones, M.T. Carter, and F.S. McClaugherty. 1962. Lime effects on soil reaction and base content of eleven soil types in Virginia. *Virginia Agric. Exp. Sta. Tech. Bull. 159.*

Noggle, J.C., J.F. Meagher, and U.S. Jones. 1986. Sulfur in the atmosphere and its effects on plant growth. pp. 251–278. In: M.A. Tabatabai (ed.), *Sulfur in Agriculture*. Amer. Soc. Agron., Madison, WI.

Nye, P.H., and D.J. Greenland. 1964. Changes in the soil after clearing tropical forest. *Plant and Soil* 21:101–112.

Odell, R.T., W.M. Walker, L.V. Boone, and M.G. Oldham. 1982. The Morrow plots. A century of learning. *Agric. Exp. Sta. Bull. 775*, Univ. Illinois, Champaign-Urbana.

Olsen, S.R., and S.A. Barber. 1977. Effect of waste application on soil phosphorus and potassium. pp. 197–215. In: L.F. Elliott and F.J. Stevenson (eds.), *Soils for management of organic wastes and waste waters*. Amer. Soc. Agron., Madison, WI.

Olson, R.A., and G.W. Rehm. 1986. Sulfur in precipitation and irrigation waters and its effects on soil and plants. pp. 279–293. In: M.A. Tabatabai (ed.), *Sulfur in agriculture*. Am. Soc. Agron., Madison, WI.

Overcash, M.R., and D. Pal. 1979. *Design of land treatment systems for industrial wastes—Theory and practice*. pp. 335–388. Ann Arbor Science, MI.

Page, A.L., T.J. Logan, and J.A. Ryan (eds.). 1987. *Land application of sludge. Food chain implications*. Lewis Pubs. Chelsea, MI.

Parfitt, R.L. 1980. Chemical properties of variable charge soils. pp. 167–194. In: B.K.G. Theng (ed.), *Soils with variable charge*. New Zealand Soc. Soil Sci., New Zealand Soil Bureau, Lower Hutt, New Zealand.

Pavan, M.A., F.T. Bingham, and P.F. Pratt. 1982. Toxicity of aluminum to coffee in ultisols and oxisols amended with $CaCO_3$. $MgCO_3$, and $CaSO_4 \cdot 2H_2O$. *Soil Sci. Soc. Amer. J.* 46:1201–1207.

Pollution from Land Use Activities Reference Group (PLUARG). 1978. Environmental management strategy for the Great Lakes System. Final report. Int. Joint Commission on the Great Lakes, Windsor, Ont., Canada.

Rao, P.S.C., P. Nkedi-Kizza, J.M. Davidson, and L.T. Ou. 1983. Retention and transformations of pesticides in relation to nonpoint source pollution from croplands. pp. 126–140. In: F.W. Schaller and G.W. Bailey (eds.), *Agricultural management and water quality*. Iowa State Univ. Press, Ames.

Ritchie, J.C., P.H. Hawks, and J.R. McHenry. 1974. Estimating soil erosion from the redistribution of fallout 137 Cs. *Soil Sci. Soc. Amer. Proc.* 38:137–139.

Ryan, J.A., H.R. Pahren, and J.B. Lucas. 1982. Controlling cadmium in the human food chain: A review and rationale based on health effects. *Envir. Res.* 28:251–302.

Salamons, W., and U.Forstner (eds.). 1988. *Chemistry and biology of solid waste. Dredged material and mine tailings*. Springer-Verlag, NY.

Sanchez, P.A. 1976. *Properties and management of soils of the tropics*. Wiley, NY.

Schindler, P.W. 1981. Surface complexes at oxide-water interfaces. pp. 1–49. In: M.A. Anderson and A.J. Rubin (eds.), *Adsorption of inorganics at solid-solution interfaces*. Ann Arbor Science, MI.

Sommers, L.E., V.V. Volk, P.M. Giordano, W.E. Sopper, and R. Bastian. 1987. Effects of soil properties on accumulation of trace elements by crops. pp. 5–24. In: A.L. Page, T.J. Logan, and J.A. Ryan (eds.), *Land application of sludge. Food chain implications*. Lewis Pubs. Chelsea, MI.

Sopper, W.E., and E.M. Seaker. 1983. A guide for revegetation of mined land in eastern United States using municipal sludge. Institute for Research on Land and Water Resources, Pennsylvania State University, University Park.

Sposito, G. 1984. *The surface chemistry of soils*. Oxford Univ. Press, NY.

Stanford, G., and S.J. Smith. 1972. Nitrogen mineralization potentials of soils. *Soil Sci. Soc. Amer. Proc.* 36:465–472.

Stevenson, F.J. 1982. *Humus chemistry*. Wiley, NY.

Stevenson, F.J. 1986. Cycles of soil. Carbon, nitrogen, phosphorus, sulfur, micronutrients. pp. 321–367. Wiley, NY.

Stewart, B.A., and B.D. Meek. 1977. Soluble salt considerations with waste application. pp. 219–232. In: L.F. Elliott and F.J. Stevenson (eds.), *Soils for management of organic wastes and wastewaters*. Amer. Soc. Agron, Madison, WI.

Stumm, W., and J.J. Morgan. 1981. *Aquatic chemistry*, 2d ed. Wiley, NY.

Tabatabai, M.A. 1985. Effects of acid rain on soils. *CRC Crit. Rev. Envir. Control.* 15:65–110.

Thomas, G.W., and W.L. Hargrove. 1984. The chemistry of soil acidity. In: F. Adams (ed.), *Soil acidity and liming*, 2d ed., 12:3–56. Amer. Soc. Agron., Madison, WI.

Traina, S.J., G. Sposito, D. Hesterberg, and U. Kafkafi. 1986. Effects of pH and organic acids on orthophosphate solubility in an acidic, montmorillonitic soil. *Soil Sci. Soc. Amer. J.* 50:45–52.

Walter, M.F., T.L. Richard, P.D. Robillard, and R. Muck. 1987. Manure management with conservation tillage. pp. 253–270. In: T.J. Logan, J.M. Davidson, J.L. Baker, and M.R. Overcash (eds.), *Effects of conservation tillage on groundwater quality. Nitrates and pesticides*. Lewis Pubs., Chelsea, MI.

Whitby, L.M., P.M. Stokes, T.C. Hutchinson, and G. Myslik. 1976. Ecological consequence of acidic and heavy metal discharge from the Sudbury smelters. *Canad. Mineral.* 14:47.

Wildung, R.E., and T.R. Garland. 1980. The relationship of microbial processes to the fate and behavior of transuranic elements in soils and plants. In: W.C. Hanson (ed.), *The transuranic elements in the environment*. ERDA Pub. Ser. TIC-22800. NTIS, Springfield, VA.

Salt-Affected Soils: Their Reclamation and Management for Crop Production

R.K. Gupta* and I.P. Abrol†

* Central Soil Salinity Research Institute, Karnal, Haryana, India.
† Indian Council of Agricultural Research, Krishi Bhawan, New Delhi 110001, India.

I. Introduction

Agricultural production in the arid and semiarid regions of the world is limited by poor water resources, limited rainfall, and the detrimental effects associated with an excess of soluble salts, constrained to a localized area or sometimes extending over the whole of the basin. In order to minimize vagaries of arid weather, bring more land under irrigation, and produce and stabilize greater yields per unit area, numerous water development projects have been commissioned all over the world. Extension of irrigation to the arid regions, however, usually had led to an increase in the area affected by shallow water tables and to intensifying and expanding the hazards of salinity. This is because irrigation water brings in additional salts and releases immobilized salts in the soil through mineral dissolution and weathering, and losing water volumes through evapotranspiration and concentrating the dissolved salts in soil solution. Fertilizers and decaying organic matter also serve as additional salt sources. Atmospheric salt depositions, though varying with location, may be an important source along the coasts. The relative significance of each source in contributing soluble salts depends on the natural drainage conditions, soil properties, water quality, soil water, and agronomic management practices followed for crop production. Whenever soils have problems of excessive soil moisture as well as natural secondary salinization/alkalization in the root zone, plant growth is either restricted or entirely prevented. However, high crop productivity in salt-affected soils can be attained if the nature and extent of salinity problems are correctly diagnosed and appropriate reclamation and management practices are adopted under favorable environmental conditions. In this review we discuss issues related to management of salt-affected soils, particularly alkali soils. Soil and plant interactions with salinity, which has been the subject of several recent publications (Shainberg and Shalhevet, 1984; American Society of Civil Engineering, 1988), is discussed only briefly.

Salt-affected soils differ from arable soils with respect to two important properties, namely, the soluble salt and the soil reaction. Any buildup in soluble salt content of soil may influence its behavior for crop production in several ways through changes in the proportions of exchangeable cations, soil reactions, the physical properties of the soils, and osmotic and specific ion toxicity effects. These changes have a bearing on the activity of plant roots and soil microbes.

II. Classification of Salt-Affected Soils

Salt-related properties of soils are subject to rapid changes under proper circumstances. Therefore, the general intent of classifying salt-affected soils is to facilitate discussions of soil management and to group soils according to crop responses (Black, 1968). On the basis of determinations made on soil samples and the influences of the two common kinds of salts (neutrals and alkali salts) on soil properties and plant growth, salt-affected soils have been separated into two broad groups, saline soils and alkali soils (Szaboles, 1974; Abrol and Bhumbla, 1978; Bhumbla and Abrol, 1979).

A. Saline Soils

Saline soils, often recognized by the presence of white salt encrustations on the surface, predominantly have chlorides and sulfates of Na, Ca, and Mg. Soils with neutral soluble salts have saturation paste $pH < 8.2$. The electrical conductivity of saturation extracts of saline soils is generally more than 4 $dS\ m^{-1}$ at 25°C. When chlorides and sulfates of Ca and Mg are the predominant salts, the sodium adsorption ratio $SAR = Na/[(Ca + Mg)/2]^{1/2}$ of the soil solution is usually less than 15. However, soil salinization with neutral soluble salts of Na invariably result in soil solution $SAR > 15$. Such soils are termed saline-sodic (U.S. Salinity Laboratory Staff, 1954). Recent experimental evidence shows that leaching of saline-sodic soils (soils containing neutral salts and high SAR solutions) with good-quality waters generally results in desalinization and simultaneous desodification (Figures 1 and 2), accomplished through the increased solubilities of naturally present $CaCO_3$ and gypsum (Dieleman, 1963; Leffelaar and Sharma, 1977; Khosla et al., 1979; Jury et al., 1979). Results of several field experiments carried out in India and Iraq on sandy loam soils suggest a limited value of amendments in the reclamation of saline-sodic soils. However, significant responses to application of amendments can be expected in soils with inherent low permeability, such as the fine-textured and swelling clay mineralogy soils, or the chemically stable, coarse-textured soils via electrolyte effect

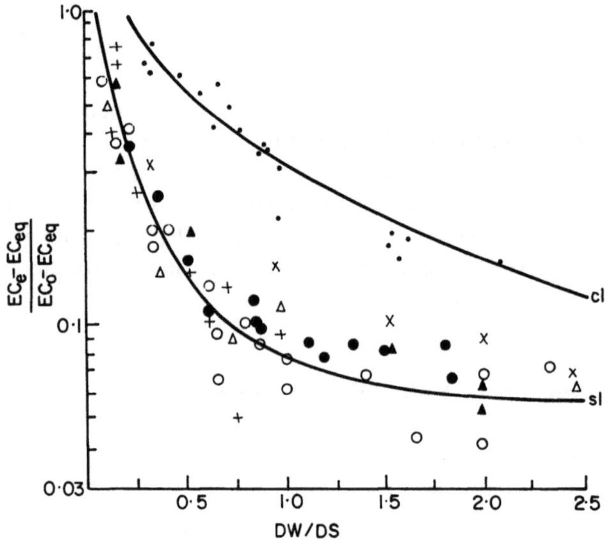

Figure 1. Salt displacement from saline-sodic (Soln. SAR >> 15, neutral salts) soils in relation to depth of leaching water per unit depth of soil (Rao et al., 1986).

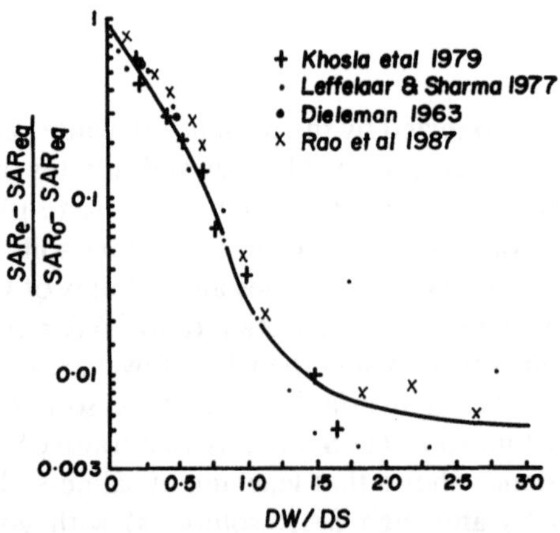

Figure 2. Desodication of saline-sodic (Soln. SAR >> 15, neutral salts) soils in relation to depth of leaching water per unit depth of soil (Rao et al., 1986).

(Oster and Shainberg, 1979). It seems that amendment use for reclamation of saline soils having high SAR in their soil solutions per se would depend primarily on the inherent infiltration characteristics of dry, loose field soils at the antecedent moisture contents and the quality and electrolyte level of the irrigation waters. Gypsum use can only be justified economically by a worthwhile yield response, usually over several seasons.

Figure 3. Relation between pH and alkalinity of aqueous carbonate systems in equilibrium with different levels of partial pressures CO_2. [After Mashhady and Rowell (1978).]

When salinity and sodicity occur together, as in many saline-sodic soils (containing neutral salts), limited evidence suggests that the effect of the two factors on plant growth is nonadditive and noninteractive, and growth is limited primarily by salinity effects (Largerwerff and Holland, 1960; Bernstein, 1962). Put another way, regardless of the sodium adsorption ratio in solution, it is the toxic excess of neutral salts in saline-sodic soils that prevent the plant from effecting normal growth (Figure 3). Many saline-sodic soils contain soluble carbonates besides the excess of neutral salts. Such soils manifest alkali soil properties. It is, therefore, only prudent that the saline-sodic soils that do not contain soluble carbonates be grouped and managed as saline soils and the rest of them grouped and managed like alkali soils. In saline-sodic soil environments the ability of the root system to tolerate salinity and in particular high levels of sodium ions is attributed partially to compensatory uptake of water and ions in the more favorable regions (Rains and Epstein, 1967; Solov'ev, 1969; Bingham and Garber, 1970; Devitt et al., 1984).

B. Alkali Soils

Alkali soils contain an excess of exchangeable sodium and have sodium carbonates. However, many a times the existence of sodium carbonates in soils goes unnoticed when extracted with water. This is because a portion of the dissolved carbonates reacts with Ca^{2+}, released from different sources (exchange sites, mineral weathering, gypsum) and precipitates immediately as $CaCO_3$ in the course of preparation of paste extracts. The high solubility of sodium salts and the condition of electroneutrality of aqueous solutions makes it imperative that remaining Na^+ charge is either balanced by sulfate ions or incorporated into the clay exchange sites. This permits the use of efflorescence crusts, pH > 8.4 and Na/Cl ratio > 1, as indicators of soils containing sodium carbonates (Nakayama, 1970; Magaritz et al., 1981; Nadler and Magaritz, 1982). When soils contain sulfates in significant amounts, a $Na/(Cl + SO_4)$ ratio > 1 may indicate the presence of carbonates of sodium.

Many soils in arid and semiarid regions have calcium carbonate in the soil profile. In the concentration range 0.03% < PC02 <1% of the soil atmosphere, calcareous soils not affected by Na^+ have a pH in the range 7.3 to 8.2 (Yaalon, 1958; Peech, 1965).

Calcareous soils containing soluble carbonates, exchangeable Na percent > 15 and saturated paste pH > 8.2, manifest problems associated with alkalinity. These soils contain soluble salts in variable quantities. With carbonate/sulfate salinization, electrical conductivity in saturated paste extracts is usually more than 4 dS m^{-1} at 25°C. The presence of soluble carbonates permits and facilitates apportioning the alkali portion of saline-sodic soils. It has been reported that an exchangeable sodium percentage (ESP) of 15 is usually associated with soil paste pH 8.2–8.3 (Szaboles, 1974; Abrol and Bhumbla, 1978). A large number of researchers have reported a direct positive relation between ESP and soil pH for calcareous alkali soils (Agarwal et al., 1982). An excess of exchangeable sodium and high pH caused by the presence of sodium bicarbonate-carbonate minerals impart these soils poor physical conditions, which adversely affect the plant growth.

Compared with the critical limiting ESP value of 15 for deterioration in soil structure (U.S. Salinity Lavoratory Staff, 1954, a considerably lower limiting value of ESP = 6 has been suggested for soils with an abundance of fine clay and lacking in soluble weatherable minerals to maintain electrolyte concentration of soil solution during leaching (Northcote and Skene, 1972; Shanmuganathan and Oades, 1983).

In nature, strongly alkaline soils invariably have high sodicity; on the other hand, a sodic soil with soil solutions of high SAR does not necessarily mean a high pH (Kelley, 1948; Beek and Breeman, 1973). In many saline-sodic soils that do not contain sodium carbonate minerals, the saturation paste pH is observed to be less than 8.2–8.3. Application of amendment is not mandatory in the management of such soils, as in the case of Na-carbonate-containing, high-pH, and high-sodicity alkali soils. From management considerations we have used the term "alkali" for soils that have high sodicity, high pH, and soluble carbonate [$(Na/Cl = SO_4) > 1.0$ as appropriate] with variable quantities of soluble salts.

Figure 4. Effect of dilution and soluble salts on pH changes of saline and alkali soils. pH was measured in water-saturated pastes in 1:2 soil-water suspensions.

The adverse effects of adsorbed Na and high pH on soil physical properties are most apparent after irrigation or rainfall, when water stagnates for long periods in low microrelief spots (Figure 4). The soil a few centimeters below the surface may be almost dry and hard in a wetting cycle. Upon drying, alkali soils become very hard and develop cracks a few centimeters wide and several centimeters deep, which close when the soil is wetted. In strongly alkaline conditions, dissolved organic matter gives the soil surface a dark black color. In alkali soils, crops show scorching and leaf burn, typical of sodium toxicity. Cultivation of alkali soil present problems because the entire field does not achieve optimum moisture conditions at one time—some spots are still very wet while others dry out. Cultivation of such land usually leaves the surface cloddy, which results in poor germination and spotty crop stands.

C. Taxonomic Classification and Nomenclature

Two soil classification schemes, the *World Soil Map Legend* (FAO-UNESCO, 1974) and *Soil Taxonomy* (Soil Survey Staff, 1975), published during the past decade, have gained a degree of international currency. Soil Taxonomy is emerging as the de facto international soil classification

Table 1. Tentative correlation of the most widely used classification systems of salt-affected soils

Subcommission on Salt-affected Soils classification: Basic grouping (1967)	Subcommission on Salt-affected Soils classification: Map of European Salt-affected Soils legend (1968)	Subcommission on Salt-affected Soils classification: Saline and Sodic Soils Map of Australia legend (1971)	Soil Map of the World (Soil Map of Europe) FAO/UNESCO project (ECA working party) classification	Australian classification (1968)	Canadian classification (1965)	French classification (1967)	USDA classification (1967)	USSR classification (1967)
Saline soils	Saline soils	Saline soils	Solonchak:	Solonchak	Saline	Sols	Salorthids	Fluffy solonchak
Alkali soils without structural B horizon	Alkali soils without structural B horizon		Orthic solonchak		subgroups	salins (excepte sous-groupe acidifie)		(nonsteppic) Crust solonchak
		Alkaline sodic soils AS1 AS2 As3	Mollic solonchak				Salorthidic Calcuiustolls Salorthidic Haplustolls	Soda solonchak (nonsteppic) Fluffy solonchak (steppic) Soda solonchak (steppic)
			Takyric solonchak					Takyrs
			Gleyic solonchak			Sols sales à alcali Sols à gley salés	Halaquepts (pp)	Meadow solonchak
	Alkali soils with structural B horizon	Nonalkaline sodic soils NS1 NS2	Solonetz:	Solonetz and solodized solonetz		Sols sodiques à horizon B et solonetz solodises		
	Solonchak-solonetz and calcareous solonetz							

Alkali soils with structural B horizon	Noncalcareous solonetz with A horizon, 15 cm	Orthic solonetz		Brown solonetz (pp) and alkali solonetz	Nadurargids Natrargids Natriboralfs Natrudalfs Natrustalfs Natrixeralfs	Desert-steppe and desert solonetz
	Solodized and/or deeply leached solonetz					
	Solonized and slightly salt-affected soils with minor structure formation	Mollic solonetz		Black, grey and brown solonetz (pp)	Natralbolls Natriborolls Natrustolls Natrixerolls	Steppe solonetz
	Deeply leached solodized soils and solods	Gleyic solonetz		Gleyed solonetz	Natraquolls Natraqualfs	Meadow solonetz
		Solodic planosols	Solod and solodic groups(pp)	Solod and solodized groups(pp)	Argiabolls(pp)	Solod

Source: Szabolcs [From *Review of Research on Salt-Affected Soils, Natural Resources Research* XV, © UNESCO 1979.]

Figure 5(a). Distribution of salt-affected soils in Asia; 1: Areas of salt-affected soils. [After Kovda et al. (1964).]

system. A brief account of the general advantages and disadvantages of the above two schemes has been given by Moore et al. (1983). Table 1, compiled by I. Szabolcs, demonstrates the correlation between the classification system elaborated by the subcommission on salt-affected soils and the other widely used classification system for salt-affected soils. Salt-affected soils such as occur extensively in Asia and Australia are shown in Figure 5.

A feature common to both the *World Soil Map* and the *Soil Taxonomy* systems is the widespread use of defined diagnostic horizons in differentiating classes. In many cases the dignostic horizons are essentially the same in both systems. Ambiguity of definitions, however, leads to difficulties in consistent identification and realization of the significance of diagnostic horizons in various soils (Isbell, 1984).

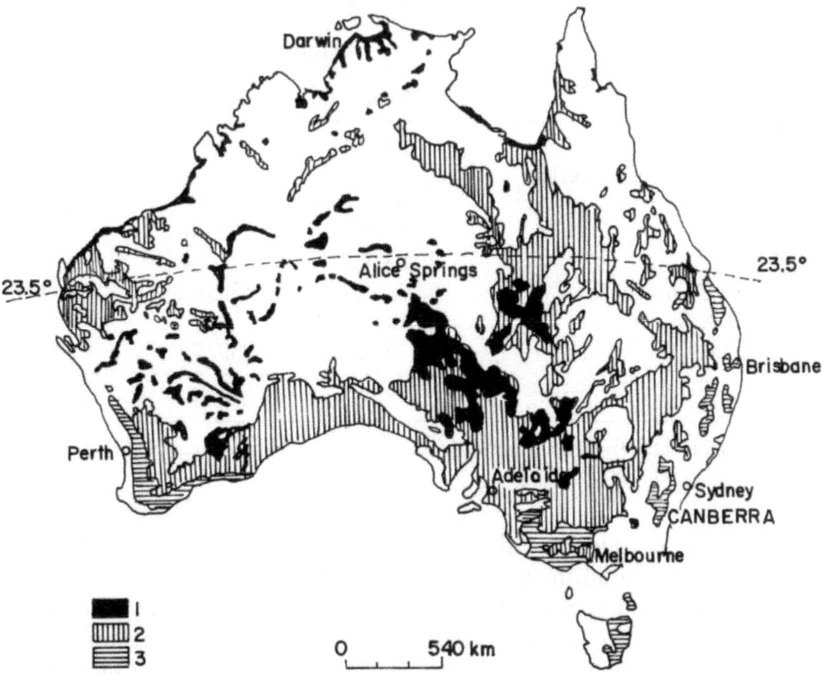

Figure 5(b). Distribution of salt-affected soils in Australia; 1: saline soils, 2: alkaline sodic soils, 3: nonalkaline Sodic Soils. [After Wright and Schuster (1972).]

Depending on the magnitude of the salinity and sodicity problems, morphometric features, and setting, including climate components, placement of salt-affected soils in different Great Group associations or their phases poses problems. In Soil Taxonomy, where salinity meets the salic subsurface natric horizons requirement, provision is made in the aridisols Great Group. Similarly, when sodicity and argillic clay requirements of a subsurface horizon are met, provision is made for a Great Group of aridisols, alfisols, and mollisols. There are apparent difficulties in grouping soils when surface horizons manifest these properties. Also, many soils that have strongly sodic argillic horizons, but do not have prismatic or columnar structure or tongues of an eluvial horizon extending more than 2.5 cm into a block-structured argillic horizon, fail to qualify as natric horizons under present definitions (Sehgal et al., 1975; Isbell and Williams, 1981). As an another example, diagnostic criteria for vertisols is based on the surface 18 or 20 cm that have been mixed as by plowing through an inversion process, development of cracks during wetting-drying cycles, etc. Under the influence of high exchangeable sodium, black cotton soils developed from basaltic material in India often cease to manifest mixing phenomena so common in other normal vertisols. As discussed earlier, there is also mounting evidence that a lower limit of exchangeable sodium percentage

Table 2. Global distribution of salt-affected soils

Continent	Area in thousand hectares		
	Saline	Alkali	Total
North America	6,191	9,564	15,755
Mexico and Central America	1,965	—	1,965
South America	69,410	59,573	129,163
Africa	53,492	26,946	80,438
South Asia	83,312	1,798	85,110
North and Central Asia	91,621	120,065	211,686
Southeast Asia	19,983	—	19,983
Australasia	17,359	339,971	357,330
Total	343,333	558,097	901,430

Source: Szabolcs (1979). [From *Review of Research on Salt-Affected Soils, Natural Resources Research* XV, © UNESCO 1979.]

at 15 is too high to identify the management constraints for soils with high shrink-swell potential. In spite of certain pitfalls, the structure of Soil Taxonomy is such that new information can be incorporated with minimal distortion. For Soil Taxonomy to develop into an effective international system of soil classification, it is probably best if nations use it in their soil survey programs and make proposals for change.

D. Global Distribution

Salt-affected soils occupy extensive areas and occur globally. In spite of the availability of many sources of information, accurate data concerning the salt-affected lands of the world are rather scarce. Widely different accounts and estimates can be found in various reports. Based on the FAO/UNESCO Soil Map of the World, the extent of salt-affected soils has been estimated by Massoud at 901.4 m ha (Table 2). Including the estimates of the extent of salt-affected soils in Europe (50.8 m ha) prepared by Szabolcs (1974), the land area facing salinity can be put at a total of 952.3 m ha.

III. Measuring Salinity, Alkalinity, and Sodicity; Some Interrelationships and Influences on Soil Properties

A. Measuring Salinity

The amount of dissolved salt in water or soil extract is usually determined in most soil testing laboratories by measuring the electrical resistance of the solution (inohms) or its reciprocal, the electrical conductance (ohms^{-1}). In the Systeme International d'Unites (SI units), the electrical conductance measured in reciprocal ohms is designated as siemens (S). Measurements of conductivity made in a cell containing two parallel platinum electrodes are usually corrected for variations in cell geometry by multiplying the results with a "cell constant." The corrected conductivity data, reported as specific conductance (EC), are conventionally normalized to a temperature of 25°C to allow valid comparison.

Although different laboratories employ a variety of soil/water ratios to obtain aqueous extracts from the soil sample, vacuum filtration of distilled water-saturated soilpaste (U.S. Salinity Laboratory Staff, 1954) remains the most widely preferred method. Whereas it is the lowest convenient soil moisture content with which to obtain soil extract for determining soluble constituents and EC, water-saturated paste relates to field soil water content and a number of soil parameters such as soil texture, sufrace area, clay content, cation-exchange capacity, etc. (Merrill et al., 1987). Other soil water ratios (1:2, 1:5, etc.), though easier to extract, are subject to errors from mineral dissolution, hydrolysis, cation exchange, and peptization. Large soil/water ratios can be used to advantage when information such as relative changes in salinity is desired and soils do not contain sparingly soluble minerals such as gypsum. At the field level, salinity measurements can also be made by using various kinds of salinity sensors [porous matrix sensors, four-electrode probes, electromagnetic (EM) induction sensors, and time-domain reflectometry (TDR)]. Advantages and limitations of each of these four kinds of salinity measurement devices have been discussed by Rhoades (1984) and will not be taken up here.

B. Alkalinity of Alkali Soils and Relationship to pH

Alkalinity (Alk) is defined here as the acid-neutralizing capacity of an aqueous carbonate system (Stumm and Morgan, 1970). In most natural waters, alkalinity equals the total concentration of cations minus the total concentration of anions other than carbonates:

$$\text{Alk} = [\text{Na}^+] + [\text{K}^+] + 2[\text{Ca}^{2+} + \text{Mg}^{2+}] - [\text{Cl}^-] - 2[\text{SO}_4{}^{2-}] \quad (1)$$

More conveniently, alkalinity as expressed in Equation (1) can be represented as

$$\text{Alk} = [\text{HCO}_3{}^-] + 2[\text{CO}_3{}^{2-}] + [\text{OH}^-] - [\text{H}^+] \quad (2)$$

The square brackets denote molar concentrations. Alkalinity can be estimated titrimetrically and is not affected by gain or loss of CO_2 or changes in temperature or pressure. Alkalinity in excess over the total amount of divalent cations is often referred to as residual alkalinity, or the equivalent term, "residual sodium carbonate" (Eaton, 1950).

$$\text{Alk. residual} = \text{Alk} - 2[\text{Ca}^{2+}] - 2[\text{Mg}^{2+}] \quad (3)$$

Equation (1) implies that any increase in alkalinity must be accompained by an equivalent increase in total cation or carbonate concentration such as due to weathering of the mineral carbonates under the influence of CO_2;

$$CaCO_3 + H_2O = Ca^{2+} + 2HCO_3^- \quad (4)$$

$$Na^+\text{-clay} + H_2O + CO_2 = H^+ \text{ clay} + Na^+ + HCO_3^- \quad (5)$$

or by the physicochemical process of hydrolysis of sodic soils. Similarly, reduction of sulfates and nitrates accompanied by oxidation of organic matter also increases alkalinity.

$$2CH_2O + SO_4^{2-} = 2HCO_3^- + H_2S \quad (6)$$

$$5CH_2O + 4NO_3^- = CO_2 + 2N_2 + 3H_2O \quad (7)$$

Precipitation of alkali earth carbonates, formation of clay minerals of the smectite and polygorskite group, and oxidation of sulfides also influence (reduce) alkalinity (Garrels and Mackenzie, 1967; Millet, 1970). Alkalinity may also change as a result of evaporation and dilution (Beek and Breeman, 1973).

Alkalinity and the equilibrium partial pressure of carbon dioxide, P_{CO_2}, adequately define the pH of an aqueous carbonate system and of alkali soils (Ponnamperuma, 1972; Mashhady and Rowell, 1978). The relation (Figure 6) is represented by the equation

$$pH = 7.82 - \log P_{CO_2} + \log \text{Alk} - 0.5\,(I^{1/2}) \quad (8)$$

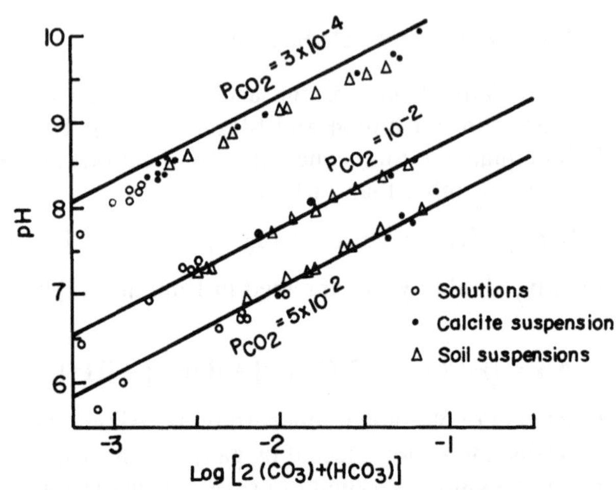

Figure 6. Relation between pH and alkalinity of aqueous carbonate systems in equilibrium with different levels of partial pressure CO_2. [Mashhady and Rowell, from "Soil alkalinity, equilibria, and alkalinity development," in *J. Soil Science* 29:65–75 (1978), Blackwell Sci. Pub. Ltd.]

where I refers to the ionic strength of the solution. The relation suggests that any increase in alkalinity or decrease in P_{CO_2} is accompanied by a rise in pH of the system. This relationship has been tested for pure systems and soils to establish that the theoretical analysis can be used for calcareous sodic soils and to confirm that it is not possible to separate pH and alkalinity ($CO_3 + HCO_3$) when considering their effects on plant growth (Whitney and Gardner, 1943; Ponnamperuma, 1972; Beek and Breeman, 1973; Mashhady and Rowell, 1978; Gupta et al., 1981). In accordance with the above relationships, the pH of soil containing an excess of $CaCO_3$ in equilibrium with the partial pressure of CO_2 in the air will vary with the calcium ion activity or the ion activity ratio $(Ca^{2+})^{1/2}/(H^+)$ according to the equations

$$CaCO_3 = Ca^{2+} + CO_3{}^{2-} \qquad pK_{sp} = 8.4 \tag{9}$$

$$CaMg(CO_3)_2 = Ca^{2+} + Mg^{2+} + 2Co_3{}^{2-} \qquad K_{sp_2} = 10^{-17} \tag{10}$$

$$(Ca^{2+}) \times PCO_2/(H^+)^2 = 10^{9.76} \tag{11}$$

$$pH - 0.5P_{Ca} = 4.88 - 0.5 \log P_{CO_2} \tag{12}$$

At the CO_2 content of atmospheric air ($P_{CO^2} = 3 \times 10^{-4}$ atm), the concentration of Ca^{2+} is 5.2×10^{-4} M (Frear and Hohnston, 1929). Accordingly, the pH of a calcareous soil in the absence of appreciable exchangeable sodium is fixed at 8.2–8.3 (Yaalon, 1958; Peech 1965).

Residual alkalinity, a constituent of alkalinity, is influenced by changes in the relative proportions of mono- and divalent cations due to cation exchange,

$$Ca^{2+}\text{-clay} + Na^{+} = 2Na^{+}\text{-clay} + Ca^{2+} \tag{13}$$

and also by dilution and evapotranspiration. Under conditions where Alk $> \Sigma$ divalent cations, as evaporation proceeds, precipitation of calcite, dolomite, CaMg $(CO_3)_2$, and sepiolite, $MgSi_3O_6(OH)_2$, would remove Ca^{2+}, Mg^{2+}, and SiO_2 and alkalinity from solutions, leaving behind monovalent cations:

$$Ca^{2} + 2HCO_3{}^{-} = CaCO_3 + 2H_2O + CO_2 \tag{14}$$

$$Mg^{2+} + 3H_4SiO_4 + 2HCO_3{}^{-} = Mg(SiO_2)_3 + 6H_2O + 2CO_2 \tag{15}$$

The divalent cations Ca^{2+} and Mg^{2+}, whose carbonates are poorly soluble, precipitate out and moderate the rise in pH by elimination of bicarbonate ions. However, Na^+ and K^+ ions tend to concentrate and increase residual alkalinity and pH of the soil solution. Hence, the presence of residual alkalinity in the starting soil solution/irrigation waters may further buildup of alkalinity during evaportranspiration, while the concentration of divalent cations in solutions would become vanishingly small in terms of the concept of chemical divide (Hardie and Eugester, 1970; Drever, 1982; Beek and Breeman, 1973).

The pH of the system would remain relatively constant while $CaCO_3$ and sepiolite were precipiating and would then increase steadily to pH values as high as 10.0 due to Na bicarbonate-carbonate minerals (Nakayama, 1970), resulting in high-SAR solutions. Beek and Breemen (1973) observed two buffer regions, the first coinciding with bulk precipitation of calcite and the second with the dissociation of HCO_3^- ions. At sufficiently high pH, alkalinity is consumed by dissociation of monomeric silica derived from silicate minerals:

$$H_4SiO_4 + CO_3^{2-} = H_3SiO_4^- + HCO_3^- \qquad (16)$$

This reaction involves silicates in the buffering process and limits the rise in pH to values between 10 and 10.4, frequently encountered in strongly alkali soils.

C. Measuring Sodicity

Determinations of sodicity and alkali status of soils are important for basing amendment requirements of alkali soils along scientific lines. It has been observed that different methods of determining cation-exchange capacity and exchangeable sodium generally yield differing results. In the literature, values of exchangeable sodium content in excess of the charge capacity of the soil (CEC) have been obtained by using the conventional NH_4OAc method (Bower, 1948; Overstreet et al., 1951; Babcock, 1960; Bower and Hatcher, 1962; Kanwar and Bhumbla, 1969). Schulz et al. (1964) reported that estimates of ESP obtained by extraction of exchangeable Na by NH_4OAc were substantially higher than the SAR or gypsum-requirement methods. The abnormal exchange behavior of alkali soils has been attributed to the occurrence of Na compounds that are relatively insoluble in water but are soluble in normal salt solutions such as NH_4OAc (Kelley, 1948), to Na fixation (Babcock, 1960), and to the presence of Na zeolites (Schulz et al., 1964). Recently it has been shown by Sposito and LeVesque (1985) that Silver Hill illitic clay specimens had a time-dependent fixation of Na^+ that did not exchange with Ca^{2+} but could be replaced by ammonium ions.

A survey of the literature on soil zeolites by Ming and Mumpton (1984) revealed that members of this group of hydrated aluminosilicates of alkali and alkaline earth cations occur widely. These minerals may be the residual zeolite phase derived directly from the rocks or phases developed in situ in strongly alkaline environments at pH above 8.6 (Pratt and Bair, 1969) directly from soil solutions. In zeolitic soils the CEC of nonzeolite exchange sites can be determined by saturating both types of exchange sites with Na and using *tert*-butyl ammonium ions during the replacement process (Barrer, 1978; Ming and Dixon, 1984).

A two-step procedure involving (1) the saturation of exchange sites with

60% ethanolic solution of 0.4 *N* NaOAc-0.1 *N* NaCl (Na : Cl, 4 : 1), pH 8.2, and (2) extraction of Na-saturated soil with 1 N $Mg(No_3)_2$, pH 7.0, solution was developed by Polemio and Rhoades (1977) to overcome limitations and eliminate sources of error in determination CEC of alkali soils. Gupta et al. (1985c) further modificed this procedure to eliminate errors due to (1) alterations in the ion sieve and steric hindrance properties of zeolites by the use of alcoholic solution and (2) release of Na from zeolitic minerals in the $Mg(No_3)_2$, pH 7.0, extracting solution. The modified procedure eliminated the use of alcohol and reduced the extraction of nonexchangeable Na from zeolitic minerals by raising the pH of the $(NO_3)_2$ solution from 7 to 8.6. Consequent to alleviating the above errors, the modified procedure seemingly is a better method for determination of CEC. The two methods of determining CEC, however, need rigorous testing on a wide range of alkali soils.

Exchangeable sodium content of alkali soils can be determined directly by leaching a portion of soil with 100 ml of 1 *N* $Mg(No_3)_2$ solution at pH 8.6 and analyzing the leachates for the sum of exchangeable plus water-soluble sodium. Estimates of soluble Na content can be obtained from the analysis of the saturation paste extracts of soils in distilled water (U.S. Salinity Laboratory Staff, 1954). Exchangeable Na can then be calculated from the difference. It may be mentioned here that erors in determination of soluble Na by analysis of saturation extracts become quite appreciable when EC exceeds 10 dS m^{-1}. In order to estimate the soluble Na in saturation extracts correctly, anion-exclusion corrections must be applied:

$$\text{Soluble Na} = \text{Na in paste extract} \times \frac{\text{Cl obtained upon exhaustive } \textit{leaching with extracting solution}}{\text{Cl in the paste extract}}$$

D. Soluble Salts and Soil pH

Calcareous soils inherently have pH around 8.0. Under natural field conditions, saturated paste pH varies from 7.1 in highly saline soils to about pH 8.2 in nonsaline, calcareous soils. Measuring pH in 1 : 2 soil-water suspensions gives pH readings of 7.6 for highly saline soils and about 8.5–8.6 for nonsaline soils. Buildup of salt reduces the pH of saline soils, but in the case of alkali soils, pH increases with increase in salinity due to the presence of sodium-bicarbonate carbonates (Figure 7). Dilution of soil-water suspensions also increases pH. The difference between the pH of a saturated paste and 1 : 2 soil-water suspensions of alkali soils has been observed to be about 1 pH unit at low salinity, and 0.2–0.4 pH unit at high salinity values. Despite higher the pH values of soil suspensions, the corresponding alkalinities in aqueous extracts have been observed to be lower by a factor of 2 or more versus that of the paste extracts (Vorob'eva et al., 1986). This pH-alkalinity paradox has been elucidated as being associated with the

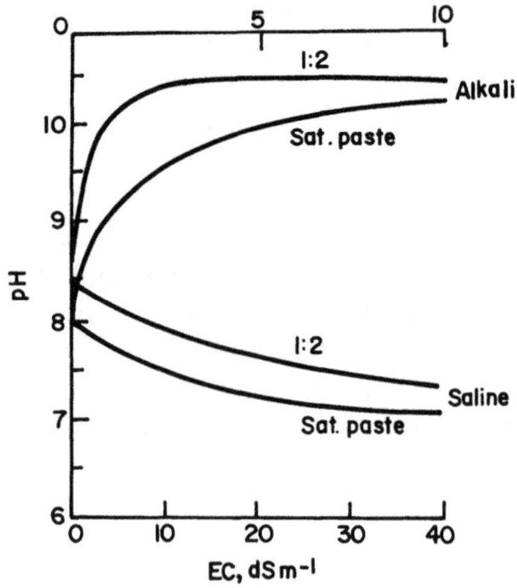

Figure 7. Effect of dilution and soluble salts on pH changes of saline and alkali soils. pH was measured in water-saturated pastes in 1:2 soil-water suspension. [From *Review of Research on Salt-Affected Soils, Natural Resources Research XV*, 1979.]

carbonate-calcium equilibria (Lindsay, 1979). For the calcium carbonate ($H_2O - CO_2 - CaCO_3$) system, the relationship between P_{CO2} and pH has been represented by the equation (Ponnamperuma, 1972):

$$pH = 5.983 - \tfrac{2}{3}\log P_{CO_2} + \tfrac{1}{3}\log \gamma_{HCO_3^-} - \tfrac{1}{3}\log \gamma_{Ca^{2+}} \tag{17}$$

$$\text{or } \log P_{CO_2} = 8.975 - 1.5\ pH + \tfrac{1}{2}\log \gamma_{HCO_3} - \tfrac{1}{2}\log \gamma_{Ca^{2+}} \tag{18}$$

where Y refers to the activity coefficient of the respective ions. The relationship between pH and alkalinity (carbonate alkalinity) can be established by substituting the $CO_3{}^{2-}$ and $HCO_3{}^-$ values in the PCO_2 term in Equation(18):

$$2CO_3{}^{2-} + HCO_3{}^- = PCO_2\left(\frac{2 \times 10^{-18.75}}{a^2H^+ \cdot \gamma CO_3{}^{-2}} + \frac{10^{-7.82}}{aH^+ \cdot \gamma HCO_3{}^-}\right) \tag{19}$$

Substitution of an equivalent expression for the P_{CO_2} term [Equation (18)] in Equation (19) and taking the logarithm yields an expression that can be used for identifying alkalinity production due to $CaCO_3$ and other anticipated sources such as sodium carbonate-bicarbonate and hydrolysis of exchangeable sodium (Vorob' eva et al., 1986). The relation suggests that an increase in PCO_2 would decrease pH and increase carbonate alkalinity. It therefore appears that the paradox might be due to a drop in P_{CO_2} with rising soil: water ratio (Whitney and Gardner, 1943) and loss of dissolved $H_2\,CO_3$ during vacuum extraction of the sample (Suarez, 1987). In spite of several limitations in the accurate measurement of soil pH, it severes as a useful parameter for meaningful interpretations, if defined operationally.

Table 3. pH Dependence on sodicity of soils

pH		Exchangeable sodium percentage		
Saturated paste	1:2 Soil-water suspensions	Chang and Dregne (1955)	Kovda (1965)	Abrol et al. (1980)
8.0–8.2	8.2–9.0	<20	<15	<20
8.2–8.4	9.0–9.4	20–30	15–25	21–35
8.4–8.6	9.4–9.6	30–45	25–30	35–50
8.6–8.8	9.6–9.8	45–60	30–40	50–65
8.8–9.0	9.8–10.0	>60	40–50	65–85
9.0–9.8	>10.0		>50	>85

E. Sodicity and Soil pH

In field conditions, a high pH value is caused by Na bicarbonate-carbonate minerals present in alkali soils (Nakayama, 1970), which precipitate Ca and Mg carbonate during evaporation and lead to an increase in the SAR of solutions. An intimate relationship between pH and sodicity (ESP) has been observed for natural alkali soil conditions. Gupta et al. (1981) developed a relation for the pH dependence of simultaneous $Na^+ - (Ca^{2+} + Mg^{2+})$ exchange (Gapon's expression) and calcite equilibria for alkali soils. The relation is expressed as

$$\text{Log ESR} = \text{pH} - 5.236 + 0.5 \log PCO_2 + \log(Na^+) + 0.5(I^{1/2}) \quad (20)$$

where ESR refers to the exchangeable sodium ratio $[Na_{ex}/(CEC\text{-}Na_{ex})]$ and I to the ionic strength of the soil solutions. An alternative expression for a similar relation [Vanselow's] (1932) expression, Equation (28)] will be discussed later.

The approximate ranges of exchangeable percentage (ESP) corresponding to the soil pH in saturated soil paste and in 1:2 soil-water suspensions as usually observed in natural field conditions are given in Table 3.

F. ESP and SAR

To overcome diffculties encountered in the routine estimation of ESP and at the same time describe the equilibrium between soil solution and adsorbed cations of soils, the U.S. salinity Laboratory Staff (1954) developed an equation relating the ratio of exchangeable Na $[ESR = Na_x/(CEC\text{-}Na_x)]$ to the ratio between Na^+ and $(Ca^{2+} + Mg^{2+})$ concentrations of equilibrium extracts, i.e., the sodium adsorption ratio,

$$\text{SAR} = \frac{\text{Na}}{\dfrac{(\text{Ca} + \text{Mg})^{1/2}}{2}} \qquad \text{ESR} = K_G \cdot \text{SAR} \quad (21)$$

where K_G is the Gapon selectivity coefficient with a value of 0.015 (mmol $L^{-1})^{-1/2}$. SAR expressed in terms of ion concerntration, (meq liter^{-1}), does not take into account the reductions in free ion concentrations and ion activities due to ion-pair formation (Sposito and Mattigod, 1977). The "true" SAR after correction for ion pairs, complexes, and salt concentration has been observed to be more than the linear 1:1 functional relationship with SAR practical (SAR_P) because of the square-root operation of the activity coefficient (γ_{Ca}) of divalent ions in the expression

$$SAR_t = \frac{\gamma_{Na}(SAR)_p}{(\gamma_{Ca})^{1/2}} \tag{22}$$

The relationship between true SAR and SAR practical is also complicated by random variation in the amounts of soluble Na and of the complexing anions HCO_3^-, Co_3^{2-}, and So_4^{2-} commonly found in alkali soils. "True SAR" (SAR_t) can be approximated from SAR_p [Equation (21)] by using a regression equation given by Sposito and Mattigod (1977):

$$SAR_t = 0.08 + 1.115(SAR)_p$$

The relationship between exchangeable sodium percentage (ESP), exchangeable sodium ratio [$ESP = Na_x/(CEC - Na_x)$], and SAR_t can be expressed by Equation (23) (Sposito, 1977):

$$ESP = \frac{100 - ESR}{1 + ESR} = \frac{100/2K_G}{1 + (SAR/2K_G)} SAR \tag{23}$$

The traditional ESR-SAR relation has been observed to be consistent with ESR-SAR predicted by the appropriately corrected form of the Gapon equation over the range $0 < SAR < 40$ (Sposito, 1977; Oster and Sposito, 1980). A limitation of the Gapon equation has been shown as due to the dependence of the magnitude of the selectivity coefficient (K_G) on the equivalent fraction of sodium (NNa) on the exchanger phase. Although the traditional K_G for $N\text{Na} > 40$ is not consistent with the thremodynamics of exchange reactions (Sposito, 1977), it has been observed that the difference in K_G traditional and the corrected thermodynamic K_G for low-CEC soils and minerals is near zero (Evangelou and Coale, 1987). The constant behavior of K_G has been suggested as being due to heterogeneity of the exchange sites and differing activity of the adsorbed cations for the low-CEC systems (Jardine and Sparks, 1984).

An unusual aspect of the SAR is that $CA^{2+} + Mg^{2+}$ ions have been lumped together in the expression [Equation (21)]. Besides the fact that ion valance influences exchange relations more than ion size, lumping together Ca^{2+} and Mg^{2+} in the SAR term has been questioned on the basis of differences in the ionic preferences of soils and clays for the two cations (Bresler et al., 1982). Exchangeable Na was appreciably higher than predicted from the SAR of the applied water, rich in magnesium (Mg/Ca ratio > 1). This is because most soils have greater selectivity for Ca than for Mg, implying that at comparable SAR values, higher ESP would be observed for Mg-Na systems than for Ca-Na systems. Calcareousness and texture of soils also influence the ionic preferences of the two divalent cations (Kovda et al., 1973; Paliwal and Gandhi, 1976; Darab, 1980; Yadav and Girdhar, 1981; Girdhar and Yadav, 1982).

High amounts of magnesium influence soil behavior through both direct and indirect effects. The direct effect of exchangeable Mg in causing more clay dispersion and lowering the hydraulic conductivity (McNeal et al., 1966a, b), usually termed "specific effect," has been reported only for the predominantly illitic soils (Emerson and Chi, 1977; Rahman and Rowell, 1979). Exchangeable Mg had an effect on hydraulic conductivity of kaolinitic and montmorillonitic soils if leaching solutions had insufficient electrolytes <10 meq liter^{-1}. Magnesium indirectly influences the dispersion flocculation of soil particles through cation exchange. The adverse effect of high exchangeable Mg in soils has been proposed as being due either to inability of the soils to release electrolytes through mineral weathering to prevent dispersion (Alperovitch et al., 1986) or to more repulsive forces as a result of the large diameter of the hydrated magnesium ion (Emerson and Chi, 1977). It has been demonstrated that exchangeable Mg reduces the dissolution weathering rates of relatively weathered noncalcareous soils and increases soil suspectibility to exchangeable Na under conditions of low electrolyte concentrations (Alperovitch et al., 1986; Kreit et al., 1982).

G. Dispersion of Soil Particles

Alkali soils frequently have poor physical properties due to high sodicity and high pH, resulting in restricted water and air movement (Acharya and Abrol, 1978; Varallyay, 1978). Alkali soils commonly lack an adequate and continuous water supply to plants due to low infiltration rates, resulting in shallow wetting zones and temporary waterlogging problems, diminished water storage in the rhizosphere due to cracking, extremely low water conductivity, and low available moisture range. Leaching alkali soils with waters having electrolyte concentrations insufficient to maintain flocculated conditions cause hydraulic conductivity reductions due to clay dispersion, movement, and consequent blockage of water-conducting pores (Frenkel et al., 1978; Pupisky and Shainberg, 1979). Recent evidence shows that dispersion of soil colloids is a determinant of a range of soil properties such as mineralogy, sodicity, pH, charge density vis-a-vis CEC, clay-organic interactions, and electrolyte levels (Collis-George and Smiles, 1963; Emerson, 1977; Quirk, 1977; Frenkel et al., 1978; Oster and Schroer, 1979; Shanmugan than and Oades, 1983; Gupta et al., 1984). Rhoades and Ingvalson (1969) concluded that dispersion rather than swelling was the operative process that lead to permeability reductions in vermicullitic soils. Also, the experimental evidence suggests that dispersion of soil aggregates within a soil would be expected to occur at a lower electrolyte concentration than that required to flocculate a clay suspension (Emerson, 1977; Quirk, 1977).

H. Permeability and Threshold Concentration of Electrolytes

The concept of threshold concentration was first developed by Quirk and Schofield (1955) and was defined as the concentration in the percolating solution that would give rise to a 10% to 15% decrease in the relative permeability at a given sodicity level. In a review, Shainberg and Letey (1984) noted that application of their basic approach has been extended to a large number of additional soils for maintaining the soil permeability at a high and stable level.

Individual salt constituents as well as total salinity of irrigation water affect the stability of soil structure and, hence, the permeability of the soils to water. It has been reported that waters with salinities less than 0.3 dS m^{-1} cause clays to swell, resulting in swelling-induced effects such as breakdown of aggregates, cursting, and reduced permeability (Quirk and Schofield, 1955; Shainberg et al., 1981b; Oster and Shainberg, 1979). High salinity levels reduce swelling, aggregate breakwdown, and soil particle dispersion, whereas high sodicity has the opposite effect. The usefulness of the permeability-concentration concept becomes immediately evident if a given sodicity level soil is to be irrigated without causing dispersion or permeability decreases; the electrolyte concentration must be above the threshold value. Therefore, both salinity and sodium adsorption ratio of the applied water have been considered together in assessing potential hazards of water quality on soil physical properties (Quirk, 1977; McNeal and Coleman, 1966; Rhoades, 1982; Rengasamy et al., 1984). There is little information on the influence of alkalinity of natural water on the threshold electrolyte concentration needed to prevent dispersion of soil particles.

I. Hydraulic Conductivity in Relation to SAR, Electrolyte Concentration, and pH

Sodium adsorption ratio is a good measure of the sodicity hazards of irrigation waters. However, during evapotranspiration the dissolved concentrations of cation and anion are altered in terms of a succession of chemical divides. The essence of the concept, as pointed out earlier, is that species present in higher relative concentrations build up in solution, and species present in low relative concentration progressively decrease during evaporation and subsequent precipitation of divalent cations. Sodium adsorption ratio, which takes into account only the cations and ignores alkali-causing anionic species, has been widely used as an index of sodicity and the related phenomenon of clay dispersion. Thus an increase in SAR of soil solution/waters provides no information on the alkalinity buildup vis-a-vis soil pH. For instance, El-Swaify (1973) reported dramatic reduction in hydraulic conductivity of a Na-dominated variable-charge-type tropical soil upon leaching with carbonate- and silica-containing waters (Figure 8). From the results of several investigations it can be concluded that disper-

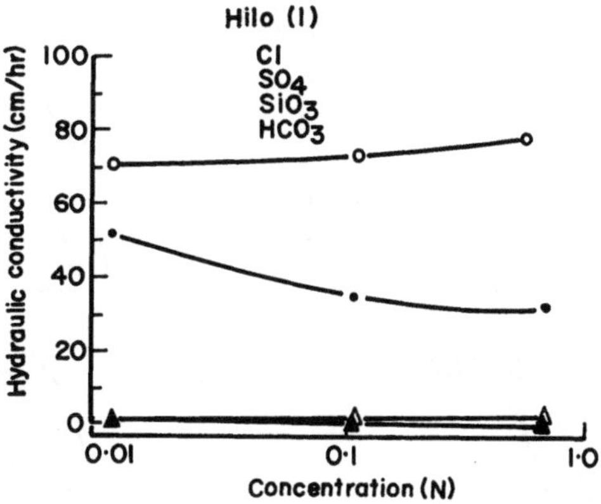

Figure 8. Effect of electrolyte concentration on the hydraulic conductivity of sodium-saturated *Hilo* soil, surface sample, 0–25 cm. [S.A. El-Swaify, in "Structural changes in tropical soils due to anions in irrigation waters," from *Soil Science*, Vol. 37: 956–958, © by Williams & Wilkins, 1973.]

sion of soil particles increases with pH (Arora and Coleman, 1979; Gupta et al., 1984; Suarez et al., 1984). In a column study with three mineralogically different soil types, Rhoades and colleagues examined the effect of pH on hydraulic conductivity at fixed electrolyte concentration and sodicity levels (Figure 9). The results make it clear that (1) greater reductions in the relative saturated hydraulic conductivity occurred at high SAR and (2) increasing pH had a significant adverse effect on the relative *K* of montmorillonitic Bonsall and kaolinitic Fallbrook soils. In the case of the vermiculitic Arlington soil with low organic carbon and higher silt content, pH changes did not cause large differences in relative *K* except at high SAR and lower levels of electrolytes. Changes in pH affect the charge on clays and organic matter and the surface charge of variable-charge minerals of iron and aluminum oxides present in alkali soils. Charge reversal of kaolinite and oxides of Fe and Al at around pH 9 (Schofield and Samson, 1954), and edge-to-face bonding, as well as bonding of oxides to negative clay surfaces at low pH values (Van Olphen, 1977), have considerable influence on dispersion and hydraulic conductivity. For similar reasons, Gupta et al. (1984) cautioned against the use of organic manures in soils *undergoing* sodication process and not preceded by amendment. Surarez et al. (1984) concluded that the adverse effect of high exchangeable sodium on particle dispersion and hydraulic conductivity is magnified by high pH of soil and water, necessitating its consideration along with previous concepts of SAR and electrolyte concentration for assigning threshold flocculation values. During reclamation of alkali soils, all three parameters—sodicity, electrolyte concentation, and pH—are in a state of flux and influence simultaneously the permeability of soils to water. For such situations Evangelou and Phillips (1987) suggested methods to predict changes in ionic preferences of the soil complex [K_{ex_v}, Equation (25)]. The pH of simultaneous clay exchange

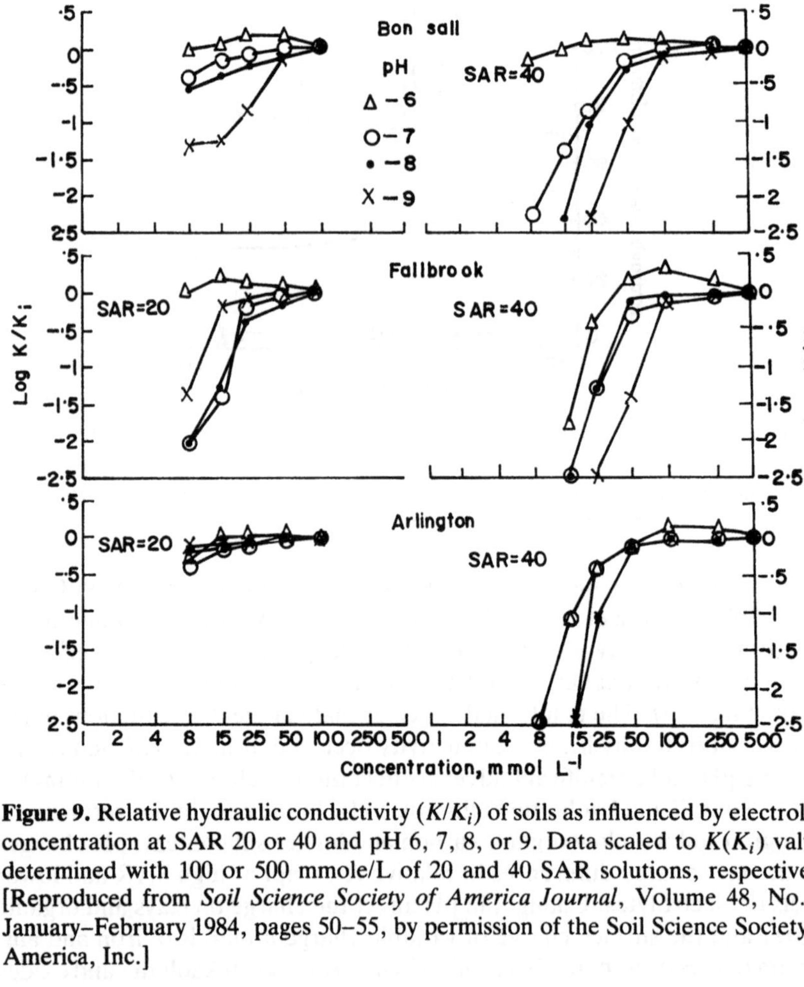

Figure 9. Relative hydraulic conductivity (K/K_i) of soils as influenced by electrolyte concentration at SAR 20 or 40 and pH 6, 7, 8, or 9. Data scaled to $K(K_i)$ values determined with 100 or 500 mmole/L of 20 and 40 SAR solutions, respectively. [Reproduced from *Soil Science Society of America Journal*, Volume 48, No. 1, January–February 1984, pages 50–55, by permission of the Soil Science Society of America, Inc.]

and calcite equilibrium can be defined by the Vanselow expression using Equation (28) (Thorstenson et al., 1979):

Reaction	Equilibrium constant	
$CaCO_3 = Ca^{2+} + CO_3^{2-}$	$\log K_c$	(24)
$Ca^{2+} + 2Na_{ex} = 2Na^+ + Ca_{ex}$	$\log K_{ex_v}$	(25)
$H^+ + CO_3^{2-} = HCO_3^-$	$\log K_2$	(26)
$CaCO_3 + H^+ + 2Na_{ex} = Ca_{ex} + 2Na^+ + HCO_3^-$	$\log K_{eq}$	(27)

The pH dependence of the system can be written in its final form as

$$pH = \log K_c + \log\left(\frac{K_{ex_v}}{K_2}\right) + \log\left[\frac{XNa_{ex}^2}{XCa_{ex} \cdot a^2Na^+ \cdot aHCO_3^-}\right] \quad (28)$$

This relation suggests that the pH of the aqueous phase will change by nearly a full unit if the mole fraction of XCa_{ex} varies 0.4 to 0.1 for a given K_{ex_v}, Na^+, and HCO_3^+ concentration. Since the pH of the aqueous phase in clay exchange and calcite equilibrium incorporates the effect of sodicity changes during reclamation, it is to be expected that soil pH provided a good basis for the gypsum response reported earlier (Loveday and Pyle, 1973; Loveday, 1974; Abrol et al., 1973) and seemed to relate with the changes in hydraulic conductivity during reclamation (Gupta et al., 1988).

IV. Reclaiming Alkali Soils

Reclamation of alkali soils require removal of part or most of the exchangeable sodium and its replacement by the more favorable calcium ions in the root zone. This can be accomplished in many ways, best dictated by local conditions, available resources, and the kinds of crops to be grown during reclamation. If the farmers can spend little for reclamation and are willing to wait for many years for good crop yields, reclamation can be accomplished, simply, by long-continued irrigation/cropping, including rice in the cropping sequence, together with incorporation of large quantities of organic manures. Incorporating organic materials such as rice husks and cereal straw has helped in increasing the water intake rates of alkali soils. For example, the effect of low-nitrogen organic materials on the infiltration rate was found to be superior and lasted longer as compared with legume straw (Williams, 1966). For quick results, however, cropping must be preceded by the application of chemical amendment followed by leaching for removal of soluble salts and other reaction products of an amendment with the alkali soils.

A. Types of Amendments

Amendments are materials that directly or indirectly, through chemical or microbial action, furnish divalent cations (usually Ca^{2+}) for replacement of exchangeable Na. Chemical amendments (Table 4) for reclaiming alkali have been broadly grouped into three categories:

1. Soluble calcium salts: $CaSO_4 \cdot 2H_2O$; $CaCl_2$, phospho-gypsum
2. Sparingly soluble calcium salts: Calcite, $CaCO_3$
3. Acids or acid formers: H_2SO_4, iron and aluminum sulfates, lime-sulfur and pyrite, etc.

B. Determining Amendment Needs

The quantity of an amendment required for an alkali soil depends on exchangeable Na content to be replaced, exchange efficiency, and depth of soil to be reclaimed. This quantity is often referred to as the gypsum requirement of soils and is usually determined by equilbrating a gypsum

Table 4. Chemical properties of various amendments

Amendment	Chemical composition	Physical description	Amt. equiv. to 1 kg of chemically pure: Gypsum (kg)	S (kg)
Gypsum	$CaSO_4$; $2H_20$	White mineral	1.0	5.38
Sulfur	S_8	Yellow element	0.19	1.00
Sulfuric acid	H_2SO_4	Corrosive liquid	0.57	3.06
Calcium carbonate	$CaCO_3$	White mineral	0.58	3.13
Calcium chloride	$CaCl_2 \cdot 2H_2$	White salt	0.85	4.59
Ferrous sulfate	$FeSO_4 \cdot 7H_2O$	Blue-green salt	1.61	8.69
Ferric sulfate	$Fe_2(SO_4)_3 \cdot 9H_2O$	Yellow-brown salt	1.09	5.85
Aluminum sulfate	$Al_2(SO_4)_3 \cdot 18H_2O$		1.29	6.94
Pyrite (30% S)	FeS_2	Yellow-back mineral	0.63	1.87

solution of known strength with the sodic soils and estimating the calcium deficiency (U.S. Salinity Laboratory Staff, 1954). This determines the Ca^{2+} required to replace exchange Na^+ plus that required to neutralize the soluble carbonates (Abrol and Dahiya, 1974) in soils.

Abrol et al. (1975) also proposed a modified procedure of estimating the gypsum requirement of alkali soils. In the modified procedure, soil samples are first leached with alcohol to remove soluble carbonates and then equilibrated with saturated gupsum solutions. Abrol et al. (1973) reviewed the available data on the mineralogy, texture, charge capacity, sodicity, and pH of a broad specturm of soils of the Indo-Gangetic plains and developed a graph to approximate the gypsum requirement from the pH of the saturated soil paste (Figure 10). Locations of the lines were determined subjectively for the predominantly micaceous soils cultivated for growing rice-wheat in rotation. In the field, the graphical method of determining gypsum requirement has proved quite helpful.

The amount of gypsum (GR) needed to replace exchangeable sodium and achieve desired sodicity per unit of land area can be calculated using the equation

$$\text{GR, (Mg ha}^{-1}) = 0.172 \times 104 \times \rho_b \times ds \times n \times (\text{CEC})\,(E_{\text{Na}_i} - E_{\text{Na}_f}) \quad (29)$$

where

ρ_b = bulk density (Mg m^{-3})

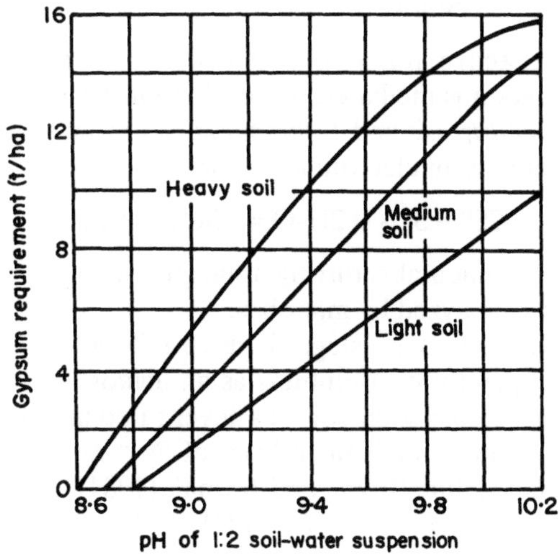

Figure 10. Graphic approximation of gypsum requirement of the illitic soils. [After Abrol et al. (1973).]

Mg = metric tons

ds = depth of soil (m)

CEC = cation exchange capacity (molc/Mg soil)

E_{Nai}, E_{naf} = initial and desired final fractions of exchangeable Na.

The Na-Ca exchange inefficiency factor (n), a measure of gypsum loss, is usually taken as 1.20 (U.S. Salinity Laboratory Staff, 1954). Equivalent quantities of different amendments needed in reference to chemically pure $CaSO_4 \cdot 2H_2O$ or elemental sulfur can be calculated using conversion factors given in Table 2.

All amendments, when used under appropriate soil conditions, have a common characteristic: they supply soluble Ca. Acid amendments react immediately with lime, naturally present in alkali soils, to produce soluble calcium. Materials such as sulfur or iron pyrite must first be oxidized to produce sulfuric acid, which in turn produces calcium sulfate, and consequently are less effective than gypsum or sulfuric acid. It is clear now that sulfuric acid is more effective than gypsum in reclaiming calcareous sodic soils (Yahia et al., 1975; Prather et al., 1978; Overstreet et al., 1951; Petrosian and Tchitchian, 1969).

Iron and aluminum sulfates ($FeSO_4 \cdot 7H_2O$ and $Al_2(SO_4)_3 \cdot 18H_2O$) hydrolyze in water to form sulfuric acid, which acts as decribed above. Oxides of iron and aluminum act as polycations, linking clay particles and promoting flocculation and soil structure stabilization.

C. Oxidation of Pyrite/Sulfur

For sulfur to be effective, it is absolutely essential to ensure its rapid oxidation. Sulfur oxidation is mediated by aerobic *Ihiobacilli*, group of chemoautotrophs. *I. Denitrifcans* and *I. thioparus* can grow anaerobically in the presence of nitrates as an electron acceptor:

$$2S^0 + 3O_2 + 2H_2O = 2SO_4^{2-} + 2H^+ \quad (30)$$

The oxidation of elemental sulfur by atmospheric oxygen is an extremely slow process until catalyzed by microbes.

Pyrite oxidation in soils is a sluggish process (McGeorge and Breazeale, 1955), and consequently its usefulness as an acidogenous amendment for reclamation of calcareous sodic soil was never rated highly. The first step in pyrite (FeS_2) decomposition is the release of Fe^{2+} and oxidation of pyrite-sulfur to elemental sulfur. Brown and Jurinak (1984) measured zero-order rate constants for formation of acid (pH), Fe^{2+}, and sulfur compounds during pyrite oxidation:

$$FeS_2 = \tfrac{1}{2}O_2 + 2H^+ = Fe^{2+}2S^0 + H_2O \quad (31)$$

$$Fe^{2+} + \tfrac{1}{4}O_2 + H^+ = Fe^{3+} + \tfrac{1}{2}H_2O \quad (32)$$

$$Fe^3 + 3H_2O = Fe\ (OH)_3 + 3H^+ \quad (33)$$

The chemical oxidation of Fe^{2+} to Fe^{3+} is an extremely slow process [Equation (32)] until accentuated by microbes. The optimum pH range for oxidation of pyrite iron (II) due to an autocatalytic mechanism has been indicated to be between 7 and 8 by Brown and Jurinak (1984). Ferric iron hydrolyzes and precipitates rapidly as ferric hydroxide at high pH [Equation (33)]. The Fe^{3+} iron can also be rapidly reduced to Fe^{2+} in the presence of pyrite/S^0:

$$FES_2 + 14Fe^{3+} + 8H_2O = 15Fe^{2+} + 2SO_4^{2-} + 16H^+ \quad (34)$$

The above schematic reactions suggest that once the sequence (Figure 11) for the formation and reduction of Fe^{3+} iron is well established, oxygenation of pyrite is of little consequence and can be oxidized by dissolved Fe^{3+} iron acting as a catalyst. It seems that in the total process of pyrite oxidation, the rate-limiting step is oxidation of Fe^{2+} by atmospheric oxygen. Microorganisms also influence the overall rate-determining step (Silverman, 1967; Stumm-Zollinger, 1972). The rate of oxygenation of Fe^{2+} in solution of pH 5 is found to be second-order with respect to hydroxy ions, therefore the rate improves 100-fold for a unit increase in pH (Stumm and Morgan, 1970). Oxygenation kinetics are also influenced by effects of ionic strengths and composition of the medium. Whereas Cu^{2+} and Co^{2+} improve oxygenation of Fe^{2+}, anions such as Cl^-, SO_4^{2-}, and phosphates significantly slow down the rates (Harmsen et al., 1954; Murakami, 1968; Quispel et al., 1952; Sung and Stumm, 1980). Chloride concentration

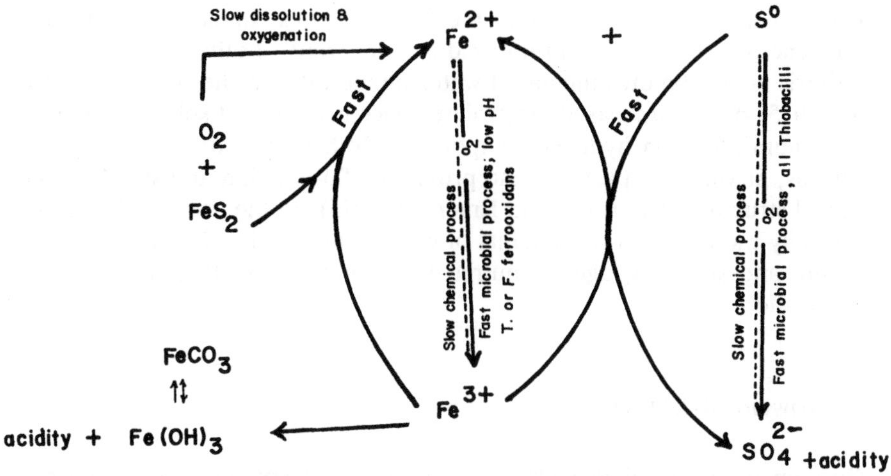

Figure 11. Schematic pathways for oxidation of pyrites in soils. [Adapted from Stumm and Morgan, *Aquatic Chemistry*, © 1970, John Wiley & Sons, Inc.]

>5 × 10 mol completely inhibits the iron-oxidizing *Ihiobacilli* (Silverman, 1967). The overall stoichiometry for complete pyrite oxidation can be represented schematically as

$$FeS_2 + \tfrac{15}{4}O_2 + {}^{5/2}H_2O = FeOOH + 2SO_4^{2-} + 4H^+ \qquad (35)$$

The above description suggests that dissolution of 1 mol of iron pyrite leads to the release of 4 equivalents of acidity: 2 equivalents each from oxidation of S and Fe^{2+}. Elemental sulfur can be oxidized by Fe^{3+} resulting from oxgenation of Fe^{2+} through abilological or biological means. In alkali soils, oxidation of S may be a rapid reaction except for reasons of formation of a protective coating of ferric oxide on the iron pyrite particle surface (Vlek et al., 1974). whereas *Ihiobacillus ferroxidans* oxidizes Fe^{2+} and S and promotes oxidation of pyrite (Unz and Lundgren, 1961), ascribing responsibility for acidification to *I. thiooxidans* requires formation of S (Temple and Delchamps, 1953). It has been reported that alkali soils lack *Ihiobacilli* (Rupela and Tauro, 1973) and consequently need inocculation with *I. thiooxidans* alone or in combination with *I. ferroxidans* (Venkatakrishanan and Abrol, 1981). They showed in in vitro studies that inoculation of pyrite with *I. thiooxidans alone or in combination with I. ferroxidans* proved inferior to inocculation with the latter microbe alone. The reasons for the poor performance of the mixed inoculum are not understood.

A large variety of other acidic materials, such ferric, ferrous, and aluminum sulfates, etc., has been tested for their relative effectiveness (Overstreet et al., 1951; Petrosian and Tchitchain, 1969; Chand et al., 1977; Miyamoto et al., 1975; Yahia et al., 1975; Prather et al., 1978). Results of field experiments carried out in the United States, India, and Armenia are generally in favor of sulfuric acid over a veriety of other amendments. Sulfuric acid promotes increased water intake rates in the soils (Yahia et al., 1975) due to increased electrolyte concentration and solubilization of aluminum and iron hydroxy compounds. Both these oxides act as polycations, linking clay particles and promoting flocculation and stabilization of soil structure (El-Swaify and Emerson, 1975). In the use of acidulants, specially the acids, as reclaimants of calcareous alkali soils, the corrosive nature and special gadgets required for handling limit their use in field application.

D. Gypsum Dissolution

Gypsum usage in alkali soils is by far the most common of all the reclaimants. The dissolution-exchange reactions, represented as

$$CaSO_4 \cdot 2H_2O(s) = CaSO_4(aq.)$$
$$CaSO_4(aq.) + 2Na_x = CaX_2 + Na_2SO_4 \text{ leachable} \qquad (36)$$

appear simple, but the actual process involves simultaneous mineral dissolution, cation exchange, and solute and water movement. The dissolution-exchange reactions remove Ca from the solution and allow additional gypsum dissolution as a linear function of exchangeable sodium content of the soils (Abrol et al., 1979; Hira and Singh, 1980; Oster and Frenkel, 1980). Dissolution also increases with SAR, ionic strength, ion-pair association, and C_{Mg}/C_{Ca} ratio (Tanji, 1969; Oster and Rhoades, 1975; Oster, 1982). Factors such as contact time, depth of mixing, charge capacity of the gypsum, particle-size distribution, gypsum content, and flow velocity also influence dissolution rates of gypsum (Kemper et al., 1975; Barton and Wilde, 1971; Keren and Shainberg, 1981; Chawla and Abrol, 1982; Keren and O'Connor, 1982; Oster, 1982; gupta et al., 1985d). With greater depth and distribution of gypsum in the sodic soil profile (low gypsum: soil ratio) the effective solubility is enhanced due to large sink capacity (Oster, 1982; Gupta et al., 1985d). At constant gypsum: soil ratio (1% gypsum rate > gypsum requirement), depth of mixing had no effect due to compensation of the increase in gypsum surface per unit volume of soil by decrease in contact time as a result of shallow mixing (Figure 12). For rapid sustained dissolution of gypsum and to avoid reduction by formation of $CaCO_3$ coatings (Keren and Kauschansky, 1981), it has been recommended to have a mix of gypsum particle sizes less than 2 mm (Chawla and Abrol, 1982). Gypsum of mixed particle sizes has the dual benefit of initial fast dissolution rates at high sodicity followed by longer sustained release of calcium.

The dissolution rate of gypsum is controlled by film diffusion and is a function of the difference between the solution concentration at saturation,

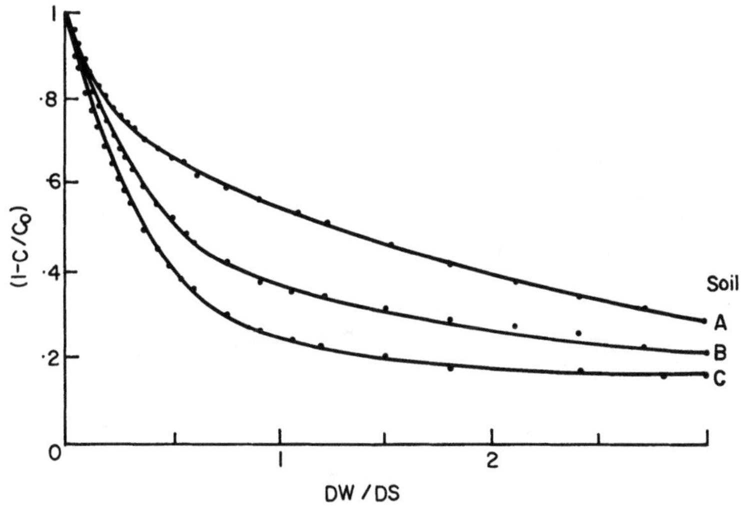

Figure 12. Scaled gypsum dissolution ($1 - C/C_0$) as a function of depth of drainage water per unit depth of soil. Data points represent three different sodicity soils (A < B < C) containing admixture of gypsum (1.0% gypsum $\geq E_{Na}$), packed to 1- and 3-cm depths in small- and big-diameter columns. [R.K. Gupta, C.P. Singh, and I.P. Abrol, in "Dissolution of gypsum in alkali soils," from *Soil Science*, Vol. 140: 382–386, © by Williams & Wilkins, 1985.]

C_s and solution concentration at a particular time, C. The mathematical relation is decribed by Kemper et al. (1975) as

$$\frac{dc}{dt} = K(C_s - C) \tag{37}$$

The solution concentration of calcium or sulfate, C, depends on the amount of gypsum, the soil water content, and the rate of Ca^{2+} removal through exchange and solute movement processes. It follows from the above description that exchange-induced dissolution rates will be fast initially, decreasing to levels equal to or less than the solubility of gypsum in water. The dissolution rate coefficient, K, which represents the first-order reaction kinetics rate constant, depends on surface area of gypsum of mixed particle sizes, gypsum content, and solution flow velocity (Kemper et al., 1975; Keren and O'Conner, 1982). Increasing the flow velocity from 12 to 673 mm h^{-1}, Keren and O'Connor concluded that despite an increase in dissolution rate coefficient, the dissolution rate decreased. This was ascribed to reduction in contact time between gypsum fragments and the flowing solution. Gupta et al. (1985c) further demonstrated that the effect of depth of mixing, contact time, sodicity, and gypsum content on dissolution rates can be integrated to yield a single curve (Figure 13) if gypsum is applied at rates equivalent to the gypsum requirement or the exchangeable sodium content of the soil. Results of the experiment presented in Figure 13 show that leaching water dissolves an identical fraction of the equivalent

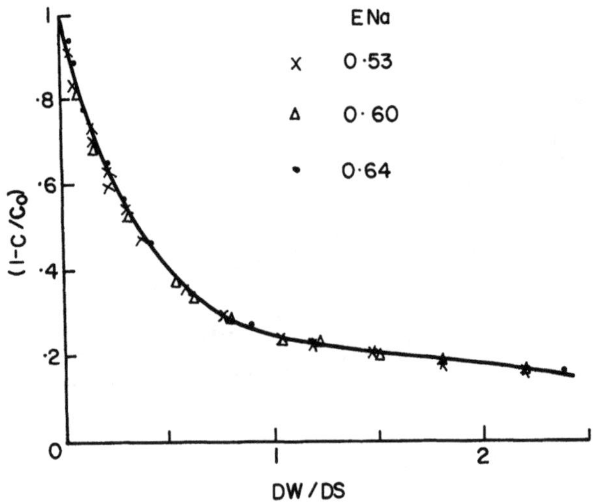

Figure 13. Scaled gypsum dissolution ($1 - C/C_0$) as a function of depth of drainage water per unit depth of soil. Data points represent three different sodicity soils containing admixture of gypsum at rates equivalent to E_{Na} packed to 1- and 3-cm depths in small and big columns. [R.K. Gupta, C.P. Singh, and I.P. Abrol, in "Dissolution of gypsum in alkali soils," from *Soil Science*, Vol. 140: 382–386, © by Williams & Wilkins, 1985.]

gypsum requirements, incorporated in the selected depths of varied sodicity soils. These results demonstrate that electrolyte concentration of the leaching solution increases with the sodicity of the soil. As a consequence, they observed near-similar percolation rates for the three different sodicity soils. Increased electrolyte concentration in the leaching water was due to an increase in the effective solubility of gypsum in the presence of an exchange phase (Abrol et al., 1979; Hira and Singh, 1980; Oster and Frenkel, 1980). Thus, these experiments demonstrate that gypsum under appropriate conditions can serve the dual purpose of a high as well as a low electrolyte-furnishing amendment. This point will be discussed further in a later section.

Equilibrium models that have been field tested have proved powerful tools for predicting sodicity profiles and water and amendment requirements in good agreement with observed values (Dutt et al., 1972; Oster and Frenkel, 1980; Tanji et al., 1972; Tanji and Deverel, 1984). It is clear from the literature that except for large flow rates, the assumption of chemical equilibrium is probably valid when soil water velocity is lower than the rate of diffusion or when sufficient surface area of gypsum is available to reach equilibrium, during leaching (Keren and O'Connor, 1982). It has been demonstrated that the dissolution reaction of gypsum is dependent on contact time and solution velocity and is not controlled by the solubility product relationship as was assumed in the equilibrium models (Tanji et al., 1972; Oster and Frenkel, 1980). It seems that the assumption of chemical equilibrium for dissolution of gypsum may be met for sodic soil conditions because of their low flow rates and sufficient surface area of gypsum of particle size less than 2 mm.

Integration of the gypsum dissolution rate Equation (37) yields Equation (38):

$$\ln(1 - \frac{C}{C_s}) = -Kt \tag{38}$$

A plot of $\ln(1 - C/C_s)$ versus time (t) should yield a straight line (Keren and Shainberg, 1981) if dissolution of gypsum is a first-order kinetic reaction with a constant dissolution rate coefficient. Keren and O'Connor showed that the term $\ln(1 - C/C_s)$ is empirically related linearly to the square root of time,

$$\ln(1 - \frac{C}{C_s}) = \alpha t^{1/2} + \beta \tag{39}$$

where α and β represent slope and intercept, respectively. A relation between t, taken as the resident time (t_b) of a flowing solution in a gypsum bed of length L, with porosity p and velocity V, was introduced by Kemper et al. (1975) as

$$t_b = \frac{Lp}{V}$$

which, when combined with Equations (39) and (38), yields

$$\mathrm{K} = (\frac{V}{L})^{1/2} [\alpha + (\frac{V}{L})^{1/2}\beta] \tag{40}$$

Equation (40) provides a way to compute the dissolution rate coefficient at various flow rates. Gobran and Miyamoto (1985) concluded that whereas the first-order reaction equation (37) is suitable up to about 50% of gypsum solubility, a second-order reaction can describe gypsum dissolution processes in the entire range.

E. Lime and Alkali Soil Reclamation

Addition of lime to alkali soils has generally been considered of doubtful value. It is usually present in many alkali soils as a natural secondary mineral. Armhein et al. (1985) concluded that dissolution kinetics of $CaCO_3$ in soil is not a simple diffusion-controlled, first-order reaction. Dissolution of $CaCO_3$ is dependent on factors such as calcite surface area, soil water composition, the chemical nature of exchangers, the temperature regime, the partial pressure of CO_2, etc.

Barren alkali soils, devoid of any vegetation, have been observed to contain little readily oxidizable carbon for support of microbial activity, needed for production of carbon dioxide and mobilization of $CaCO_3$. Application of gypsum followed by cropping increased the urease (UA) and dehydrogenase (DHA) activity, measures of biological activity, by about threefold (Rao and Ghai, 1985). Growing of trees, 12 years after, markedly improved the biological activity and reclaimed the alkali soil as compared with continuously cropped and grassed plots (Table 5). Reclamation was thus apparently through more CO_2 production and mobilization of $CaCO_3$, in favorable moisture and temperature regimes that prevailed under a forest canopy. The calcite content during the corresponding 12-year period decreased by 1.0%, 1.5%, and nearly 2.0% with cereal cropping, grasses, and agroforestry land use, respectively. Displacement of exchangeable Na by Ca from $CaCO_3$, though not very effective in soils having pH > 8.0, is greatly affected by factors listed previously (Turner and Clark, 1956; Cole 1957; Yaalon, 1958). Addition of organic manures, submergence (ponded water conditions), and growing rice help reduce soil pH and Na saturation and improve water intake rates (see Figure 11) of alkali soils (McNeal et al., 1966a; Chhabra and Abrol, 1977; Gupta et al., 1988). Using the equilibrium model and piston movement of soil solutions it has been shown that the presence of lime in sodic soils amended with gypsum reduces gypsum requirement by 9% for P_{CO2} of 10 kPa 10% CO_2 (Oster and Frenkel, 1980; Hoffman, 1980). Though there is no reliable estimate of the amount of calcite dissolved in alkali soils during rice culture, it is fairly clear that the benefit of submergence could be enhanced by emphasizing factors that are responsible for greater CO_2 production.

The presence of lime in soils has been suggested to stabilize the soil aggregates and prevent clay dispersion (Rimmer and Greenland, 1976; Emerson, 1983; Gupta et al., 1984). In less sodic soils (ESP < 20), lime prevents clay dispersion due to an increase in electrolyte concentration through the dissolution mechanism (Shainberg et al., 1981a; Shainberg and Gal, 1982). At high ESP, calcite does not provide enough electrolytes in soil solution to prevent the dispersive effect of exchangeable Na (Shainberg and Gal, 1982), though it reduces clay dispersion considerably (Gupta et al., 1984). Since the specific surface area of calcite is highly variable for soils (Holford and Mattingley, 1975) and P_{CO2} content is important in the regulation of the kinetics of dissolution of calcite (Tanji and Whittig, 1985), it seems difficult to ascertain a threshold electrolyte concentration for flocculation of calcareous sodic soils. Coatings of sesquioxide alter the surface reactions of $CaCO_3$, further complicating the determination of flocculation concentrations required to prevent dispersion of calcareous sodic soils (Rengasamy et al., 1984).

F. Role of Rice Culture and Organic Manures in Reclamation of Alkali Soils

Rice culture enjoys a favored place in any cropping sequence recommended for adoption during reclamation of alkali soils. Rice is preferred to be grown owing to its high tolerance to soil sodicity (Abrol and Bhumbla,

Table 5. Physicochemical properties and enzymatic activities of alkali soils (0–30 cm) after 12 years of different land use

Land use	pH	ESP	$CaCO_3$ (%)	Organic Carbon (%)	Kjeldahl N (kg ha^{-1})	Available N (kg ha^{-1})	Urea hydrolyzed (mu g^{-1} soil/h) urease	TPF[a] mu g^{-1} soil/day dehydrogenase
Original soil	10.10	87.4	2.11	0.11	0.03	21.00	3.90	10.60
Gypsum + cropping	7.95	18.3	1.07	0.28	0.05	82.25	9.95	35.45
Grasses	8.00	28.4	0.52	0.53	0.06	91.00	22.80	83.57
Trees	7.00	2.3	0.17	1.47	0.14	157.55	55.55	110.00

[a]Trihydrophenyl formazan.

Source: Rao and Ghai (1985).

1979). The requirement of ponded water conditions for optimum rice growth promotes buildup of P_{CO2} and leaching of salts resulting from exchange of sodium by calcium. Several research investigations point out that the pace of reclamation of alkali soils, as evidenced from increased permeability and reductions in sodicity, is considerably enhanced under rice culture. Increased hydraulic conductivity resulting from rice culture has been attributed either to the physical action of plant roots in facilitating water movement (Goertzen and Bower, 1958; Chhabra and Abrol, 1977) or to the removal of entrapped air from the larger conducting pores (McNeal et al., 1966b). Results of a column study show that more carbon dioxide in rice culture causes reduction in soil pH and exchangeable Na content through mobilization of carbonates (Gupta et al., 1988). The physical presence of active roots influences chemical reclamation of alkali soils more through CO_2 production and mobilization of calcium carbonate, while at the same time seems to yield a certain hydraulic conductivity. In these studies, entrapment effects were obviated by flushing the soil columns with carbon dioxide. The influence of FYM on HC, however, was observed as a chemical action of CO_2 on calcite naturally present in alkali soil. It was earlier concluded that the benefits of flooding alkali soils can be accentuated through raising P_{CO2} by incorporating organic manures and growing rice (Ponnamperuma, 1972). From the results of field experiments conducted at the Central Soil Salinity Research Institute, Karnal, India, it can be concluded that the process of reclamation of alkali soils should begin in summer. This is because the CO_2 content in soil air is greater in summer than in winter (Tanji and Whittig, 1985), a consequence of more microbial activity at higher temperatures. Greater carbon dioxide production, under appropriate temprature and soil moisture regimes, helps in greater mobilization of soil calcium carbonate and removal of Na ions from the exchange phase.

G. Amendment Application Methods

The gypsum requirement equation (29) implies that reduction in sodicity and soil depth to be reclaimed depend on the types of crops to be grown in a given cropping sequence. Therefore, the concept of an appropriate profile of sodicity for which to aim in reclamation seems to depend on the nature of the crop sequence and time scale. The experimental evidence for specification of appropriate sodicity profiles for different crops is only limited (Miyamoto et al., 1975).

The researchers in India (Central Soil Salinity Research Institute, 1979) have indicated that alkali soils can be brought under cultivation with a minimum of inputs by including rice in the cropping sequence. The sodicity profile resulting upon amending the surface 15 cm of soil has proved quite efficient for the rice-wheat crop sequence on experiment stations and on farmers' fields. Field studies and backup laboratory investigations have shown that mixing the gypsum requirement in the desired reclamation depth proved more beneficial than mixing with deeper depths. Khosla et al. (1973) found that increasing the depth of mixing gypsum reduces its reclamation efficiency due to a dilution effect (large soil: gypsum ratio). In areas with surface crusting and infiltration problems, shallow mixing of the gypsum in the surface layer is considered best (Loveday, 1984). Mechanical disturbance of the profile and incorporating gypsum in the target zone is perhaps the best practice for reclamation of solonetzic soils with impervious sodic B horizons (Rasmussen et al., 1972; Toogood and Cairns, 1978).

Acidents such as pyrites and sulfur, which must first be oxidized by soil microorganisms before they are effective, should be surface-applied or mixed in the surface few centimeters of soils. Ponded water or saturated soil moisture conditions are unfavorable to oxidation of these materials and would further slow down their chemical action. Sulfuric acid, according to Miyamoto et al. (1975) should be surface-applied directly (sprinkled) for reasons of better distribution, less destruction of soil aggregates, and more efficient leaching of salts. Elemental S can be applied as a water suspension containing 55% to 60% S using conventional fluid fertilizer equipment (Thorup, 1972).

H. Reclamation Efficiency of Chemical Methods

Besides the external factors, such as amount of amendment, depth of mixing, flow rates, etc., physical properties of the amendment, such as particle size, also influence reclamation efficiency. Chawla and Abrol (1982) have shown that better results can be achieved if gypsum ground to pass through a 2-mm sieve is used for reclamation.

Use of high-electrolyte amendments such as $CaCl_2 \cdot 2H_2O$ has often been advocated to speed up reclamation of alkali soils. High-electrolyte percolating solutions rich in calcium increase soil permeability but decrease the percentage of applied Ca that exchanges for adsorbed Na (exchange efficiency) due to continuous reduction in the sodicity of the soils being reclaimed. The efficiency of exchange is usually much greater at high ESP values (Chaudhry and Warkentin, 1968) and declines to about 20% to 40% at ESP levels below 15 (Greene and Ford, 1983; Manin et al., 1982; Loveday, 1976). The low efficiency of exchange is due to (1) the slowness of exchange of Na inside the structural elements, and (2) the fact that part of the added calcium exchanges with Mg. At low sodicity, amendments such as gypsum (Doering and Willis, 1975; Rhoades, 1982), which produce leaching solutions with low electrolyte, likely prove more efficient. Rhoades and his collaborators (Prather et al., 1978) integrated the beneficial aspects of the two types of amendments into a dual amendment concept. In the dual amendment concept of alkali soils reclamation, for three-fourths of the amendment requirement, gypsum is mixed in the top soil depth; the remaining one-fourth of the amendment need is applied to the surface as a highly soluble calcium salt such as $CaCl_2 \cdot 2H_2O$. They found dual amendment application more effective in increasing soil permeability and speeding reclamation than gypsum incorporated alone. When gypsum was added at rates equivalent to the exchangeable Na content of the reclamation depth and uniformly mixed, gypsum was observed to be as effective as dual amendment (Gupta and Singh, 1988). Soil column profiles of sodicity (Figure 14) and pH and hydraulic conductivity as observed during leaching and at the end of the experiment evidently remonstrate with the dual amendment concept except when deeper soil layers are to be reclaimed. Results of the two studies differed mainly because in the experimental setup of Prather, three-fourths of the amendment gypsum need of 55-cm-long soil column was incorporated into the top 5 cm of soil depth only. Leaching losses of calcium from the surface-concentrated $CaCl_2$ versus internally incorporated $CaCl_2$ and gypsum followed the order: $CaCl_2$ mixed > $CaCl_2$ − surface > $CaSO_4$ mixed + $CaCl_2 \approx CaSO_4$ mixed (Gupta and Singh, 1988). Surface-zone applied $CaCl_2$ is considerably more effective than equimolar amounts incorporated through the entire target zone. The higher efficiency of gypsum (mixed) compared with $CaCl_2$ was due to its continued solvation and exchange in many more effective steps over more parcels of water flux and not being all swept forward with the wetting front. Due to the high solubility of $CaCl_2$, surface placement results in strong enough Ca concentration to effect reclamation to a level that is considerably higher than for mixed $CaCl_2$ but less than the reclamation effected by gypsum. Results of soil column studies have shown that gypsum was better than $CaCl_2$ in maintaining a high hydraulic conductivity for a chemically stable soil that did not release salt into the soil solution (Shainberg et al., 1982). In the experimental studies by Gupta and Singh, calculated reclamation efficiencies of 84%, 71%, and 59% for gypsum mixed and dual amendment, $CaCl_2$ surface, and mixed placement treatments, respectively. Keren and O'Connor (1982) reported a reclamation efficiency of 81% for gypsum at a soil water flow velocity of 1.16 cm ha^{-1}. The reclamation efficiency of $CaCl_2$ applied as solution had earlier been reported to range from 77% to 63% (Alperovitch and Shainberg, 1973). Reclamation with $CaCl_2$ is more efficient with surface application than mixing it in the target zone (Magdoff and Bresler, 1973; Gupta and Singh, 1988).

I. Water Requirement for Reclamation

Predicting the water requirement for gypsum dissolution and leaching of soluble salts to achieve a desired reduction in sodicity has been a focal point of research for those concerned with reclamation of alkali soils. For predicting water requirement for gypsum dissolution and to leach reaction products, Hira et al. (1981) developed an equation:

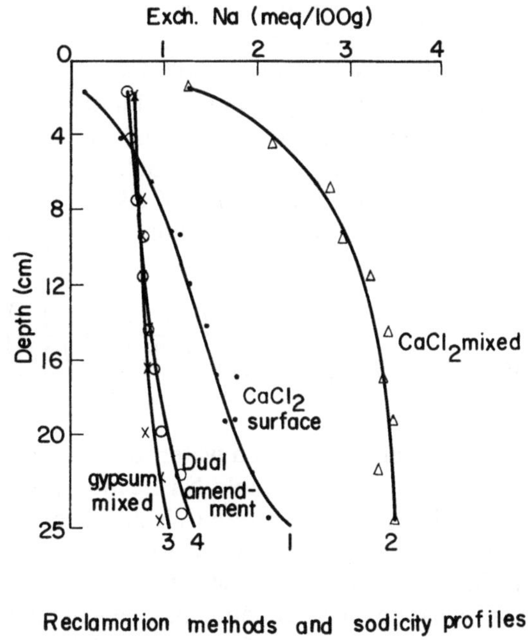

Figure 14. Reclamation methods and sodicity profiles of soil columns containing admixture of amendments, $CaSO_4$ and $CaCl_2$ furnishing equimolar concentrations of Ca, equivalent to the exchangeable Na contents of the soils. [R.K. Gupta and C.P. Singh, in "A comparative evaluation of reclamation efficiency of surface concentrated versus internally incorporated $CaCl_2$ and gypsum amendments in an alkali soil," from *Soil Science*, Vol. 146: in press; © by Williams & Wilkins, 1988.]

$$Z^{-1/3} = 1 - KE_{Na_i}\left(\frac{2}{3\rho_b D_0 m_0} I\right)^{1/2} \tag{41}$$

where

$Z \quad = \dfrac{m_0 - m}{m_0}$

m = amount of gypsum dissolved (g = equiv.)

m_o = initial amount of gypsum applied per unit surface area of soil

D_o = initial gypsum particle diameter (cm)

ρ_b = bulk density of gypsum (g cm^{-3})

E_{Na_i} = initial exchangeable Na per unit surface area of soil (g − equiv.)

I = depth of irrigation water (cm)

K = empirical constant (cm^{-2})

They calculated the depth of irrigation waters as 6.8 cm, required to dissolve 99% of applied gypsum of particle size < 0.26 mm. To leach the soluble salts from 0 to 30 cm soil depth, an additional 15 cm of water would be required. Since the process of gypsum dissolution, exchange, and salt and water movement occur simultaneously and are in a state of flux, leaching requirement for sodicity reduction should provide a realistic estimate of the water required for reclamation of alkali soils. For predicting water requirement for desired sodicity reduction, the empirical equation of Jury et al. (1979) is of particular interest:

$$\mathrm{PV} = ESF_i - ESF_f + 0.003\,\frac{\left[\left(\frac{1}{ESF_f}\right) - \left(\frac{1}{ESF_i}\right)\right]}{\beta} \tag{42}$$

where PV is the pore volume of applied water for leaching, ESF refers to the initial (i) and final (f) exchangeable Na fraction (exchangeable Na/CEC). The term $\beta = \theta C_i/(\rho_b\ \mathrm{CEC})$, where C_i is the effective concentration of $(Ca^{2+} + Mg^{2+})$ in the leaching solution at 1 pore volume, ρ_b is soil bulk density, and θ is the average volumetric water content.

The difficulty in using the above equation has been in finding a verifiable value for divalent ion concentration, C_i, as inferred from monitoring the drainage. In presence of Na-exchanger phase, gypsum solubility (C_i) after passage of 1.0 PV of drainage has been obtained in the range of 18–22 meq liter^{-1} (Keren and O'Conner, 1982; Gupta and Singh, 1988), depending on the amounts and fineness of added gypsum and flow velocity. Using a near-constant C_i, a value of 22 meq liter^{-1} of Ca^{2+} in the effluents that crossed the bottom plane or the interface between the Na-soil-gypsum layer and the Na-soil layer after 1 PV of drainage, the estimated irrigation water requirement for desodification were found to be somewhat lower than the observed value (Gupta and Singh, 1988).

V. Crop Management in Alkali Soils

A. Alkali Soils and Plant Growth

Alkali soils have poor soil-water-air relations, which adversely affect plant growth and makes them difficult to work with in moist and dry conditions. In alkali soils having high exchangeable sodium percentage >50% and low CO_2 partial pressure <0.01 atmosphere the calcium concentrations are below the limiting value of 0.2 meq liter^{-1} needed for plant growth (Wallace, 1966) and the carbonate values are quite high and interfere with plant growth and nutrition (Cruz-Romero and Coleman, 1974) of some of the essential nutrients. Accumulation of elements in plant parts in toxic amounts result in injury, reduced growth, and even plant mortality. Elements commonly observed in toxic concentrations in alkali soils include sodium, molybdenum, boron, and at times selenium. In certain crops, toxicity of bicarbonate ions has also been reported.

Under field conditions, plant growth is adversely affected by a combination of the above factors, the extent depending on the nature of the crop grown, the level of exchangeable sodium, and the agronomic management.

B. Relative Tolerance of Crops and Grasses to Alkali Soil Environments

The relative performance of various crops in fields at different soil sodicity levels has been compiled from some of the experiments conducted at the Central Soil Salinity Research Institute, Karnal, India. Sodicity tolerance ratings of different crops have been fixed according to crop hazard as inferred by 50% reduction in relative yields.

Choice of crops to be grown during reclamation of alkali soils is a very important consideration for acceptable yields. Growing crops tolerant to sodicity/alkalinity can ensure reasonably good returns in the initial phases of reclamation. The relative tolerance of crops, field tested for alkali soil conditions, has been indicated in Table 6.

Besides the sodicity-tolerant crops, grasses can also be grown on alkali soils. Grasses such as *Diplachne fusca* (Karnal grass), *Chloris gayana* (Rhodes grass), and *Brachiaria mutica* (Para grass) have been reported as highly tolerant to alkali soil environments (Kumar and Abrol, 1986). Relative performance of different grasses to alkali soil conditions is illustrated

Table 6. Relative crop tolerance to alkalinity/sodicity of soils

ESP Range[a]	Crops	References
10–15	Safflower, mash, peas, lentil pigeon-pea, urd bean	Singh et al., 1981; Abrol and Bhumbla, 1979; Singh and Abrol, 1983; Central Soil Salinity Research Institute, 1979.
16–20	Bengal gram, soybeans	Singh and Abrol, Central Soil Salinity Research Institute, 1987
20–25	Groundnut, cowpea, onion, pearl millet	Singh and Abrol, 1985b; Singh et al., 1980
25–30	Linseed, garlic, guar	Singh et al., 1980
30–50	Raya, wheat, sunflower	Chhabra et al., 1979; Singh et al., 1979
50–60	Barley, Sesbania	Abrol and Bhumbla, 1979
60–70	Rice	Abrol and Bhumbla, 1979

[a] Relative crop yields are only 50% of the maximum in respective sodicity ranges.

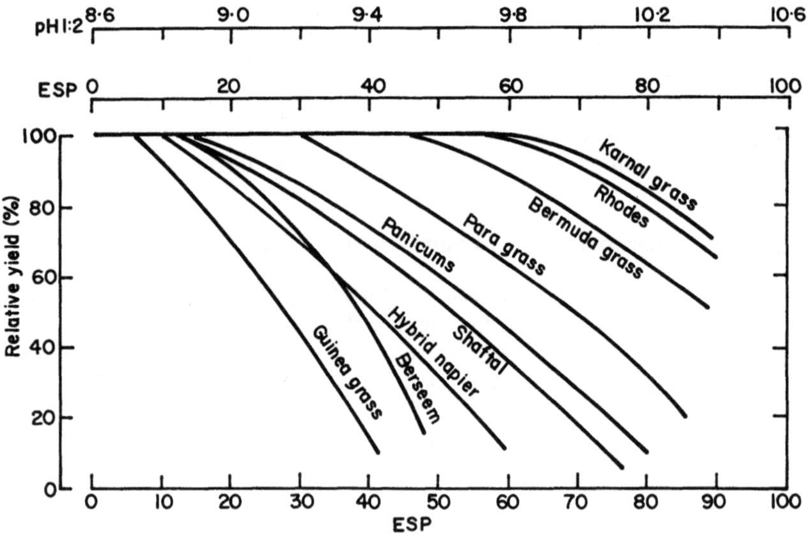

Figure 15. Relative tolerance of grasses in alkali soils. [After Kumar and Abrol (1986).]

in Figure 15. It may be pointed out that whereas Karnal and Para grasses like wet moisture regimes, others need provision of surface drainage to avoid reductions in yields.

C. Nutrient Management in Alkali Soils

1. Nitrogen

Because of low fertility of alkali soils, most crops suffer from inadequate nitrogen supply and need to be supplemented through chemical fertilizers. Urea fertilizer is one of the most widely used nitrogen sources for crops. When urea is hydrolyzed to ammonia and CO_2 by the enzyme urease, the process has the most commonly expressed disadvantage of loss of N via ammonia volatilization. Urease activity has been shown to be positively related to organic carbon, total nitrogen, and silt plus clay contents (Zantua et al., 1977; Rao and Ghai, 1985). Studies have also been conducted to determine the effect of various soil properties on volatilization of ammonia from soil. Increased soil pH has been shown to positively influence NH_3 volatilization from urea and other NH_3 sources (Fenn and Kissel, 1973). Results of some recent studies show that the potential for high NH_3 volatilization from surface-applied urea could be better evaluated from the low hydrogen ion-buffering capacity of a soil rather than the initial soil pH (Reynolds and Wolf, 1987; Freney et al., 1985). Initial soil pH, however, is an important property of alkali soils because of its influence on the infiltration rates of soils and also on the pH and alkalinity of the flood water under rice culture.

Work at the Central Soil Salinity Research Institute, Karnal, India, suggests that a given crop suquence in alkali soils generally needs 20 to 25% more nitrogen than the same cropping pattern on a normal calcareous soil (Central Soil Salinity Research Institute, 1979). Higher nitrogen requirements of crops grown in alkali soil during the reclamation phases was due primarily to more N losses through ammonia volatilization (Rao and Batra, 1983). Nitrification inhibitors and urease inhibitors, except phenylphosphorodiamidate (PPD), had little effect on urea hydrolysis and ammonia volatilization from slow-release and other urea fertilizers during rice culture on alkali soil (Rao, 1987; Rao and Ghai, 1986a, 1986b; Rao and Batra, 1984). In wheat, urease inhibitors such as hydroquinone (HQ) and PPD, however, proved quite effective in terms of grain yield and nitrogen recovery (Rao and Ghai, 1986a). Losses of ammonia were higher at the field moisture range and in unreclaimed soils. In alkali soils N losses have been reported as high as 87% in an upland Berber soil, pH 10.5, of North Sudan (Jewitt, 1942) but may range generally between 30% and 65% depending on the level of applied nitrogen and prevailing soil moisture conditions in alkali soils (International Rice Research Institute, 1976; Bhardwaj and Abrol, 1978; Rao and Batra, 1983).

Ammonia volatilization losses follow first-order reaction rate kinetics (Vlek and Stumpe, 1978; Fleisher and Hagin, 1981) with a half-life (t) of about 62 to 65 at field capacity and about 10.5 days for waterlogged conditions in alkali soils (Rao and Batra, 1983). Rates of volatilization and the amounts of N losses could be substantially reduced (by about 90%) if the nitrogenous fertilizers were placed at a depth of 6 to 7 cm for upland soils (Rao and Batra, 1983) and at about 10 cm depth for rice culture (International Rice Research Institute, 1977; Rao, 1987). N loss could be reduced further if the ammonia pool resulting from urea hydrolysis was partly substituted by green manures or organic manures. Ghai et al. (1988) reported that *Sesbania aculeata*, if grown for 45 days and green manured, enriches alkali soils by contributing 122 kg N ha^{-1} for rice crop.

Rao and Batra (1983) reported that N losses were completed by the fourth day in upland soil, but were least at 40 kg N ha^{-1}. Complementary anions of the ammoniacal fertilizers such as are in AS AN, AC, and DAP did not influence N losses through volatilization in laboratory and field experiments. Application of N through two sources, ammonium sulfate and urea, resulted in similar yields in both rice and wheat (Singh, 1987). Because of the minimum nitrogen losses at 40 kg N ha^{-1}, three or four equal splits of the recommended 120 kg N ha^{-1} for rice and wheat is an efficient practice in alkali soils of the Indo-Gangetic plains (Singh, 1987; Rao and Ghai, 1988b). Rao (1987) found an excellent negative correlation between the response of rice in terms of N uptake and the ammoniacal nitrogen level in the flood water two days after fertilization in alkali soils. This may prove to be a valuable index for future evaluation of fertilizer sources and practices for rice.

2. Phosphorus and Potassium

Alkali soils have been reported to contain high amounts of available phosphorus (Singh and Nijhawan, 1943; Chhabra et al., 1981). Recently, it has been observed that the amounts of water-soluble P increased with alkalinity vis-a-vis soil pH in all major alkali benchmark soil series of the Indo-Gangetic plains (Gupta et al., 1989). Based on measurements in clay suspensions, Pratt and Thorne (1948) had earlier shown that solution P concentrations in alkali soils were high at high sodicity and high pH.

Recently Gupta et al. (1989) have shown that level of electrolyte, sodium saturation, pH, and the presence of alkalinity causing anionic species together or in combination enhances release of P from calcareous sodic soils. Chhabra et al. (1981) reported that Olsen's extractable P decreased in field soils as the level of gypsum application was raised. They also observed that the leaching losses of P in alkali soils could be substantially or completely prevented if gypsum requirement is mixed in the target zone. We observed that strongly alkaline calcareous sodic soils have the bulk of soil P in Ca P (54%) and in residual inorganic P forms (28%). High values of residual P forms are usually anticipated with moderately and strongly weathered soils wherein inorganic forms are randomly dispersed in the matrices of iron oxide coatings and concretions. Soils in the Indo-Gangetic plains have experienced relatively little chemical weathering. Data on depth distribution of P forms in the soil profile and the mineralogy of alluvial soils suggest that residual P forms consist of "included apatite," possibly in plagioclase feldspares, chlorite, and quartz.

Likely presence of included apatite in plagioclase feldspars may accelerate their weathering rate and also the rate and pattern of inorganic transformations in alkali soils. Compared with an uncultivated virgin alkali soil, cultivation decreased the total inorganic P forms and increased the organic P fractions in the soil over the time scale. Of the total inorganic P fractions, reductions were observed primarily in the calcium-bound and residual P forms (Gupta et al., 1989).

Results of a long-term fertility trial in a typical Natrustalf ammended with gypsum indicated that rice-wheat crops grown in sequence would need no P and K fertilization during the initial six years of reclamation (Chhabra, 1985). Response to P fertilization in iron pyrite-treated alkali soils, however, has been observed in coordinated field trials in India, (Coordinated Research Project of ICAR, 1988). In a pot experiment, wherein an alkali soil was used with three phosphatic fertilizer sources (monocalcium phosphate, diammonium phosphate, and nitrophosphate), DAP proved best at low sodicities, ESP <30, but MCP was reported a better fertilizer source for wheat crop at higher sodicity values (Gupta et al., 1985b).

Potassium availability in alkali soils generally has been reported as adequate. Predominance of micaceous minerals in soils of many arid and semiarid regions (Jackson et al., 1948; Sidhu and Gilkes, 1977; Kapoor et al., 1981a,b), Na-K exchange in biotite, and dissolution of muscovite structural units release large amounts of K in alkaline sodium soil environments (Pal, 1985). In the micaceous soils of the Indo-Gangetic plains, crop response to applied potassium fertilizers has not been observed even after nearly 15 years of rice-wheat cropping (Dargan and Chillar, 1973; Chhabra, 1985).

3. Zinc

Calcareous soils having an inherent pH around 8.0 are known for their responsiveness to zinc fertilization, though these soils contain sufficient total Zn (Kanwar and Randhawa, 1974; Takkar and Nayyar, 1981; Singh and Abrol, 1986). Large yield gains upon zinc fertilization of alkali soils have often been ascribed to the role of zinc in improving the tolerance of crops to a sodic environment (Shukla and Prasad, 1974). Apart from the ameliorative role of Zn in plant growth, the differences between effective and ineffective amounts of Zn seem to vary widely depending on the chemical nature and pH of alkali soils (Takkar and Singh, 1978; Sharma et al., 1982; Singh et al., 1983). Whereas soil pH has been indicated to chiefly govern Zn solubility, the level of sodication was of little consequence (Singh et al., 1983; Singh and Abrol, 1985b).

In normal soils the relation between Zn solubility and pH has been negative, such that an increase in soil pH decreased the availability of zinc (Shaw and Dean, 1952). In strongly alkali soils, available Zn concentrations have been observed to increase with sodicity even more than in the reclaimed soils (Singh et al., 1983; Takkar and Nayyar, 1981). The kinds and amounts of amendment used for reclamation of alkali soils also influenced Zn concentrations in soil solutions (Singh et al., 1983) due to differential reductions in soil pH. The relation for available zinc versus pH as observed with varying amounts of different chemical amendments is shown in Figure 16. These experimental results suggest little influence of the chemical nature of the exchange complex except at high application rates of $FeSO_4$, which led to sharp reductions in soil pH. Alkali soils have a total of 60 to 80 μg Zn/g soil (Takkar and Randhawa, 1978) associated with the various Zn forms in decreasing order, with mineral amorphous and semicrystalline Zn forms, respectively. Zinc fertilization at the rate of 10 kg $ZnSO_4 \cdot 7H_2O$ ha^{-1} on a regular basis to a rice-wheat sequence for nearly three years led to a build up in various Zn forms (Singh and Abrol, 1986) in the order amorphous Zn > crystalline Zn > complexed Zn residual

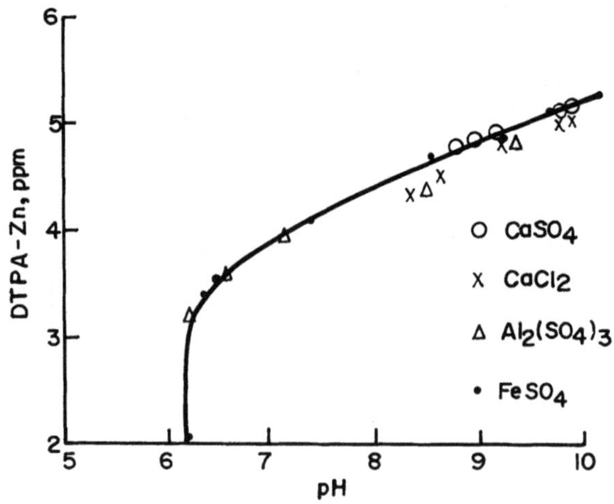

Figure 16. Effect of graded levels of four chemical membranes in pH of an alkali soil and the DTPA-extractable zinc status of amended soils. [Singh et al., Factors affecting DTPA-extractable zinc in sodic soils, *Soil Science*, 136, © by Williams & Wilkins, 1983.]

mineral Zn > exchangeable Zn. Equimolar Zn fertilizer concentration, when applied only once at the beginning of the experiment, showed that the amorphous and crystalline Zn forms decreased and were transformed to residual mineral Zn forms in due course of time. Singh and Abrol (1986) reported that exchangeable and amorphous Zn contribute mainly to Zn uptake by plants. In strongly alkaline soils where Fe^{3+} activity is controlled by amorphous ferric hydroxide, the activities of the metal ion Zn in soil solutions seems in the range supported by ferrite mineral ($ZnFe_2O_4$) (Lindsay, 1979). The information on likely changes in Zn forms brought out during reclamation of alkali soils has been summed up in Figure 17, a Zn distribution diagram (Gupta et al., 1987). This figure shows that Zn fixation increases with increasing pH as a result of precipitation reactions and entrapment of Zn within interlayer wedge zones of illite (Reddy and Perkins, 1974). Specifically sorbed zinc forms (zone III, Figure 17), which here include Zn from dissolution of precipitates plus specifically sorbed Zn bound to octehadral-OH in layer silicates (Hodgson, 1963) and to the amorphous oxides, increased with reduction in pH of alkali soils. Thus, whereas precipitation reactions govern Zn solubility at high pH, reactions responsible for zone III turn out to be important during reclamation.

Increase in solution Zn concentration at high pH was suggested to increase mobilization of Zn by soluble dispersed organic matter (Jeffrey and Uren, 1983; Jahiruddin et al., 1985; Saeed and Fox, 1977). Gypsum application reduced sodicity and soil pH, flocculates the dispersed organic matter, coprecipitating the associated organically complexed Zn, and in-

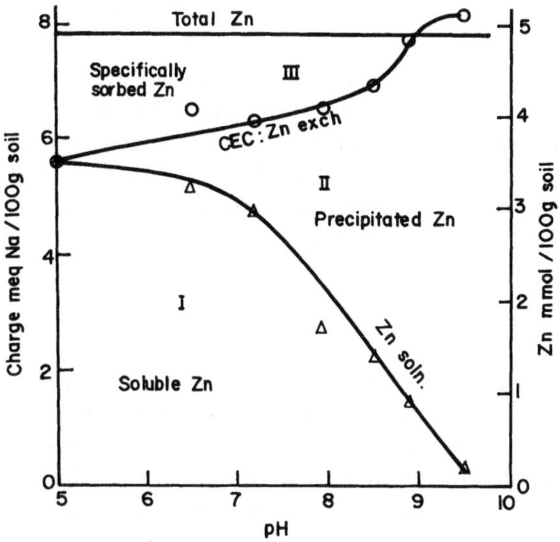

Figure 17. Influence of pH changes on the amounts of zinc forms in the alkali soil. Zinc distribution diagram.

creases the specifically sorbed Zn forms, well known for their low desorbability. Thus, whereas poor soil physical properties and retrogradation of precipitates limit adequate supply of Zn to crops at high pH, coprecipitation and specific adsorption reactions seem to mitigate the benefit of increased Zn solubility at lower pH values during desodication. Since a single, heavy dose of Zn fertilizer leads to transformation of Zn to stable mineral-bound Zn forms (Singh and Abrol, 1986), the regular application of Zn in small doses to each of the crops grown in a sequence has been shown to be a more efficient nutrient management practice.

Nutrients such as boron and molybdenum are not likely to be limiting for plant nutrition in alkali soils. In fact, with increasing sodicity and soil pH, concentration of these elements increases in the soil solution, and at higher concentrations they could prove toxic. However, once the alkali soils are amended with gypsum and leached, concentration of these nutrients in soil solution drops to within safe limits and remains no longer toxic to plants.

D. Water Management

1. Drainage

Provision of adequate surface and subsurface drainage is extremely important in reclamation of alkali soils. Since most farmers begin reclamation in monsoon season and grow rice immediately afterwards, as much rain water should be stored in the rice fields as will not affect crop growth adversely.

In many alkali soil areas, high water table is the major cause of formation of such soils. Lowering the water table in such cases is absolutely essential. Fortunately, the quality of ground waters in many areas is such that waters can be used for irrigation. For this reason, installation of tublewells and use of the pumped water for irrigation has proved an effective and economical way of controlling the ground water table in the Indo-Gangetic plains. Detailed drainage requirements for an area/catchment, however, must be worked out and implemented for long-term success of reclamation programs.

2. Irrigation

The optimum depth of irrigation and the interval between two irrigations depends on such factors as the atmospheric evaporativity, proliferation and depth of root penetration, capacity of the soil to store and transmit water, and the nature of plant responses to soil water stress. For crops other than rice, irrigation management presents major difficulties in obtaining optimum crop yields in alkali soils. Excess exchangeable sodium profoundly influences soil water behavior through dispersion of soil aggregates and the clogging action of dispersed particles. Alkali soils have low infiltration rates, reduced capacity to store water in a form available to plants, and poor ability to transmit water to the growing roots. All these constraints warrant water application at controlled rates, smaller depth, and more frequently.

3. Cultural Practices

Cultural practices can often be modified to suit adverse soil conditions to obtain optimum crop yields. Crop stands in alkali soils can be often improved by increasing the seed rate or the number of transplants per hill in the case of rice. Increased seed rates compensate for reduced germination and mortality of seedlings during initial seedling development stage (Central Soil Salinity Research Institute, 1979). In rice, older seedlings are observed as being somewhat more tolerant of alkali condition than younger seedlings. Planting 40 to 45-day-old rice seedlings has an edge over planting 25 to 30 day-old seedlings.

Similarly, planting older seedlings to establish tree species has proved beneficial (Central Soil Salinity Research Institute, 1979). For planting trees in alkali soils, the proper method of site preparation is important. Since most tree species have deep root systems and most alkali soils tend to have a concretion pan, the auger-hole technique has been found promising for establishing tree seedlings in highly alkali soils (Gill, 1987). The technique consists of making a hole, about 15 cm in diameter and 100 to 150 cm deep, with a power-driven auger, and filling back the hole with a mixture of original sodic soil, 2 to 3 kg of gypsum, and 7 to 8 kg of farmyard manure. With the auger-hole techique, a very small fraction of the total soil volume (profile) is reclaimed for establishing saplings of multipurpose tree species on alkali soils.

VI. Reclamation of Saline Soils

A. Leaching Soluble Salts and Controlling Salinity

Crop yields in a region are directly influenced by the distribution of soil salinity and water table depths. Reclamation of saline soils therefore means reducing soil salinity to acceptable levels by leaching and alleviating waterlogging caused by irrigation and preventing subsequent salinization. This is usually accomplished by placing drains in the soil beneath crops to remove excess irrigation water that would lead to raising of the underlying water table if not removed and to secondary salinization. The amount of water used for reclamation of saline soils to acceptable salinity levels depends significantly on initial soil salinity, the quality of irrigation water, the soil depth to be reclaimed, the water application technique, etc. The rule of thumb is that nearly 80% of the soluble salts from a 50-cm-deep soil profile can be removed by 50 cm of water. Leaching curves for sandy loam soils from India and Iraq and clay loam soils from the United States have been presented earlier (Figures 1 and 2). It has been shown that intermittent ponding and controlled water application modes (sprinklers) perform better at low evaporation rates (Carter and Fanning, 1964), because a large percentage of water flows through the fine pores. However, Minhas and Khosla (1986) have demonstrated that intermittent ponding does not lead to any saving in water for leaching of salt under high evaporative demand conditions. Excellent reviews on salinity control and crop management in saline soils have recently appeared in book form (Shainberg and Shalhevent, 1984; American Society of Civil Engineering, 1988), and consequently we shall deal with these topics only briefly.

In permanent reclamation plans for saline areas, availability of adequate outlets for disposition of saline drainage waters is of paramount importance. Disposal of drainage water can be a matter of serious concern, as has been illustrated by the as-yet-unresolved drainage problems of the San Joaquin Valley of California, where the problem is further complicated by removal of water with a higher selenium concentration than in the applied irrigation water.

A major process of salinity control depends on movement of water through the soil profile (leaching) to remove excess salts from the root zone. It is prudent that leaching and drainage for salinity control must minimize: (1) the flow of water through the profile to reduce dissolution of precipitated/soil minerals in the profile, and (2) reduce the drainage volumes. Reductions in drainage volumes can be achieved through water economy and by placing drains at depths such that spatial and temporal variability of soil water and soil salinity distributions are favorable for optimum crop returns (Rolston et al., 1985), and only unavoidable water is collected from shallow aquifers (Van Schilfgaarde, 1984). Field observations from Pakistan, India, and the United States suggest that the water table depths required to prevent an adverse effect on the yields of sorghum, cotton, toria, Brassica, and several other crops requiring minimal supplemental irrigation varied between 0.6 to 1.0 m (Doering et al., 1982; Oosterbaan, 1982; Gupta and Khosla, 1982). Limited evidence suggests that drains placed at depths more than 1.1 to 1.5 m (usually the zone of calcite concentrations in the Indo-Gangetic plains) may be drawing on the aquifer more than necessary (Rao et al., 1986a;b Van Schilfgaarde, 1984). Reuse of drainage water for irrigation of crops and in afforestation programs seems an immediate management practice to reduce this volume. Other disposal avenues for the effluents, however, have to be explored for a permanent solution of the salinity problem.

B. Agronomic Management of Saline Soils

In saline soils, maintenance of crop productivity at optimum rather than at maximum levels requires consideration of salt distribution within root zones influenced by the water extraction pattern of the crop, the method of water application, soil profile modifications (chiseling, deep plowing), changing the land configuration of special planting practices or irrigation management, mulching and rain water leaching, and adoption of an appropriate crop rotation involving salt-tolerant cultivars (U.S. Salinity Laboratory Staff, 1954).

Crop tolerance to salinity varies widely between glycophytes (which tolerate only low concentrations of salt) and halophytes (which tolerate relatively higher salt concentrations). Genotypic variation in salinity tolerance has been documented in sugar cane, maize, rice, barley, and several other crops. Several other factors, such as temperature, humidity, light intensity, stage of growth, moisture, and soil fertility, also influence plant response to salinity. Because of the above factors, it is difficult to establish a uniform scale of salt tolerance. Tolerance ratings of crops are useful, however, in providing guidelines to relative crop responses in saline environments with different agronomic management conditions. An exhaustive review of published reports led Maas (1988) to compile a list on the relative salt tolerance of cereals, fiber, vegetables and fruits, grasses, and forage crops. A quantitative salt tolerance rating of some of the important crops has been given in Table 7 for quick, relative comparisons among crops plants.

Salt tolerance of crops in Table 7 has been expressed in terms of two essential coefficients: (1) threshold soil salinity (*ct*), electrical conductivity of saturated paste extract without any yield reduction; and (2) slope (*S*), the percentage yield decrease per unit of salinity increase beyond the

Table 7. Salt tolerance of some selected crops, grasses, and trees: Electrical conductivity of saturated soil extract

Crop	Salt tolerance threshold (dS m^{-1})	50% Yield (dS m^{-1})	50% Emergence (dS m^{-1})
Barley	8.0	18.0	16–24
Cotton	7.7	17.0	15
Sugar beet	7.0	15.0	6–12
Sorghum	6.8	15.0	13
Wheat	6.0	13.0	14–16
Soybean	5.0	7.5	—
Beet, red	4.0	9.6	14
Date palm	4.0	16.0	—
Spinach	2.0	8.5	—
Peanut	3.2	5.0	—
Sugar cane	1.7	9.8	—
Tomato	0.5	7.6	—
Safflower	—	14.0	—
Cowpea	1.3	9.1	16
Corn	1.7	5.9	21–24
Lettuce	1.3	5.2	11
Onion	1.2	4.2	5.6–7.5
Rice	3.0	7.2	18
Bermuda grass	6.9	14.8	—
Rye Grass	5.6	12.1	—
Asparagus	4.1	29.0	—
Alfalfa	2.0	9.0	—
Sesbania	2.3	9.3	—
Berseem	1.5	9.5	—
Squash, Zucchini	4.7	9.9	—
Jojoba (*Simmondsia chinensis*)	—	—	Tolerant
Guayule	15	19.0	—

Source: Maas (1988).

threshold. In equation form, Mass and Hoffman (1977) represented salt tolerance by relative yield, YR:

$$YR = 100 - S(C - Ct)$$

where C refers to average root zone salinity. Slope values for the different crops can be calculated from the data in Table 7 using the expression slope $(S) = 50/(EC_{50\%\ \mathrm{yield}} - EC_{\mathrm{threshold}})$. The threshold and slope coefficient for the salt tolerance model, determined for steady-state salinity conditions, probably never near reality in practical field conditions, yet have found field application in validation of crop-water production function models for saline water use (Letey et al., 1985). Modeling efforts and experimental results suggest that salt tolerance coefficients determined under steady-state conditions are applicable to field conditions where the salinity is not uniform with soil depth (Hoffman, 1986).

VII. Conclusions

On a global basis, salt-affected soils occupy an estimated 952.2 m ha of land, constituting nearly 7% of total land area or nearly 33% of the area of potential arable lands of the world. Until about 1950, the increase in world food demand was met by an increase in the cultivated area. But subsequently the food-production growth rates were barely sufficient to keep pace with population growth, particularly in the poor, populous countires (Barr, 1981). The pressure of food shortages is compelling the developing countries to bring new lands under crop occupation. Reclamation of salt-affected soils will provide a unique opportunity for increasing bioproduction and alleviating pressure on traditionally cultivated land.

Considerations of soil management and crop responses permit classification of salt-affected soils into saline and alkali/sodic soils. In *Soil Taxonomy*, placement of salt-affected soils in different Great Groups or their phases at times poses problems. But the new information emerging from extensive use of *Soil Taxonomy* by nations can be incorporated into its structure. An increase in the extent and range of our knowledge on the physical, chemical, and mineralogical properties of salt-affected soils has greatly improved explanations on many aspects of the physical and chemical behavior of soils and consequently the scope for modification of reclamation practices to suit the socioeconomic structures of the countries. Whereas individual farmers can initiate programs for reclamation of alkali soils, the problem for saline soils has to be tackled on a watershed basis because all of the area may not need drainage provision. Amendment needs for reclamation of alkali soils should take into consideration the crops to be grown in sequence and their sodicity/salinity tolerance. Use of reclamation models likely will prove very useful. Management of salt-affected soils needs special considerations of irrigation and other agronomic practices. If ground water cannot be used for irrigation, adequate provision of surface and subsurface drainage is extremely important in reclamation. Among other things, drainage design should take into account the presence of any salt-bearing subsoil horizon and drainage requirement of crops, particularly in the monsoon/rainy season.

Contrary to the general belief, many salt-affected soils are fairly well supplied with plant nutrients, for example, the coastal saline soils in the humid tropics for rice (Ponnamperuma, 1984) and the alkali soils for P and K. Crop cultivation can be possible on these soils without costly fertilizers.

Micronutrients such as zinc, however, play a very key role in crop production in alkali soils. It seems that substitution of part of the inorganic requirement of crops by organic forms can greatly improve N use efficiency.

In many Asian countries, green manure crop is incorporated in wet moisture regimes during the rainy season. The practice of organic matter addition has generally proved beneficial in crop production. But the role of organic manures needs further evaluation for situations of soils undergoing sodication processes due to use of sodic water, water having residual alkalinity, and also in conditions where the water table is close to the soil surface.

VIII. Perspective

Although considerable progress has been made in the past toward understanding the problems associated with management of salt-affected soils, yet the rehabilitation of such lands is proceeding at a snail's pace. This is because salt-affected soil farmers have poor resource endowment, which is not readily enhanced by the fragile and uncertain infrastructure. Reclamantion of salt-affected soils requires additional agriculture inputs such as amendments, water, and infrastructure for drainage. Requirements of extra resources for practicing agriculture on such lands limit the pace of reclamation on an extensive scale.

Prognosis concerning development of irrigation predicts about 400 m ha of irrigated lands for the first part of the twenty-first century. New prospects for increasiong and stabilizing yields in arid and semiarid regions by extension of irrigation may be severely marred if adequate care is not taken to prevent hazards of secondary salinization/alkalinization, which so far seem to have outpaced all our reclamation efforts. This suggests the need for adoption of extensive land use strategies such as in forestry/farm forestry, etc., for speedy rehabilitation of these areas with only moderate and phased-out use of agricultural inputs. Results of forestry trials conducted at the Central Soil Salinity Research Institute, Karnal, India, have demonstrated the great potential of farm forestry in easing out several of the salinity problems resulting from mismanagement of irrigation and poor natural drainage. Some matters, such as the role of trees and crop combinations in control of salinity and water table, drainage volume, reduction, influence of microclimate changes on crop tolerance, use of low-quality water and the role of trees on bioamelioration processes still evade us. With new, quantitative understanding of the interaction of water quality and conditions of use in different soil types and climates, tangible prospects of substantial increase in bioproductivity of salt-affected soils can be realized in the near future. In view of the ever-increasing human and animal population and the greater demands for food, fodder, fiber, and fuelwood, there can be no complacency in our efforts to restore the salt-affected lands to agriculture production. Divergent socioeconomic conditions of farmers call for reclamation technologies that amalgamate the concept of root zone salinity/sodicity with a range of soil, water, and agronomic management practices. Reclamation technologies should include considerations of salt tolerance of crops and also realize the role of the microbial community, appropriate to different farming systems.

Acknowledgments

We are grateful to Professor Kenneth K. Tanji. Department of Land, Air and Water Resources, University of California, Davis, for providing us with several reports and reprints of original papers. We also thank Ms. Charlynne Merrill for her extra efforts in typing the final manuscript.

References

Abrol, I.P., and D.R. Bhumbla. 1978. Some comments on terminology relating to salt-affected soils. pp. 6–19 to 6–27. In: *Proc. Dryland-Saline-Seep Control, Edmonton, Canada, June 1978.*

Abrol, I.P., and D.R. Bhumbla. 1979. Crop responses to differential gypsum applications in a highly sodic soil and the tolerance of several crops to exchangeable sodium under field conditions. *Soil Sci.* 127:79–85.

Abrol, I.P., and I.S. Dahiya. 1974. Flow associated precipitation reaction in saline-sodic soils and their significance. *Geoderma* 11:1–9.

Abrol, I.P., D.R. Bhumbla, and K.S. Dargan. 1973. Reclaiming alkali soils. *Tech. Bull. No. 2*, Central Soil Salinity Research Institute, Karnal, India.

Abrol, I.P., R. Chhabra, and R.K. Gupta. 1980. A fresh look at the diagnostic criteria for sodic soils. pp. 142–147. In: *Proc. Int. Symp. Salt-Affected Soils, Karnal, India.*

Abrol, I.P., I.S. Dahiya, and D.R. Bhumbla. 1975. On the method of determining gypsum requirement of soils. *Soil Sci.* 120:30–36.

Abrol, I.P., R.K. Gupta, and S.B. Singh. 1979. Note on the solubility of gypsum and sodic soil reclamation. *J. Indian Soc. Soil Sci.* 27:481–483.

Acharya, C.L., and I.P.Abrol. 1978. Exchangeable sodium and soil water behavior under field conditions. *Soil Sci.* 125:310–319.

Agarwal, R.R., J.S.P. Yadav, and R.N. Gupta. 1982. *Saline and alkali soils of India.* ICAR, New Delhi.

Alperovitch, N., and I. Shainberg. 1973. Reclamation of alkali soils with $CaCl_2$ solutions. pp. 431–440. In: A. Hadas (ed.), *Physical aspects of soil, water and salts in ecosystems*. Springer-Verlag, Berlin.

Alperovitch, N., I. Shainberg, and J.D. Rhoades. 1986. Effect of mineral weathering on the response of sodic soils to exchangeable magnesium. *Soil Sci. Soc. Amer. J.* 50:901–904.

American Society of Civil Engineering, Irrigation and Drainage Division. 1988. *Agriculture salinity assessment and management manual.* ASCE Pub. (in press).

Armhein, C., J.J. Jurinak, and W.M. Moore. 1985. Kinetics of calcite dissolution as affected by carbon dioxide partial pressure. *Soil Sci. Soc. Amer. J.* 49:1393–1398.

Arora, H.S., and N.T. Coleman. 1979. The influence of electrolyte concentration on flocculation of clay suspensions. *Soil Sci.* 127:134–139.

Babcock, K.L. 1960. Evidence for sodium fixation in salt-affected soils. *Soil Sci. Soc. Amer. J.* 24:85–86.

Barr, T.N. 1981. The world food situation and global prospects. *Science* 214:1087–1095.

Barrer, R.M. 1978. Cation-exchange equilibria in zeolites and feldspathoids. In: L.D. Sand and F.A. Mumpton (eds.), *Natural zeolites—occurrence, properties.* Pergamon, NY.

Barton, F.M., and M.M. Wilde. 1971. Dissolution rates of polycrystalline samples of gypsum and orthorhombic forms of calcium sulfate by rotating disc method. *Trans. Faraday Soc.* 67:3590–3597.

Beek, G.G.E.M., and N. Breeman. 1973. The alkalinity of alkali soils. *J. Soil Sci.* 24:129–136.

Bernstein, L. 1962. Salt-affected soils and plants. pp. 139–174. In: *Problems of the arid zones, Proc. UNESCO Symp. Paris.*

Bhardwaj, K.K.R., and I.P. Abrol. 1978. Nitrogen management in alkali soils. pp. 83–86. In: S.P. Sen (ed.), *Nitrogen assimilation and crop productivity.* Associated Publ. Co., New Delhi.

Bhumbla, D.R., and I.P. Abrol. 1979. Saline and sodic soils. pp. 719–738. In: *Soils and Rice, Symp.*, IRRI, Los Banos, Laguna, Phillipines, Sept. 20–23, 1977.

Bingham, F.T., and M.J. Garber. 1970. Zonal salinization of the root system with NaCl and boron in relation to growth and water uptake of corn plants. *Soil Sci. Soc. Amer. J.* 34:122–126.

Black, C.A. (ed.). 1968. *Soil-plant relationships.* Wiley, NY.

Bower, C.A. 1948. Rapid tests for soluble and exchangeable sodium in saline and alkali soils. *J. Amer. Soc. Agron.* 40:1100–1105.

Bower, C.A., and J.T. Hatcher. 1962. Characterization of salt-affected soils with respect to sodium. *Soil Sci.* 93:275–280.

Bresler, E., B.L. McNeal, and D.L. Carter. 1982. *Saline and sodic soils, principles-dynamics-modelling.* Springer-Verlag, Berlin, Heidelberg, New York.

Brown, A.C., and J.J. Jurinak. 1984. Non-biological pyrite oxidation in alkali soils. *Agron. Abst.* p. 175.

Carter, D.L., and C.D. Fanning. 1964. Combining surface mulches and periodic water applications for reclaiming saline soils. *Soil Sci. Soc. Amer. J.* 28:564–567.

Central Soil Salinity Research Institute. 1979. *A decade of research.* CSSRI, Karnel, India.

Centeral Soil Salinity Research Institute. 1986, 1987. *Annual reports.* Karnal, India.

Chand, M., I.P. Abrol, and D.R. Bhumbla. 1977. A comparison of the effect of eight amendments on soil properties and crop growth in a highly sodic soil. *Indian J. Agric. Sci.* 47:348–354.

Chang, C.W., and H.E. Dregne. 1955. Effect of exchangeable sodium on soil properties, on growth and cation content of alfalfa and cotton. *Soil Sci. Soc. Amer. J.* 19:29–35.

Chaudry, G.H., and B.P. Warkentin. 1968. Studies on exchange of sodium from soils by leaching with calcium sulphate. *Soil Sci.* 105:190–197.

Chawla, K.L., and I.P. Abrol. 1982. Effect of gypsum fineness on the reclamation of sodic soils. *Agric. Water Mgt.* 5:41–50.

Chhabra, R. 1985. Crop response to phosphorus and potassium fertilization of a sodic soil. *Agron. J.* 77:699–702.
Chhabra, R., and I.P. Abrol. 1977. Reclaiming effect of rice grown in a sodic soil. *Soil Sci.* 124:49–55.
Chhabra, R., I.P. Abrol, and M.V. Singh. 1981. Dynamics of phosphorus during reclamation of sodic soils. *Soil Sci.* 132:319–324.
Chhabra, R., S.B. Singh, and I.P. Abrol. 1979. Effect of exchangeable sodium percentage on the growth, yield and chemical composition of sunflower (*Helianthus annuus*). *Soil Sci.* 127:242–247.
Cole, C.V. 1957. Hydrogen and calcium relationships of calcareous soils. *Soil Sci.* 83:141–150.
Collis-George, N., and D.E. Smiles. 1963. An examination of cation balance and moisture characteristics of determining the stability of soil aggregates. *J. Soil Sci.* 14:21–32.
Coordinated Research Project of ICAR, *Salt-affected soils and saline water use*, 1986–1988. *Annual reports*. Karnal, India.
Cruz-Romero, G., and N.T. Coleman. 1974. Reactions among calcium carbonate, carbon dioxide and sodium adsorbents. *Soil Sci. Soc. Amer. J.* 38:738–742.
Cruz-Romero, G., and N.T. Coleman. 1975. Reactions responsible for high pH in Na saturated soils and clays. *J. Soil Sci.* 26:169–175.
Darab, K. 1980. *Magnesium in solenetz soils*. pp. 92–101. In: *Int. Symp. Salt-Affected Soils*, Karnal, India.
Dargan, K.S., and R.K. Chillar. 1973. A technique for sugarbeet and vegetables in saline-alkaline soils. *Indian Farming*. 22:13–16.
Devitt, D., W.M. Jarrell, W.A. Jury, O.R. Lunt, and L.H. Stolzy. 1984. Wheat response to sodium uptake under zonal saline-sodic conditions. *Soil Sci. Soc. Amer. J.* 48:86–92.
Dieleman, P.J. 1963. *Reclamation of salt affected soils in Iraq*. Inst. Land Reclamation and Improvement, ILRI Pub. 11, Wageningen.
Doering, E.J., and W.O. Willis. 1975. Chemical reclamation for sodic strip-mine spoils. ARS-NC-20, U.S. Dept. Agric., Washington, DC.
Doering, E.J., L.C. Benz, and G.T. Reichman. 1982. Shallow-water table concept for drainage design in semi-arid and sub-humid regions. pp. 42–49. In: *Proc. 4th Natl. Drainage Symp. Amer. Soc. Agric. Eng.*
Drever, J.O. 1982. *The geochemistry of natural waters*. Prentice-Hall, Englewood Cliffs, NJ.
Dutt, G.R., R.W. Terkeltoub, and R.S. Rauschkloub. 1972. Prediction of gypsum and leaching requirements for sodium-affected soils. *Soil Sci.* 114:93–103.
Eaton, F.M. 1950. Significance of carbonates in irrigation waters. *Soil Sci.* 69:123–133.
El-Swaify, S.A., 1973. Structural changes in tropical soils due to anions in irrigation waters. *Soil Sci.* 37:956–958.
El-Swaify, S.A., and W.W. Emerson. 1975. Changes in physical properties of soil clays due to precipitated aluminum and ion hydroxides: 1. Swelling and aggregate stabiltiy after drying. *Soil Sci. Soc. Amer. J.* 39:106–1063.
Emerson, W.W. 1977. Interparticle bonding. pp. 477–498. In: *Soils: An Australian viewpoint*. Division of Soils, CSIRO, Melbourne, Academic Press, London.
Emerson, W.W. 1983. Physical properties and structure. pp. 78–104. In: J.S. Rus-

sel and E.L. Greacen (eds.), *Soil factors in crop production in a semi-arid environment*. Univ. Queensland Press, Brisbane, Australia.

Emerson, W.W., and C.L. Chi. 1977. Exchangeable calcium, magnesium and sodium and the dispersion of illite. II. Dispersion of illites in water. *Aust. J. Soil Res.* 15:225–265.

Evangelou, V.P., and E.J. Coale. 1987. Dependence of the Gapon coefficient on exchangeable sodium for mineralogically different soils. *Soil Sci. Soc. Amer. J.* 51:68–72.

Evangelou V.P., and R.E. Phillips. 1987. Sensitivity analysis on the comparison between the Gapon and Vanselow exchange coefficients. *Soil Sci. Amer. J.* 1473–1479.

FAO–UNESCO. 1974. *Soil map of the world.* 1:5000000. Volume 1. Legend. UNESCO, Paris.

Fenn, L.B., and D.E. Kissel. 1973. Ammonia volatilization from surface applications of ammonium compounds on calcareous soils: I. General theroy. *Soil Sci. Soc. Amer. J.* 37:855–859.

Fleisher, Z., and J. Hagin. 1981. Lowering ammonia volatilization losses from urea application by activation of nitrification process. *Fertilizer Res.* 2:101–107.

Frear, G.L., and J. Johnston. 1929. The solubility of calcium carbonate (calcite) in certain aqueous solutions at 25°C. *J. Amer. Chem. Soc.* 51:2082–2093.

Freney, J.R., R. Leuning, J.R. Simpson, O.T. Denmead, and W.A. Muirhead. 1985. Estimating ammonia volatilization from flooded rice fields by simplified techiques. *Soil Sci. Soc. Amer. J.* 49:1049–1054.

Frenkel, H., J.O. Goertzen, and J.D. Rhoades. 1978. Effect of clay type and content, exchangeable sodium percentage and electrolyte concentration on clay dispersion and soil hydraulic conductivity. *Soil Sci. Soc. Amer. J.* 42:32–39.

Garrels, R.M., and F.T. Mackenzie. 1967. Equilibrium concepts in natural water system. *Adv. Chem. Ser.* 67:222–242.

Ghai, S.K., D.L.N. Rao, and L. Batra. 1988. Nitrogen contribution to wetland rice by green manuring with *Sesbania* spp. in an alkaline soil. *Biol. Fertilizer Soils* 6:22–25.

Gill, H.S. 1987. Doctoral dissertation, Kurukshetr Univ., Kurukshetra, India.

Girdhar, I.K., and J.S.P. Yadav. 1982. Effect of magnesium rich waters on soil properties, yield and chemical composition of wheat. *Soil Sci.* 134:348–352.

Gobran, G.R., and S. Miyamoto. 1985. Dissolution rate of gypsum in aqueous salt solutions. *Soil Sci.* 140:89–94.

Goertzen, J.O., and C.A. Bower. 1958. Carbon dioxide from plant roots as a factor in the replacement of adsorbed sodium in calcareous soils. *Soil Sci. Soc. Amer. J.* 22:36–37.

Greene, R.S.B., and G.W. Ford. 1983. The effect of gypsum on cation exchange and leaching in two red duplex wheat soils. *Aust. J. Soil Res.* 21:187–193.

Gupta, R.K., and B.K. Khosla. 1982. Seasonal variations in salt and water content profiles in shallow and saline ground watertables. *Indian J. Agric. Sci.* 52:506–510.

Gupta, R.K., and R.R. Singh. 1988. A comparative evaluation of the reclamation efficiency of surface concentrated versus internally incorporated $CaCl_2$ and gypsum amendments in an alkali soil. *Soil Sci.* 146.

Gupta, R.K., D.K. Bhumbla, and I.P. Abrol. 1984. Effect of soil pH, organic matter and calcium carbonate on the dispersion behavior of alkali soils. *Soil Sci.* 137:245–251.

Gupta, R.K., D.R. Bhumbla, and I.P. Abrol. 1985a. Release of exchangeable sodium from an alkali soil upon amendment application. Role of variable charge and cation hydrolysis. *Soil Sci.* 139:312–312.

Gupta, R.K., R. Chhabra, and I.P. Abrol. 1981. Relationship between pH and exchangeable sodium in a sodic soil. *Soil Sci.* 131:215–219.

Gupta, A.P., S.S. Khana, and N.K. Tomar. 1985a. Effect of sodicity on the utilization of phosphatic fertilizers by wheat. *Soil Sci.* 139: 47–52.

Gupta, R.K., C.P. Singh, and I.P. Abrol. 1985b. Determination of cation exchange capacity and exchangeable sodium in alkali soils. *Soil Sci.* 139:326–332.

Gupta, R.K., C.P. Singh, and I.P. Abrol. 1985c. Dissolution of gypsum in alkali soils. *Soil Sci.* 140:382–386.

Gupta, R.K., R.R. Singh, and I.P. Abrol. 1988. Influence of simultaneous changes in sodicity and pH on the hydraulic conductivity of an alkali soil under rice culture. *Soil Sci.* 146:395–402.

Gupta, R.K., R.R. Singh, and K.K. Tanji. 1989. Phosphorus release in sodium-dominated soils. Communicated.

Gupta, R.K., Van dem Sef Elshout, and I.P. Abrol. 1987. Effect of pH on Zn adsorption precipitation reactions in an alkali soil. *Soil Sci.* 143:198–204.

Hardie, L.A., and H.P. Eugester. 1970. The evolution of closed-basin brines. pp. 273–290. *Mineral. Soc. Amer. Spec. Publ. No. 3.*

Harmsen, G.W., A. Quispel, and D. Otzen. 1954. Observations on the formation and oxidation of pyrites in the soil. *Plant Soil* 5:324–347.

Hira, G.S., and N.T. Singh. 1980. Irrigation water requirement for dissolution of gypsum in sodic soils. *Soil Sci. Soc. Amer. J.* 44:353–358.

Hira, G.S., M.S. Bajwa, and N.T. Singh. 1981. Prediction of water requirement for dissolution of gypsum in sodic soils. *Soil Sci.* 131:353–358.

Hodgson, J.E. 1963. Chemistry of the micronutrient elements in soils. *Adv. Agron.* 15:119–159.

Hoffman, G.J. 1980. Guidelines for the reclamation of salt affected soils. Second Inter-American Conf. Salinity and Water Management Technology, Juarez, Mexico.

Hoffman, G.J. 1986. Yields of irrigated crops under saline conditions. In: *Conf. Proc. Decision criteria for residuals management in agriculture*. Univ. California, April, 1986.

Holford, I.C.R., and G.E.G. Mattingley. 1975. Surface area of calcium carbonate in soils. *Geoderma* 13:247–255.

Internatinal Rice Research Institute. 1976, 1977. *Annual reports.* Los Banos, Phillipines.

Isbell, R.F. 1984. Soil classification problems in the tropics and sub-tropics. pp. 17–26. In: *Int. Workshop on Soils—Research in Tropics*. Sept. 12–16, 1983. Townsville, Queensland, Australia.

Isbell, R.F., and J. Williams. 1981. Dry soils in Australia: characteristics and classification. pp. 124–150. In: F.H. Beinroth and A. Osman (eds). *Proc. Third Int. Soil Classification Workshop.* ACSAD, Damascus, Syria.

Jackson, M.L., S.A. Twez, A.L. Willis, G.A. Bourbeau, and R.P. Dennington. 1948. Weathering sequence in clay size minerals in soils and sediments. I. Fundamental generalizations. *J. Phys. Colloid Chem.* 52:1237–1260.

Jahiruddin, M., N.T. Livesey, and M.S. Cressner. 1985. Observations on the effect of soil pH upon zinc absorption by soils. *Commun. Soil. Sci. Plant Anal.* 16:909–922.

Jardine, P.M., and D.L. Sparks. 1984. Potassium-calcium exchange in multireactive soil system. II. Thermodynamics. *Soil Sci. Soc. Amer. J.* 48:45–50.

Jeffery, J.J., and N.C. Uren. 1983. Copper and zinc species in the soil solution and the effects of soil pH. *Aust. J. Soil Res.* 21:479–488.

Jewitt, T.N. 1942. Loss of ammonia from ammonium sulphate applied to alkaline soils. *Soil Sci.* 54:401–409.

Jury, W.A., W.W. Jarrell, and D. Devitt. 1979. Reclamation of saline-sodic soils by leaching. *Soil Sci. Soc. Amer. J.* 43:1100–1106.

Kanwar, J.S., and D.R. Bhumbla. 1969. Physico-chemical characteristics of sodic soils of the Punjab and Haryana and their amelioration by use of gypsum. *Agrokem. Talajt.* 18(Suppl.):325–330.

Kanwar, J.S., and N.S. Randhawa. 1974. *Micronutrient research in soil and plant in India. A review*. ICAR Publ., New Delhi, India.

Kapoor, B.S., H.B. Singh, and S.C. Goswami. 1981a. Analcime in a sodic profile. *J. Indian Soc. Soil Sci.* 28:513–515.

Kapoor, B.S., H.B. Singh, S.C. Goswami, I.P. Abrol, G.P. Bhargava, and D.K. Pal. 1981b. Weathering of micaceous minerals in some salt affected soils. *J. Indian Soc. Soil Sci.* 29:486–492.

Kelley, W.P. 1948. *Cation exchange in soils*. Waverly Press, Baltimore, MD.

Kemper, W.D., J. Olsen, and C.J. DeMooy. 1975. Dissolution rate of gypsum in flowing water. *Soil Sci. Soc. Amer. J.* 39:458–463.

Keren, R., and P. Kauschansky. 1981. Coating of calcium carbonate on gypsum particle. *Soil Sci. Soc. Amer. J.* 45:1242–1244.

Keren, R., and G.A. O'Connor. 1982. Gypsum dissolution and sodic soil reclamation as affected by water flow velocity. *Soil Sci. Soc. Amer. J.* 46:726–732.

Keren, R., and I. Shainberg. 1981. Effect of dissolution rate on the efficiency of industrial and mineral gypsum in improving infiltration of a sodic soil. *Soil Sci. Soc. Amer. J.* 45:103–107.

Khosla, B.K., K.S. Dargan, I.P. Abrol, and D.R. Bhumbla. 1973. Effect of depth of mixing gypsum on soil properties and yield of barley, rice and wheat grown on a saline sodic soil. *Indian J. Agric. Sci.* 43:1024–1031.

Khosla, B.K., R.K. Gupta, and I.P. Abrol. 1979. Salt leaching and effect of gypsum application in a saline-sodic soil. *Agric. Water Mgt.* 2:193–202.

Kovda, V.A. 1965. Alkaline soda-saline soils. *Agrokem. Talajt.* 14(Suppl.): 15–48.

Kovda, V.A., C. van den Berg, and R.M. Hagan. 1973. *Irrigation, drainage and salinity*. FAO/UNESCO, Hutchinson, London.

Kreit, J.F., I. Shainberg, and A.J. Herbillion. 1982. Hydrolysis and decomposition of hectorite in dilute salt solutions. *Clays Clay Mineral* 30:233–241.

Kumar, A., and I.P. Abrol. 1986. *Grasses in alkali soils. Tech. Bull. No. 11*, Central Soil Salinity Research Inst., Karnal, India.

Lagerwerff, J.V., and J.P. Holland. 1960. Growth and mineral content of carrots and beans as related to varying osmotic and ionic composition effects in saline-sodic sand culture. *Agron. J.* 52:603–608.

Leffelaar, P.A., and R. Sharma. 1977. Leaching of a highly saline-sodic soil. *J. Hydrol.* 32:203–218.

Letey, J., A. Dinar, and K.C. Knapp. 1985. Crop-water production function model

for saline irrigation waters. *Soil Sci. Soc. Amer. J.* 49:1005–1009.
Lindsay, W.L. 1979. Chemical equilibrium in soil. Wiley, NY.
Loveday, J. 1974. Recognition of gypsum responsive soils. *Aust. J. Soil Res.* 12:87–96.
Loveday, J. 1976. Relative significance of electrolyte and cation exchange effects when gypsum is applied to a sodic clay soil. *Aust. J. Soil. Res.* 14:361–371.
Loveday, J. 1984. Amendments for reclaiming sodic soils. pp. 220–237. In: I. Shainberg and J. Shalhavet (eds.), *Soil salinity under irrigation processes and management*. Springer-Verlag, Berlin, Heidelberg.
Loveday, J., and J.C. Pyle. 1973. The Emerson dispersion test and its relationship to hydraulic conductivity. *Tech. Paper 15*, CSIRO Aust. Div. Soils.
Maas, E.V. 1988. Crop salt tolerance. In: *Agriculture salinity assessment and management manual.* ASCE Publ. (in press).
Maas, E.V., and G.J. Hoffman. 1977. Crop salt-tolerance and current assessment. *J. Irrig. Drainage Div., ASCE* 103(2):115–134.
Magaritz, M., A. Nadler, H. Koyumdjisky, and J. Dan. 1981. Using Na/Cl ratio to trace sources of salinity in a semi-arid zone. *Water Resources Res.* 17:602–608.
Magdoff, F., and E. Bresler. 1973. Evaluation of methods of reclaiming sodic soils with $CaCl_2$. pp. 431–440. In: A. Hadas (eds.), *Physical aspects of soil water salts in ecosystems, Vol. 4.* Springer-Verlag, Berlin, Heidelberg.
Manin, M., A. Pissarra, and J.W. Van Hoom. 1982. Drainage and desalinization of heavy clay soil in Portugal. *Agric. Water Mgt.* 5:227–240.
Mashhady, A.S., and D.L. Rowell. 1978. Soil alkalinity. Equilibria and alkalinity development. *J. Soil Sci.* 29:65–75.
McGeorge, W.T., and E.L. Breazeale. 1955. The value of pyrite and pyrrhotite as soil conditioner. *Rep.* 12A, Agric. Exp. Sta. Univ. Arizona, Tucson.
McNeal, B.L., and N.T. Coleman. 1966. Effect of solution composition on soil hydraulic conductivity. *Soil Sci. Soc. Amer. J.* 30:308–312.
McNeal, B.L., W.A. Norvell, and N.T. Coleman. 1966a. Effect of soluction composition on soil hydraulic conductivity and on the swelling of extracted soil clays. *Soil Sci. Soc. Amer. J.* 30:308–317.
McNeal, B.L., G.A. Pearson, J.T. Hatcher, and C.A. Bower. 1966b. Effect of rice culture on the reclamation of sodic soils. *Agron. J.* 58:238–240.
Merrill, S.D., J.R. Deutsch, and M.W. Pole. 1987. Saturation percentage. In: D.R. Williams and G.E. Schuman (eds.), *Reclaiming mine soils and overburn in the Western United States*. Soil Conerv. Soc. Amer. Ankeny, IA.
Millet, G. 1970. *Geology of clays*. Springer-Verlag, Berlin, Heidelberg.
Ming, D.W., and J.B. Dixon. 1984. Quantitative determination of soil zeolites using a cation exchange capacity method. *Agron. Abstr.*, p. 274.
Ming, D.W., and F.A. Mumpton. 1984. Zeolites in soils. *Agron. Abstr.*, p. 274.
Minhas, P.S., and B.K. Khosla. 1986. Solute displacement in a silt loam soil as affected by dry method of water application under different evaporation rates. *Agric. Water Mgt.* 12:64–74.
Miyamoto, S., J. Ryan, and J.L. Strochlein. 1975. Sulfuric acid and leaching requirements for reclaiming sodium affected calcareous soils. *Plant Soil* 43:573–585.
Moore, A.W., R.F. Isbell, and K.H. Northkote. 1983. Classification of Australian soils. pp. 253–266. In *An Australian viewpoint*. Div. Soils, CSIRO, Melbourne, Academic Press, London.

Murakami, H. 1968. Some problems on the oxidation of oxidizable sulfur. *J. Sci. Soil Manure* (Jap.) 39:116–120.

Nadler, A., and M. Magaritz. 1982. The effect of sodium carbonate presence on the soil extract's chemical composition. *Soil Sci. Soc. Amer. J.* 46:1333–1336.

Nakayama, F.S. 1970. Hydrolysis of $CaCO_3$, Na_2CO_3 and $NaHCO_3$ and their combinations in the presence and absence of external CO_2 source. *Soil Sci.* 109:391–398.

Northcote, K.H., and J.K.M. Skene. 1972. *Australian soils with saline and sodic properties*. Soil Pub. 27, CSIRO, Melbourne, Australia.

Oosterbaan, O.J. 1982. *Annual report*. Int. Inst. Land Reclamation, Wageningen, pp. 50–54.

Oster, J.D. 1982. Gypsum usage in irrigated agriculture; a review. *Fertilizer Res.* 3:73–89.

Oster, J.D., and H. Frenkel. 1980. The chemistry of the reclamation of sodic soils with gypsum and lime. *Soil Sci. Soc. Amer. J.* 44:41–45.

Oster, J.D., and J.D. Rhoades. 1975. Calculated drainage water compositions and salt burdens resulting from irrigation with river waters in Western United States. *J. Envir. Qual.* 4:73–79.

Oster, J.D., and F.W. Schroer. 1979. Infiltration as influenced by irrigation water quality. *Soil Sci. Soc. Amer. J.* 43:444–447.

Oster, J.D., and I. Shainberg. 1979. Exchangeable cation hydrolysis on soil weathering as affected by exchangeable sodium. *Soil Sci. Soc. Amer. J.* 43:70–75.

Oster, J.D., and G. Sposito. 1980. The Gapon coefficient and the exchangeable sodium percentage-sodium adsorption ratio relation. *Soil Sci. Soc. Amer. J.* 44:258–260.

Overstreet, R., J.C. Martin, and H.M. King. 1951. Gypsum, sulfur, sulfuric acid for reclaiming an alkali soil of the Fresno series. *Hilgardia* 21:113–126.

Pal, D.K. 1985. Potassium release from muscovite and biotite under alkaline conditions. *Pedologie* 35:133–136.

Paliwal, K.V., and A.P. Gandhi. 1976. Effect of salinity, Ca: Mg ratio in irrigation water and soil texture on the predictabiltiy of exchangeable sodium percentage. *Soil Sci.* 122:85–90.

Peech, M. 1965. Hydrogen ion activity. In: C.A. Black (ed.), *Methods of soil analysis, Part II*. ASA Mono. 9., Madium, WI.

Petrosian, G.P., and A.I. Techitchian. 1969. Soda-saline soils of the Ararat Plain and methods of their reclamation. *Agrokem. Talajt.* 18(Suppl.):121–138.

Polemio, M., and J.D. Rhoades. 1977. Determining cation exchange capacity: A new procedure for calcareous and gypsiferous soils. *Soil Sci. Soc. Amer. J.* 41:524–527.

Ponnamperuma, F.N. 1972. The chemistry of submerged soils. *Adv. Agron.* 24:29–96.

Ponnamperuma, F.N. 1984. Role of cultivar tolerance in increasing rice production in saline lands. pp. 255–271. In: R.C.Staples and G.H. Taenniessen (eds.), *Salinity tolerance in plants. Strategies for crop improvement*. Wiley, NY.

Prather, R.J., J.D. Goertzen, J.D. Rhoades, and H.Frenkel. 1978. Efficient amendment use in sodic soil reclamation. *Soil Sci. Soc. Amer. J.* 42:782–786.

Pratt, P.F., and F.L. Bair. 1969. Sodium hazard of bicarbonate irrigation water. *Soil Sci. Soc. Amer. J.* 33:880–883.

Pratt, P.F., and D.W. Thorne. 1948. Solubility and physiological availability of phosphate in sodium and calcium systems. *Soil Sci. Soc. Amer. J.* 13:213–217.

Pupisky, H., and I. Shainberg. 1979. Salt effects on the hydraulic conductivity of a sandy soil. *Soil Sci. Soc. Am. J.* 43:429–433.

Quirk, J.P. 1977. Chemistry of saline soils and their physical properties pp. 79–90. In: T.Talsma and J.R.Philip (eds.), *Salinity and water use*. Aust. Acad. Sci., Canberra.

Quirk, J.P., and R.K. Schofield. 1955. The effect of electrolyte concentration on soil permeability. *J. Soil Sci.* 6:163–178.

Quispel, A., G.W. Harmsen, and D. Otzen. 1952. Contribution to the chemical and bacteriological oxidation of pyrite in soil. *Plant Soil* 4:43–55.

Rahman, A.W., and D.L. Rowell. 1979. The influence of Mg in saline and sodic soils: A specific effect or a problem of cation exchange. *J. Soil Sci.* 30:535–546.

Rains, D.W., and E. Epstein. 1967. Preferential absorption of potassium by leaf tissue of the mangrove, *Avicennia marina*: An aspect of halophyte competence in coping with salt. *Aust. J. Biol. Sci.* 20:847–857.

Rao, D.L.N. 1987. Slow release urea fertilizers—Effect on flood water chemistry, ammonia volatilization and rice growth in an alkali soil. *Fertilizer Res.* 13:209–212.

Rao, D.L.N., and L. Batra. 1983. Ammonia volatilization from applied nitrogen in alkali soils. *Plant Soil* 70:219–228.

Rao, D.L.N., and L. Batra. 1984. Evaluation of conventional, slow release and nitrification inhibitor treated fertilizers for rice in an alkali soil. *IRRN* 9:26–27.

Rao, D.L.N., and S.K. Ghai. 1985. Urease and dehydrogenase activity of alkali and reclaimed soils. *Aust. J. Soil Res.* 23:661–665.

Rao, D.L.N., and S.K. Ghai. 1986a. Urease inhibitors: Effect on wheat growth in an alkali soil. *Soil Biol. Biochem.* 18:255–258.

Rao, D.L.N., and S.K. Ghai. 1986b. Effect of phenylphosphorodiamidate on urea hydrolysis, ammonia volatilization and rice growth in an alkali soil. *Plant Soil* 94:313–320.

Rao, K.V.G.K., O.P. Singh, R.K. Gupta, S.K. Kamra, R.S. Pandey, P.S. Kumbhare, and I.P. Abrol. 1986. *Drainage investigations for salinity control in Haryana*. Tech. Bull. 10, Central Soil Salinity Res. Inst., Karnal, India.

Rasmussen, W.W., D.P. Moore, and L.A. Alban. 1972. Improvement of a solonetzic (slick spot) soil by deep ploughing, subsoiling and amendments. *Soil Sci. Soc. Amer. J.* 36:137–142.

Reddy, M.R., and H.F. Perkins. 1974. Fixation of zinc by clay minerals. *Soil Sci. Soc. Amer. J.* 38:229–231.

Rengasamy, P., R.S.B. Greene, and G.W. Ford. 1984. The role of clay fraction in the particle arrangement and stability of soil aggregates. A review. *Clay Res.* 3:53–67.

Reynolds, C.M., and D.C. Wolf. 1987. Influence of urease activity and soil properties on ammonia volatilization from urea. *Soil Sci.* 143:418–425.

Rhoades, J.D. 1982. Reclamation and management of salt-affected soils after drainage. pp. 123–187. In: *Proc. 1st Annual Western Provincial Conf. on rationalization of water and soil research and management*, Lethbridge, Alberta, Canada.

Rhoades, J.D. 1984. Principles and methods of monitoring salinity. pp. 143–167. In: I. Shainberg and J. Shalhevet (eds.), *Soil salinity under irrigation, processes and management*. Springer-Verlag, Berlin, Heidelberg.

Rhoades, J.D., and R.D. Ingvalson. 1969. Macroscopic swelling and hydraulic conductivity properties of four vermicullitic soils. *Soil Sci. Soc. Amer. J.* 33:364–367.

Rimmer, D.L., and D.J.Greenland. 1976. Effect of $CaCO_3$ on the swelling behavior of a soil clay. *J. Soil Sci.* 27:129–139.

Rolston, D.E., J.W. Biggar, R.J. Miller, D.R. Nielsen and F.E.Broadbent. 1985. Crop response to temporally and spatially variable soil-salinity profiles. pp. 93–101. In: *Soil and plant interactions with salinity*. Kearney Found. Five-Year Rep., 1980–85.

Rupela, O.P., and P. Tauro. 1973. Isolation and characterization of *Thiobacillus* from alkali soils. *Soil Biol. Biochem.* 5:891–897.

Saeed, M., and R.L. Fox. 1977. Relationship between suspension pH and Zn solubility in acid and calcareous soils. *Soil Sci.* 124:199–203.

Schofield, R.K., and H.R. Samson. 1954. Flocculation of kaolinite due to the attraction of oppositely charged crystal faces. *Discuss. Faraday Soc.* 18:135–145.

Schulz, R.K., R. Overstreet, and I. Barshad. 1964. Some unusual ionic exchange properties of sodium in certain salt affected soils. *Soil Sci.* 99:161–165.

Sehgal, J.L., G.F. Hall, and G.P. Bhargava. 1975. An appraisal of the problems in classifying saline-sodic soils of the Indo-Gangetic plains in N.W. India. *Geoderma* 14:75–91.

Shainberg, I., and M. Gal. 1982. The effect of lime on the response of soils to sodic conditions. *J. Soil Sci.* 33:489–498.

Shainberg, I., and J. Letey. 1984. Response of soils to sodic and saline conditions. *Hilgardia* 52(2):1–57.

Shainberg, I., R. Keren, and H. Frenkel. 1982. Response of sodic soils to gypsum and calcium chloride application. *Soil Sci. Soc. Amer. J.* 46:116–117.

Shainberg, I., and J. Shalhavet (eds.), 1984. *Soil salinity under irrigation processes and management*. Springer-Verlag, Heidelberg, Berlin.

Shainberg, I., J.D. Rhoades, and R.J. Prather. 1981a. Effect of low electrolyte concentration on clay dispersion and hydraulic conductivity of a sodic soil. *Soil Sci. Soc. Amer. J.* 45:273–277.

Shainberg, I., J.D. Rhoades, D.L. Suarez, and R.J. Prather. 1981b. Effect of mineral weathering on clay dispersion and hydraulic conductivity of sodic soils. *Soil Sci. Soc. Amer. J.* 45:287–291.

Shanmuganathan, R.T., and J.M. Oades. 1983. Modification of soil physical properties by addition of calcium compounds. *Aust. J. Soil Res.* 21:285–300.

Sharma, B.D., P.N. Takkar, and U.S. Sadana. 1982. Evaluating levels and methods of zinc application to rice in sodic soils. *Fertilizer Res.* 3:161–167.

Shaw, E., and L.A. Dean. 1952. Use of dithiozone as an extractant to estimate the zinc untrient status of soils. *Soil Sci.* 73:341–347.

Shukla, U.C., and K.G. Parsad. 1974. Ameliorative role of zinc on maize growth under alkali soil conditions. *Agron. J.* 66:804–806.

Sidhu, P.S., and R.J. Gilkes. 1977. Mineralogy of soil developed on alluvium in the Indo-Gangetic plain (India). *Soil Sci. Soc. Amer. J.* 41:1194–1201.

Silverman, M.P. 1967. Mechanism of bacterial pyrite oxidation. *J. Bacteriol.* 94:1046–1057.

Singh, G. 1987. Nitrogen management of rice-wheat sequence in alkali soils. *Indian J. Agron.* 32:387–391.

Singh, M.V., and I.P. Abrol. 1985b. Solubility and adsorption of zinc in a sodic soil. *Soil Sci.* 140:406–411.

Singh, M.V., and I.P. Abrol. 1986. Transformation and movement of zinc in an alkali soil and their influence on the yield and uptake of zinc by rice and wheat crops. *Plant Soil* 94:445–449.

Singh, M.V., R.R. Chhabra, and I.P. Abrol. 1983. Factors affecting DTPA extractable zinc in sodic soils. *Soil Sci.* 36:359–366.

Singh, O., and S.D. Nijhawan. 1943. Availability of phosphates in alkaline and calcareous soils. *Indian J. Agric. Sci.* 130:131–141.

Singh, S.B., and I.P. Abrol. 1983. Influence of exchangeable sodium on the yield of pea and its chemical composition and nutrient uptake. *Indian J. Agric. Sci.* 53:685–689.

Singh, S.B., R. Chhabra, and I.P. Abrol. 1979. Effect of exchangeable sodium on the yield and chemicla composition of raya (*Brassica juncea* Linn.). *Agron. J.* 71:767–771.

Singh, S.B., R. Chhabra, and I.P. Abrol. 1980. Effect of soil sodicity on the yield and chemical composition of cowpea grown for fodder. *Indian J. Agric. Sci.* 50:870–874.

Singh, S.B., R. Chhabra, and I.P. Abrol. 1981. Effect of exchangeable sodium on the yield, chemical composition and oil content of safflower and linseed. *Indian J. Agric. Sci.* 51:885–891.

Soil Survey Staff. 1975. *Soil taxonomy: a basic system of soil classification for making and interpreting soil surveys.* U.S. Dept. Agric. Handbook No. 436. U.S. Govt. Printing Office, Washington, D.C.

Solov'ev, V.A. 1969. Plant growth and water and mineral element supply under conditions of NaCl salinization. *Soviet Plant Physiol.* 16:722–727.

Sposito, G. 1977. The Gapon and Van Selow selectivity coefficients. *Soil Sci. Soc. Amer. J.* 41:1205–1206.

Sposito, G., and C.S. LeVesque. 1985. Sodium-calcium-magnesium exchange on Wyoming beutonite in the presenc of adsorbed sodium. *Soil Sci. Soc. Amer. J.* 49:1153–1159.

Sposito, G., and S.W. Mattigod. 1977. On the chemical foundation of the sodium adsorption ratio. *Soil Sci. Soc. of Amer. J.* 41:32–329.

Stumm, W., and J.J. Morgan. 1970. *Aquatic chemistry*. Wiley-Interscience, NY.

Stumm-Zollinger, E. 1972. Bacterial mediation in the oxidation of pyrite. *Arch. Mikrobiol.* 83:110–115.

Suarez, D.L. 1987. Prediction of pH errors in soil water extracts due to degassing. *Soil Sci. Soc. Amer. J.* 51:64–67.

Suarez, D.L., J.D. Rhoades, R. Lavado, and C.M. Grieve. 1984. Effect of pH on saturated hydraulic conductivity and soil dispersion. *Soil Sci. Soc. Amer. J.* 48:50–55.

Sung, W., and W. Stumm. 1980. Kinetics and product of ferrous iron oxygenation in aqueous systems. *Environ. Sci. Technol.* 14:561–570.

Szabolcs, I. 1974. *Salt affected soils in Europe.* Martinus Nijhoff, The Hague, Research Inst. Soil Sci. Agric. Chem., Acad. Sci., Budapest.

Szabolcs, I. 1979. Review of research on salt-affected soils. *Natural Resources Res.* 15:137. UNESCO, Paris.

Takkar, P.N., and V.K. Nayyar. 1981. Effect of gypsum and zinc in rice nutrition on sodic soil. *Exp. Agric.* 17:49–55.

Takkar, P.N., and N.S. Randhawa. 1978. Micronutrients in Indian agriculture.

Fertilizer New 23:1–26.

Takkar, P.N., and T. Singh. 1978. Zinc nutrition of rice as influenced by rates of gypsum and zinc fertilization of alkali soils. *Agron. J.* 70:447–450.

Tanji, K.K. 1969. Solubility of gypsum in aqueous electrolytes as affected by ion association and ionic strengths up to 0.15 cm and at 25°C. *J. Envir. Sci. Technol.* 3:656–661.

Tanji, K.K., and S.J. Deverel. 1984. Simulation modelling for reclamation of sodic soils. pp. 238–251. In: I. Shainberg and J. Shalhevet (eds.), *Soil salinity under irrigation, processes and management*. Springer-Verlag, Heidelberg, Berlin.

Tanji, K.K., and L.D. Whittig. 1985. Carbonate and sulfate chemistry and mineralogy in salt-affected soils. pp. 73–84. In: *Soil and plant interactions with salinity*. Kearney Found. Five-Year Rep., 1980–85.

Tanji, K.K., L.D. Doneen, G.V. Ferry, and R.S. Ayers. 1972. Computer simulation analysis on reclamation of salt-affected soils in San Joaquin Valley, California. *Soil Sci. Soc. Amer. J.* 36:127–133.

Temple, K.L., and E.W. Delchamps. 1953. Autotrophic bacteria and the formation of acid in bituminous coal mines. *Appl. Microbiol.* 1:255–258.

Thorstenson, D.C., D.W. Fisher, and M.G. Croft. 1979. The geochemistry of the Fox Hills-Basal Hill creek aquifer in southwestern North Dakota and northwestern South Dakota. *Water Resources Res.* 15:1479–1498.

Thorup, J.T.1972. Soil sulfur application. *Sulfur Inst. J.* 8:16–17.

Toogood, J.A., and R.R. Cairns (eds.), 1978. *Solonetzic soils technology and management*. Bull. B-78-1, Univ. Alberta, Edmonton.

Turner, R.C., and J. Clark. 1956. The pH of calcareous soil. *Soil Sci.* 82:337–341.

Unz, R.F., and D.G. Lundgren. 1961. A comparative nutritional study of three chemoautotrophic bacteria: *Ferrobacillus ferroxidans*, *Thiobacillus ferroxidans*, and *Thiobacillus thioxidans*. *Soil Sci.* 92:302–313.

U.S.Salinity Laboratory Staff. 1954. *Diagnosis and improvement of saline and alkali soils*. Handbook 60, U.S. Government Printing Office, Washington, DC.

Van Olphen, H. 1977. *An introduction to clay colloid chemistry*, 2d. Interscience, New York.

Van Schilfgaarde, J. 1984. Drainage design for salinity control. pp. 190–197. In: I. Shainberg and J.Shalhevet (eds.), *Soil salinity under irrigation, processes and management*. Springer-Verlag, Berlin, Heidelberg.

Varallyay, G. 1978. Moisture status and flow phenomena in salt-affected soils. In: *Proc. Indo-Hungarian Seminar, managment salt-affected soils*, Karnal, India.

Venkatakrishannan, S., and I.P. Abrol. 1981. Amelioration of a sodic soil through *Thiobacilli* inoculation and pyrite application. *Indian Soc. Soil Sci.* 29:526–529.

Vlek, P.L.G., and J.M. Stumpe. 1978. Effect of solution chemistry and environmental conditions on ammonia volatilization losses from aqueous system. *Soil Sci. Soc. Amer. J.* 42:416–421.

Vlek, P.L.G., Th.J.M. Blom, J. Beek, and W.L. Lindsay. 1974. Determination of the solubility product of various iron hydroxides and jarosite by the chelation method. *Soil Sci. Soc. Amer. J.* 38:429–432.

Vorob'eva, L.P., L.P. Vyrodova, and E.E. Goronenkova. 1986. Evaluating alkalinity of calcareous soils. *Moscow Univ. Soil Bull.* 41:12–17.

Wallace, A. 1966. *Current topics in plant nutrition*. pp. 115–117. Lithographed by Edwards Bros., Ann Arbor, MI.

Whitney, R.S., and R. Gardner. 1943. The effect of carbon dioxide on soil reac-

tions. *Soil Sci.* 55:127–141.
Williams, W.A. 1966. Management of nonleguminous green manures and crop residues to improve infiltration rates of an irrigated soil. *Soil Sci. Soc. Amer. J.* 30:631–634.
Yaalon, D.H. 1958. Studies on the effect of saline irrigation water on calcareous soils. II. The behavior of calcium carbonate. *Bull. Res. Council Isr.* 7G:115–122.
Yadav, J.S.P., and I.K. Girdhar. 1981. Effect of different magnesium: Calcium ratios and sodium adsorption ratio values of leaching water on the properties of calcareous versus noncalcareous soils. *Soil Sci.* 131:194–198.
Yahia, T.A., S. Miyamoto, and J.L. Strochlein. 1975. Effect of surface applied sulfuric acid on water penetration into dry calcareous and sodic soil. *Soil Sci. Soc. Amer. J.* 39:1201–1204.
Zantua, M.I., L.C. Dumenil, and J.M. Bremner. 1977. Relationships between soil urease activity and other soil properties. *Soil Sci. Soc. Amer. J.* 41:350–352.

Biological Degradation of Soil

G.K. Sims*

I. Introduction

Soil microorganisms play key roles in cycling of nutrients, decomposition of wastes and residues, and detoxification of pollutant compounds in the environment. Factors that affect these organisms, and/or their ability to mediate anthropic functions, have received attention in recent years. The purpose of this chapter is to present a review of the present literature that

*Department of Agronomy, The Ohio State University, Columbus, OH 43210, U.S.A.

addresses the potential for biological degradation of soil. Biological degradation of soil as described herein refers to the impairment or elimination of one or more "significant" populations of microorganisms in soil, often with a resulting change in biogeochemical processing within the associated ecosystem. "Significant" microorganisms are those for which an ecologically significant role is understood.

The ability of soil microbial populations to function properly is of critical importance to the health and well-being of humankind. Cycling of nutrient elements by microorganisms facilitates continued production of plant- and animal-derived food. Depending on the overall balance of various biotic and abiotic processes affecting nutrient elements, specific microbial transformations may become either beneficial or detrimental to human health. For example, the availability of nutrients to plants (and therefore people) and the quality of surface and ground waters may be affected by changes in the relative rates of competing reactions involving a particular chemical species. Activities of soil microorganisms may affect environmental quality not only through overproduction of some chemical species, but also through failure to detoxify pollutants, or through undesirable proliferation of biomass (eutrophication).

Because of the extreme importance of microbially mediated processes in soil, it is important to understand the long-term effects of various practices on biological processes in soils. The precise nature of microbial communities in soil has thus far been inextricable. Presently, most cells of microorganisms observed in soil cannot be recovered in culture (Olsen and Bakken, 1987). The activities of soil microbial populations are so interwoven that the sum of the activities of all the populations within a community produce a kind of moire accent effect, which is difficult to unravel. For example, it may not be possible to separate direct effects of a treatment on a particular organism from indirect effects resulting from inhibition or stimulation of a second organism with which the first organism has any relationship other than neutralism. For such reasons, it has been difficult to interpret the effects of particular treatments on soil microorganisms or their functions, or even to understand the real cause for effects that have been clearly demonstrated. The reader is therefore invited to question the existing knowledge base, and to propose new means by which biological degradation of soils can be better understood.

II. Microbial Communities in Soil

A. The Cast of Characters

The terrestrial environment contains a very large number of ecological niches, which may be separated spatially or temporally, and therefore may permit occupancy by organisms competing for common resources. As a

Table 1. Density and biomass values of different groups of soil organisms in cultivated temperate soils[a]

Group of organisms	Dry biomass[b] $(g\ m^2)$[f]	Dry biomass[c] (kg ha)	Abundance[d] $(No.\ m^2)$	Live biomass[d] $(g\ m^2)$	Abundance[e] (No. g)	Live biomass[e] (kg ha)
Bacteria	32.76	36.9	3×10^{14}	300	10^4–10^9	300–3000
Actinomycetes	84–144	0.2			10^5–10^8	
Fungi		454.0		400	2×10^4–10^6[g]	500–5000
Algae					10^2–5×10^4	7–300
Protozoa		1.0	5×10^8	38	10^4–10^5	50–200
Nematodes		2.0	10^7	12		
Earthworms		12.0	10^5	132		
Enchytraeidae		4.0				
Molluscs		5.0	*h*	h		
Acari		1.0	2×10^5	3		
Collembola		2.0	2×10^4	5		
Diptera		3.0	*h*	h		
Other arthropods		6.0	2×10^h	36[h]		

[a]From Tiedje and Dazzo (1982).

[b]From Clark and Paul (1970).

[c]Lynch and Poole (1979).

[d]From Richards (1974) and Macfadyen (1963).

[e]From Alexander (1977).

[f]Range over nonwinter months for this Saskatchewan, Canada, site; most precisely measured data are presented.

[g]Shown as number of propagules/g, which is not a uniform indicator of fungal density; the more accepted approach is hyphal length, for which the typical range is 10 to 100 m/g. The biomass estimate is based on the length value.

[h]Organisms of these groups are grouped as other fauna in this case.

Source: Tiedje and Dazzo (1982)

result, a typical community of soil microorganisms may be characterized by robust genetic diversity and therefore extensive physiological capabilities. The major groups of microorganisms typically observed in soil include representatives of the eubacteria (including cyanobacteria and the actinomycete line), archaebacteria, fungi, algae, protozoa, viruses, and probably viroids, and prions. The relative contributions of major taxa to the total microbial biomass of soils is given in Table 1. For the purpose of this review, we will limit our discussion primarily to bacteria and fungi, with some discussion of soil fauna. Considerably less is known of the impact of soil degradation of the ecology of other groups in terrestrial ecosystems. Descriptions of dominant genera and their characteristics may be found in Alexander (1977) and Teidje and Dazzo (1982).

Within each of the above groups of organisms there exist common features that appear to be successful adaptations for occupation of certain ecological niches. For example, the eubacteria, with the exception of the actinomycete line, exist as single cells, representing the simplest forms of cellular organisms. Soil bacteria are for the most part considered saprophytic or parasitic, with a few autotrophic and symbiotic forms. Due to small size in relation to competing and predatory organisms in other groups, bacteria may exploit resources unavailable to larger organisms, and may escape predation by larger forms through sheltering in microsites of appropriate physical dimensions. Bacteria may also be rendered inaccessable to smaller organisms, such as the ectoparasite *Bdellovibrio* in the presence of soil materials (Roper and Marshall, 1978). Small size and single-cell organization may pose disadvantages for successful exploitation of resources contained in large objects, such as plant residues. A particular residue particle may be a rich source of carbon but deficient in nitrogen or some other nutrient needed by the organism. A single-cell organism is not able to transport nutrients from the soil matrix to the site of growth within the residue particle. Filamentous morphology observed among fungi and actinomycetes may therefore be much better suited for pioneering such substrates. This may in part explain why fungi are often dominant, early colonizers of lignified plant residues.

Single-cell growth habit is also noted among protozoa, which may obtain carbon saprophytically or by predation upon smaller forms, such as bacteria. The growth habit of predatory forms makes sense in light of the intrinsic property of mobility through the soil matrix that is imparted by the relatively small size of these organisms. However, in order to predate bacteria by engulfing them, protozoa must generally be larger than bacteria they attack, which explains the ability of bacteria to escape predation by entering small soil pores. Such escape is more difficult to explain in the case of small parasites, such as *Bdellovibrio*.

Many bacteria, including motile forms, occur in colonies in soil, and may thus be much less independent than it would otherwise seem. Colonial growth habit allows more efficient utilization of resources. For example, degradation of polymeric substrates usually involves an initial depolymerization step by extracellular enzymes. Products released from insoluble substrates might be rapidly diluted and lost to other organisms when population densities are low. Colonial growth might also afford other advantages, such as protection against dessication, toxicants, or ultraviolet radiation. Soil bacteria are commonly motile, and probably chemotactic, suggesting that it is possible for these organisms to actively pursue or avoid substances that occur heterogeneously in soil microsites, diffusing away to the bulk solution. Chemotactic capability requires a considerable commitment of genetic material and expenditure of metabolic energy (MacNab, 1987). Very little is known of the ecological significance of chemotaxis within the soil ecosystem (Chet and Mitchell, 1976; MacNab, 1987), yet it seems likely that such a genetically and metabolically expensive adaptation would not have been retained unless it offered some major advantage to the organism (MacNab, 1987). Is it possible that microorganisms move chemotactically toward substrates, and away from toxicants in the soil? If so, what would be the implications of chemotaxis with regard to detoxification of pollutants, or introduction of toxicants into the soil? The ecological significance of chemotaxis in soil is presently unknown, but may become an important area of research in microbial ecology.

The advantages of filamentous growth associated with many soil fungi are obvious, such as the ability to exploit a larger volume of soil, or to transport nutrients from hyphae in enriched microsites to hyphae in depleted microsites. Filamentous growth offers protection from being washed through soil by percolation of water through soil, and affords advantages of multicellular organization, such a specialization of certain cells, more effective genetic exchange strategies, etc.

The morphological adaptations described above are important factors in the susceptibility and opportunistic response of soil microorganisms to soil degradation. Various physiological adaptations, described below, are also important determinants in the response of soil organisms to anthropogenic perturbances.

B. Relevant Aspects of Mircrobial Metabolism

1. Microbial Energetics

Excellent reviews are available on various aspects of microbial physiology, including derivation of energy (Gottshalk, 1986; Stanier et al., 1976). No attempt will be made here to review the process of microbial energetics; rather, the presentation will be limited to critical aspects of the process that indicate potential susceptibility to environmental perturbances.

The universal currency of energy in biological systems is adenosine 5′-triphosphate (ATP). A high Gibbs free-energy yield is associated with hydrolysis of each of two phosphate bonds found in the ATP molecule

(−7.3 kcal mol^{-1} for hydrolysis of ATP to ADP). Coupling reactions to ATP hydrolysis typically shifts the equilibrium ratio of products to reactants by about 10^8. Cells generate ATP via two basic mechanisms: substrate-level phosphorylation, where a phosphate group (with a high free energy of hydrolysis) is transfered from a phosphorylated intermediate to ADP; and oxidative phosphorylation, in which ATP is produced during the transfer of electrons from NADH or $FADH_2$ to O_2 (or some other electron acceptor) via a series of electron carriers. An important contrast of the two mechanisms is the need for a functioning electron-transport chain and a terminal electron acceptor (generally a compound other than the substrate) for oxidative phosphorylation.

The soil community contains individuals with obligate, optional, or no reliance upon oxidative phosphorylation. Obligately aerobic organisms function only in a respiratory mode, using oxygen (or in some cases, nitrate) as a terminal electron acceptor. These organisms are not competitive for substrates in the absence of appropriate electron acceptors. Other organisms are able to switch to a fermentative mode of energy transduction, in many cases using the substrate itself (or a comparable organic compound) as both an electron source and sink. Generally, the average oxidation state of the products of fermentations is the same as the oxidation state of the substrates. Since roughly half of the substrate may be used as an electron acceptor, fermentation is less productive of ATP than oxidative phosphorylation. Therefore, in the presence of electron acceptors, respiratory metabolism is more competitive than fermentation. Fermentation becomes less favorable as the availability of feermentable substrates (compounds that are neither highly oxidized nor reduced) declines, and potentially toxic products of fermentations (commonly simply organic acids) accumulate. Obligate anaerobes, including some forms with electron-transport systems adapted for sulfate or other electron acceptors, are competitive only under strictly anaerobic conditions.

It is apparent from the discussion above that in some environments, the availability of electron acceptors (including oxygen, nitrate, and sulfate) and fermentable substrates may become as important as the availability of carbon for heterotrophic activities in soil. Thus factors that affect the availability of these materials will also affect the turnover of carbon, which fuels other microbial transformations in soil.

2. Oxygenases

Several biological reactions of environmental significance are catalyzed by oxygenases (enzymes requiring molecular oxygen as a substrate). These enzymes incorporate one (monooxygenases) or two (dioxygenases) oxygen atoms into each molecule of product formed. Oxygenases are involved in the degradation of aliphatic and aromatic molecules (including lignins, hydrocarbons, and pesticides) and ammonium oxidation by chemoautotrophic bacteria (Schmidt, 1982). Oxygenases are also involved directly in the removal of halogens from xenobiotic molecules. Since O_2 is a required substrate for oxygenases, these reactions will not take place in the absence of O_2 regardless of any ability of the organism to function anaerobically. For example, a psuedomonad may have the ability to oxidize an aromatic compound via an oxygenase activity, and the ability to use nitrate as an electron acceptor (denitrification) in the absence of oxygen, but may not oxidize the aromatic compound under denitrifying conditions. Therefore, the availability of oxygen to the microbial community affects not only respiratory activities, but also specific reactions involving O_2.

3. Dependence of Enzyme Catalysis and Nutrient Availability on pH

The ratio of protonated to deprotonated forms of an enzyme and its substrates affects the activity of the the enzyme (Ainsworth, 1977). Enzymes are generally active only over a narrow range of pH, and exhibit some optimum pH for maximum activity. Microorganisms contain membrane-bound proteins that are directly affected by the pH of the external environment. Also, pH affects the solubility of many essential nutrient elements, with a resulting indirect effect on the growth of microorganisms. It follows that factors which affect the pH of soil solution markedly affect the activities of soil microorganisms.

C. Microbially Mediated Processes in Soil

Our interest in biological degradation of soil has little to do with an inherent regard for the safety of microorganisms themselves, or even the threat of extinction of any particular microorganism. Rather, scientists are concerned with potential impairment of biogeochemical processes mediated by soil microorganisms. Among the most significant processes affected by soil microorganisms are the cycling of nutrients, including carbon, nitrogen, sulfur, or phosphorus; detoxification of pollutant molecules and complex wastes (including inorganic and organic pollutants); and suppression of pathogenic organisms. Numerous texts have been written to describe the cycling of nutrients by soil microorganisms (Alexander, 1977; Richards, 1987; Smith, 1982; Stevenson, 1986). Therefore, we introduce only those aspects of nutrient cycling that highlight specific features suggesting the significance of biological processing of nutrients and pollutants in soil, and the nature of vulnerability of these processes to soil degradation.

1. Carbon Cycling

Soil microbial communities are generally regarded as primarily heterotrophic, requiring inputs of carbon from without the microbial community. For most terrestrial ecosystems, it appears that carbon and energy inputs are derived primarily from higher plants rather than from autotrophic microorganisms, although such microorganisms are indeed present in the soil. Carbon entering the soil ecosystem as plant carbon generally leaves the system as simple organic gases, particularly carbon dioxide and methane. Retention of plant carbon by soil microorganisms may involve considerable residence in biomass or humic materials, but is seldom, if ever, permanent. This fact is evidenced by the dissipation of plant materials shortly after introduction into the soil enviroment.

Cycling of plant carbon appears to be one of the least sensitive features of the soil microbial community. This should be expected, in light of the dependence of most soil microorganisms on plant carbon. A rather dramatic environmental insult must be affected before cycling of plant derived carbon is impared. For example, as many as 90% of soil fungi may be capable fo utilizing cellulose. Elimination of any one particular fungus is unlikely to markedly affect decomposition fo celluose. Moreover, there are numerous bacteria capable of utilizing cellulose (*Cellulomonas, Pseudomonas, Vibrio, Bacillus*), which may in part replace the role of the fungi in the event of major changes in fungal populations. Despite the improbability of a specific impairment of cellulose degradation in soil, this condition has been observed in the field (Sparrow and Sparrow, 1988), suggesting that it is difficult to extrapolate functioning of soil communities from estimates of available biochemical potential (e.g. fraction of the total population with a particular biochemical trait). Similar scenarios could be developed for other aspects of the carbon cycle.

2. Nitrogen Cycling

Nitrogen is among the most important nutrients for living systems and is often the limiting factor in productivity of ecosystems. It is also one of the nutrients most affected by the activities of soil microorgansims. For this reason, many studies have investigated the effects of land degradation on various aspects of the nitrogen cycle. Of the major components of the N cycle (mineralization, immobilization, nitrification, denitrification, and nitrogen fixation), nitrogen fixation and nitrification appear to be the most sensitive to enviromental perturbances. Factors contributing to the sensitivity of the nitrification process include intolerance of many nitrifiers (particularly ammonium oxidizers) to acidity, and the strong dependence of nitrification on molecular oxygen. Nitrifiers (e.g. *Nitrosomonas* and *Nitrobacter*) exhibit unusually high pH minima and optima for growth (Table 2). Because of the involvement of oxygenases in autotrophic nitrification, wet conditions leading to anoxia generally impede nitrification. Symbiotic nitrogen fixation is a very complicated biological phenomenon, involving numerous coordinated processes that must take place for establishment of effective root nodules or other analogous structures. Certain pollutant molecules might interfere with induction of nodulation genes by aromatic

Table 2. Optimal conditions of pH for growth of select microorganisms

Organism	Minimum	pH optimum	Maximum
Enterobacter spp.	4.4	6.0–7.0	9.0[a]
Pseudomonas spp.	5.6	6.6–7.0	8.0[a]
Nitrosomonas spp.	7.0	8.0–8.8	9.4[a]
Nitrobacter spp.	6.6	7.6–8.6	10.0[a]
Thiobacillus spp.	1.0	2.0–2.8	6.0[a]
Micrococcus spp.	5.0	7.2	9.0[b]

[a]Data from Atlas (1984b).
[b]G. Sims, Unpublished data.

compounds (mostly flavinoids) produced by the plant, or other chemical signals necessary of establishment of nodules.

3. Phosphorus Cycling

Pollution from phosphorus, introduced in detergents or as agricultural amendments, has been implicated as a causitive agent in the eutrophication of fresh-water lakes. This has been the most commonly reported effect of land management on phosphorus cycling. These effects are discussed elsewhere within this volume (Logan, 1989). Though very little information is available on the effects of land degradation on other aspects of phosphorus cycling, one may conclude by inference that certain perturbations may alter phosphorus availability through effects mycorrhizal fungi, which have been identified as contributers to phosphorus nutrition of many plants. The effects on mycorrhizae of degradative processes, such as strip mining and soil erosion, will be discussed in following sections.

4. Other Cycles in Soils

Cycling of sulfur, iron, manganese, and other elements of specialized interest has been reviewed extensively (Alexander, 1977; Atlas and Bartha, 1987; Stevenson, 1986). Iron and manganese undergo redox reactions in soil as a function of the availability of oxygen. Reductive processes involving these metals have generally been attributed to indirect action of microorganisms (e.g., electron donors are thought to be reduced fermentation products), although there has been some evidence suggesting the use of these metals as terminal electron acceptors (Tugel et al., 1986). Oxidation of iron and manganese may occur spontaneously (depending on the mineral form) in the absence of microbial activity when reduced environments are exposed to oxygen (Alexander, 1977). Sulfur may also be subject to redox reactions in soil. As in the case of iron and manganese, sulfur (or sulfate) reduction has generally been attributed to microbial activity, whereas abiotic reoxidation may occur when reduced soils are aerated. As one would expect, factors that influence the availability of oxygen in soil (especially drainage) also affect cycling of iron, manganese, or sulfur. Among the most devastating forms of land degradation known is the process of acid mine drainage, which results primarily from biological and chemical oxidation of pyrite (FeS_2) associated with materials exposed during strip mining. Large quantities of sulfuric acid are released, resulting in acidification of waterways, soils, and sediments receiving runoff. Land degradation resulting from acid mine drainage is discussed elsewhere in this volume (Logan, 1989).

III. Aspects of Soil Biology Affected by Land Degradation

Information on the effects of anthropic activities on soil biology has been limited primarily to studies on population biology of bacteria, fungi, and micro- or mesofauna, and the effects of various treatments on functioning of communities of soil microorganisms and/or soil fauna. With few exceptions, investigation of functionality has been limited to components of the carbon or nitrogen cycle. Very few experiments have demonstrated mechanisms for observed effects. Most experiments have been of short duration, or have consisted of point measurements taken 5 to 10 years after the initial treatment was made. Few studies can be found that demonstrate the kinetics of adverse effects or the process of recovery of the soil community.

A. Indicators of Biological Degradation of Soil

1. Population Biology and Diversity

Among the most common measurements of biological degradation of soil are measurements of the effects of some treatment on courts of organisms within a particular taxonomic group. Bacteria and fungi have been enumerated primarily with viable counting techniques, usually employing plating on nutrient-rich agar media. The validity of viable counts as predictors of whole populations of bacteria or fungi remains controversial for reasons too numerous to catalog here (Olsen and Bakken, 1987). Unfortunately, there are no perfect alternatives to viable counts. Other methods of quantifying soil organisms are also somewhat inadequate due to problems in either the mechanics of obtaining or the logic of interpreting numerical data obtained. Suffice it to say that due to the inadequacies of available methodology, there have been inherent problems with most data on population biology of soils; thus results and interpretation must be carefully considered.

An extension of population biology that has received some attention is taxonomic, or physiological, diversity. Although very few attempts have been made to examine diversity of soil microorganisms in detail, the concepts involved are pervasive in the literature and merit some individual attention. *Diversity* may be used to describe the manner in which species or operational taxa are assembled within a community. The concept of species diversity was proposed by Fisher et al. (1943) and has been used to describe communities of animals, plants, and microorganisms. Diversity is often expressed by a mathematical index, such as the McIntosh uniformity index (McIntosh, 1967), which measures the uniformity of a community, the Simpson dominance index (Simpson, 1949), which assesses the degree of dominance or heterogeneity within a community, and the Shannon index (Shannon, 1948; Shannon and Weaver, 1949), which is used to provide general measurement of both species richness and equitability.

The most common application of diversity indices has been in the demonstration of effects of perturbations or environmental insults on communities within a particular econiche. It has been suggested that species diversity is a measure of entropy within a community and therefore would suggest the amount of energy required to maintain organization of the community (Atlas, 1984a, 1984b). It has been held that diversity confers stability to ecosystems, though this concept has often been challenged. This suggests that communities possessing more genetic information should be better able to cope with environmental insults. A measurement of diversity, such as one of the aforementioned indices, might therefore be expected to reflect the ability of a community to withstand perturbations, and might also provide an indication of environmental stress, which could be expected to reduce the diversity of expressed genetic information. In practice, there seems to be little evidence that diversity indices will allow predictions of fragile terrestrial econiches, with the possible exceptions of extreme environments, such as cold regions, which we might have expected to be sensitive a priori. However, in many cases, environmental insults have produced rather striking effects on diversity indices of particular types of habitats. It is yet very difficult to ascertain the utility of this kind of measurement for assessing the degree of biological degradation of land, since very few studies have been performed, perhaps because of the rather labor-intensive nature of these experiments. Because of the controversial nature of diversity measurements, and the broad implications possible if successful applications are found, citations to diversity measurements are included in many sections of this chapter.

2. Nutrient Cycling

The most common assessments available for the funtional state of damaged terrestrial ecosystems have been studies on nutrient cycling processes. Most common have been assays of "soil enzymes," components of the nitrogen cycle (especially nitrification, mineralization, and nitrogen fixation), decomposition of cellulose and/or wood, and respiration measurements. Soil enzyme activities and respiration are not direct measurements of nutrient turnover, but have been construed as indicators of the functional status of the soil community. Peculiarities of these measurements will be discussed as they are introduced.

3. Accumulation of Pollutant Molecules

Impairment of the functioning of microbial communities can result in accumulation of toxic substances that would otherwise have been transformed to some innocuous species. For example, reductive dechlorination of toxic organic compounds may be inhibited in the presence of excess sulfate, presumably resulting in accumulation of chlorinated hydrocarbons in sulfate-rich anaerobic environments (Gibson and Suflita, 1986). Certain microorganisms, including the nitrite-oxidizing chemoautotroph, *Nitrobacter*, are extremely sensitive to minute concentrations of ammonia. If ammonium concentration and soil pH are high, the functioning of *Nitrobacter* may be impaired, resulting in the accumulation of toxic quantities of nitrite (produced through ammonium oxidation). Accumulation of microbially produced nitrite has been associated with soils receiving large quantities of animal wastes (Alexander, 1971)

4. Change in Redox Status

Changes in oxidation-reduction (redox) status are commonly associated with practices that result in excess water in soils. The rate of diffusion of oxygen through water-filled soil pore space is about one ten-thousandth the rate through air-filled pores, resulting in a much slower rate of oxygen supply in wet soils. When chemical and biological processes consume more oxygen than can be supplied, the result is a shift toward anerobic metabolism, and accumulation of reducing substances (Kaspar and Teidje, 1982). This is reflected in a reduced redox potential as measured by platinum electrode, and a preponderance of reduced species [particularly Fe (II), Mn (II), sulfide, and methane] in solution or headspace. A reducing condition in soil is a good indicator of poor oxygen supply, and is commonly associated with tillage practices and other land management techniques that result in compaction, ponding, and other conditions resulting in reduced oxygen supply. Severity of the condition may be reflected in the composition of reduced products accumulating. For example, the evolution of appreciable quantities of methane indicates the presence of extremely reduced sites within the soil matrix.

IV. Effects of Toxic Substances on Microorganisms

Numerous reports are available on the effects of various toxic substances on the structure and function of microbial communities. Usually, the effects of a particular toxicant are temporary, and the duration of effects is

Table 3. Duration of selected perturbances of microbial communities or processes in soils

Perturbance	Duration	Reference
Disposal of retorted oil shale	>6 yr	Segal and Manicinelli (1987)
Crude oil Spill	>10 yr	Sparrow and Sparrow (1988)
Clearcutting of forests	300 yr[a]	McFee and Stone (1965)
Pesticide application	4–6 wk	Chandra (1964)
	2–16 wk	Biedebeck et al. (1987)
Pyridine spill	3–4 wk	Brand and Sims (1987)
Strip mining	50–100 yr	Atlas (1987)

[a] Estimated time required for equilibrium of forest floor biomass. Microbial community structure may stabilize in significantly less time.

determined by the persistence of the toxicant and the ability of the microorganisms to adapt to its presence. In warm climates, the effects on microorganisms of most organic toxicants, such as fuel materials, solvents, and pesticides, are usually of short duration, since most organics are eventually detoxified in nature. Contamination of soils by toxic metals often results in long-term effects on microbial communities (Table 3). Soils of the arctic region are particularly sensitive to environmental insults, and long-term damage to community structure and function may result from contamination of even readily degraded materials. What follows is a discussion of observed effects of various toxic substances on the growth and activities of soil microorganisms.

A. Pesticides

Of all the effects of chemical inputs on soil biology, the effects of pesticide use are the most extensively documented. Since pesticides, including fumigants, fungicides, herbicides, and insecticides, are used in agriculture because they are toxic to some organism, and because many metabolic processes are common to all cellular organisms, it is not surprizing that pesticides often display toxic effects on nontarget organisms, including microorganisms. Several reviews are available on the effects of pesticides on nontarget organisms (Parr, 1974; Thompson and Edwards, 1974; Jones, 1956; Fletcher, 1960; Bollen, 1961; Kreutzer, 1963, 1965; Martin, 1964, 1966; Alexander, 1969; Helling et al., 1971; Lal and Lal, 1988). The amount of information available on such effects is staggering, yet it appears we still know little about what really happens to soil microorganisms when pesticides are applied. Most pesticides appear to have some effects on soil organisms, although often urealistically high concentrations have been used to induce effects. Generally, data consist of measurements of microbial transformations, or population counts for particular groups of interest. What follows is a discussion of some of the effects that have been seen on various groups of organisms, and processes that they mediate.

1. Effects of Pesticides on Microbial Growth and Biomass

Microbial biomass was reduced by exposure to extremely high rates (50 mg kg^{-1} to 50 g kg^{-1}) of BHC, although these rates are unrealistic for most situations (Anan'eva et al., 1986). Treatment of soil with a commercial formulation of paraquat (1,1′-dimethyl-4,4′-bipyridinium) resulted in proliferation of bacteria and fungi during a 14-day period (Smith and Mayfield, 1977). The results were confounded by the observation that treatment reduced CO_2 evolution, cellulose degradation, and nitrogenase activity. It is interesting to note that pesticides may affect particular predators more than their prey, and therefore result in increased numbers of prey upon pesticide application. Treatment of soil with glyphosate [N-(phosphonomethyl)glycine] or diquat (6,7-dihydrodipyrido [1,2-*a*:2′,1′-*c*]-pyrazinediium) + paraquat increased the the incidence of take-all disease of wheat (Mekwatanakarn and Sivasthamparam, 1987). Inoculation of soil with untreated soil resulted in suppression of the disease, suggesting that the herbicides were negatively affecting biocontrol of *Gaeumannomyces graminis* var. *tritici* by other soil microorganisms.

2. Effects of Pesticides on Nutrient Cycling

Most numerous are accounts of pesticide effects on varius aspects of the nitrogen cycle. The most extensively studied appear to be nitrification and symbiotic nitrogen fixation, which are mediated by a much less diverse group than the other processes studied. These processes are expected to be sensitive to toxicants and environmental perturbances because of the relatively few species represented, and in the case of symbiotic nitrogen fixation, the possibility of aromatic pesticides interfering with chemical signals involved in the communication between the host plant and the symbiont prior to and during the infection process. Of all the processes studied, nitrification appears to be the most sensitive to pesticide application.

Application of amitrole (3-amino-1,2,4-triazole), 2,3,6-TBA (2,3,6-trichlorobenzoic acid), 2,4-DB [4-(2,4-dichlorophenoxy)butyric acid], and diallate (S-2,3-dichloroallyl diisopropylthiocarbamate) at normal field rates resulted in inhibition of nitrification for at least eight weeks (Chandra, 1964; Domsch and Paul, 1974). Other pesticides, such as atrazine (6-chloro-N^2-ethyl-N^4-isopropyl-1,3,5-triazine-2,4-diamine) (Hauke-Pacewiczowa, 1971; Setty et al., 1970), bromacil (5-bromo-3-*sec*-butyl-6-methyluracil)(Pancholy and Lynd, 1969), picloram (4-amino-3,5,6-trichloropyridine-2-carboxylic acid)(Dubey, 1969), and simazine (6-chloro-N^2,N^4-diethy-1,3,5-triazine-2,4-diamine(Domsch and Paul, 1974), also inhibited nitrification, although the effects were generally of short duration (Domsch and Paul, 1974; Bartha et al., 1967; Voets et al., 1974). In some cases, degradation products were also inhibitory to nitrification (Corke and Thompson, 1970). Examples of no effect (Domsch and Paul, 1974; Deshmukh and Shrikhande, 1974; van Schreven et al., 1070; Thorneburg and Tweedy, 1973), and even stimulation (Joshi et al., 1976; Zavarzin and Belyaeva, 1966) of nitrification have also been reported for most of the pesticides described above. Similar information is available on scores of other pesticides (Lal and Lal, 1988). Inhibition of nitrification is generally not considered a problem on site, where nitrification results in increased mobility, and susceptibility of N fertilizers to leaching losses. The impact of off-site effects is generally unkown, but is probably wise to avoid.

Nodulation, nitrogen fixation, and growth of various legumes can be inhibited by pesticides (Daitloff, 1970; Faizah et al., 1980; Garcia and Jordan, 1967, 1969;). Again, examples no effect (Dunigan et al., 1972; Stovold and Evans, 1980) and even stimulatory effects (Mallik and Tesfai, 1985; Grossbard, 1970; Hossain and Alexander, 1984; Chamber and Montes, 1982; Jones and Giddens, 1984; Ostwal and Guar, 1971) of pesticide application on nitrogen fixation have been reported. The effects of pesticides on nonsymbiotic nitrogen fixation by heterotrophic and photosynthetic bacteria have also been investigated. Again, both stimulatory (Nayak and Rao, 1982; Tu, 1981) and inhibitory (Nayak and Rao, 1980; Vlassak et al., 1976) effects have been reported.

Potential effects on other aspects of the nitrogen cycle have been reported. The processes of denitrification, and especially ammonification, are mediated by extremely diverse groups of microorganisms. Some 23 genera are reported to possess the capacity to use nitrate as a terminal electron acceptor (denitrification), and most heterotrophic organisms capable of assimilating carbon-bound nitrogen will excrete excess N as ammonium (ammonification or mineralization). Even though these groups are quite diverse, pesticide effects on their activities have been reported. As expected, microbial responses have been positive (Cervelli and Rolston, 1983; Ishizawa and Matsuguchi, 1966), negative (Rolston and Cervelli, 1980; Bollag and Kurek, 1980), or lacking (Bollag and Kurek, 1980; Bollag and Henninger, 1976).

The most common indicator of the effects of pesticides on microbial activity, and cycling of carbon, is respiration. Since most of the soil microbial community is heterotrophic, it is expected that only broad-scale effects of pesticides on soil microorganisms would result in changes in respiration levels. As expected, soil respiration is much less sensitive than nitrification to application of pesticides (Parr, 1974). Moreover, even large concentrations of many pesticides were not inhibitory (Bartha et al., 1967). Pesticides designed for antimicrobial action, such as fungicides, generally exhibited the most pronounced effects on respiration (Chandra and Bollen, 1961).

B. Toxic Organic and Inorganic Pollutants

Numerous organic compounds other than pesticides enter soils as a result of industrial activities, including chemical synthesis, mining of coal, and processing of petroleum, coal tars, oil shales, and other energy-related materials. Organic contaminants resulting from these activities occur at widely ranging concentrations, from near saturation of the soil pore space in the case of oil or fuel spills to $<10^{-9}$ kg ha^{-1}. Attempts to determine the effects of such compounds have employed a variety of research techiques. Generally, much more dramatic effects on microbial communities and their activities have been observed for toxic organics than for pesticides, which generally have undergone extensive screening for ecotoxicological effects prior to release.

1. Effects on Microbial Communities

Spillage of crude oil on arctic and subarctic soils resulted in decreased bacterial diversity and increased proliferation of heterotrophic bacteria (Sexstone et al., 1978a, 1978b). Increased microbial activity was attributed to residues formed by increased plant mortality and increased soil temperature resulting from darkening of the soil surface (Sexstone and Atlas, 1977; Sparrow et al., 1978), although decomposition of plant cellulose was thought to be inhibited by crude oil (Linkins et al., 1978; Parkinson et al., 1975). The effects of crude oil spills (and other environmental insults) are extremely persistent in cold-region soils. Microbial biomass, estimated by ATP content, as well as cellulose and wood decomposition rates, were significantly lower in treated plots 10 years after the initial application of crude oil (Sparrow and Sparrow, 1988). Impared activities and reduced microbial biomass as well as the persistence of petroleum residues in these soils suggested that the effects of the initial environmental insult had changed little in 10 years and were expected to persist for many more years. It should be noted that effects of the oil spill were temporally variable (Figure 1), suggesting a need for periodic sampling to establish the extent of damage. Partial remediation of crude oil spills has been achieved by the addition of fertilizer materials. Application of fertilizer containing urea and phosphate (27:27:0) resulted in increased bacterial numbers and stimulation of microbial degradation of *n*-alkanes in the the petroleum (Jobson et al.,1974). Inoculation with oil-utilizing bacteria had little effect on oil degradation.

Addition of retorted oil shales to surface soils resulted in reduction in populations of fungi, and biomass (indicated by ATP measurements), whereas no effect was seen on bacterial populations (Hersman and Klein, 1979). Bacterial diversity (cluster analysis) and species richness were suppressed in soils containing oil shales (Segal and Mancinelli, 1987). Sites that had been revegetated appeared less affected by shale disposal, and

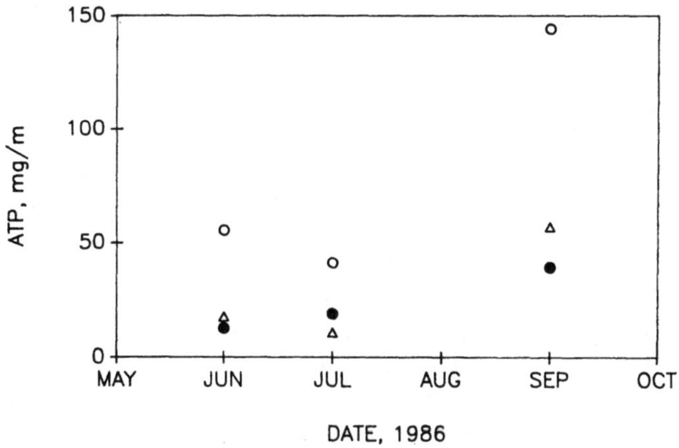

Figure 1. Temporal variability in the effects of oil spillage on ATP (biomass) content of tundra soil. Open circles are for untreated soil, filled circles are for soils treated in winter, open triangles are for soils treated in summer. Data are adapted from Sparrow and Sparrow (1988).

increased in bacterial diversity and species richness with increasing duration of vegetative cover (3 to 6 years). Based on physiological diversity measurements, Metzger et al. (1986) suggested that bacteria in the rhizosphere of range plants were less susceptible to the effects of oil shales than those in the bulk soil. Treatment of soil with pyridine, a major toxicant present in oil-shale retort water (Pereira et al., 1983), resulted in decreased bacterial diversity but did not affect bacterial abundance or biochemical estimators of biological activity (Brand and Sims, 1987).

2. Effects on Nutrient Cycling

Spillage of crude oil and disposal of retorted oil shales on land have been shown to suppress cellulose degradation, suggesting an inpairment of carbon cycling in these systems (Segal and Mancinelli, 1987; Sparrow and Sparrow, 1988; Linkins et al., 1978; Parkinson et al., 1975), although the results have not always been consistent with populations and activities of heterotrophic organisms (Sexstone and Atlas, 1977; Sparrow et al., 1978). Generally, utilization of oil by bacteria in spills has been extremely limited, particularly in cold-region soils (Sexstone et al., 1978b), and little degradation of non-petroleum-based organic carbon occurred over an extended period of time (Sparrow and Sparrow, 1988). The results of these investigations suggest that oil spills may induce long-term effects on carbon cycling in soils, particularly those of the arctic and subarctic regions. Such spills represent the most dramatic form of environmental insult, where soil pores may be completely filled with oil, resulting in biological damage not only from toxicity, but also from disruption of membranes or other physical effects of submersion in a nonpolar solvent matrix. Ecosystems may be much more tolerant of low concentrations of crude oil or its constituents.

Total nitrogen in soils treated with crude oil was significantly higher than untreated soil 10 years after the initial application (Sparrow and Sparrow, 1988). The results suggest that inputs/outputs were higher for treated plots, possibly due to differential effects on biological processes leading to nitrogen loss. This idea was supported by the occurrence of almost all the soil inorganic nitrogen as ammonium, which is much stable in the environment than nitrate. The authors also suggested increased N fixation in the treated plots as a possible explanation, although this seems unlikely. Inhibition of nitrification rather than stimulation of N fixation seemed likely, since both processes appear to be sensitive to organic pollutants. For example, nitrogen fixation was inhibited by the addition of retorted oil shales to soils (Hersman and Klein, 1979).

Toxicity of metals to soil microorganisms and microfauna is well documented. Metal contamination (largely Zn and Cu from a brass mill) resulted in a 20% reduction in the rate of forest litter decomposition (Bengtsson et al., 1988). Data strongly indicated reduced activity of soil animals as a major factor in the inhibition of carbon cycling in the presence of metals. Enrichment of soil with heavy metals (Cu, Ni, Zn, Cd) resulted in decreased microbial biomass and ATP concentration more than 20 years after metal inputs were terminated (Brookes and McGrath, 1984). Heterotrophic and autotrophic nitrogen fixation, nitrification, and dehydrogenase activity were inhibited as well (Brookes et al., 1984, 1986). Mineralization of organic nitrogen and adenylate energy charge associated with whole soils were unaffected by metal enrichment (Brookes et al., 1984; Brookes and McGrath, 1987). Incidence of mycorrhizal root tips was significantly decreased in soils naturally enriched in Cu, Pb, and Zn (Bell et al., 1988). The presence of toxic metals in sewage sludges apparently inhibited N_2 fixation in soybeans for up to eight years after application of high-metal sludge materials (Heckman et al., 1987). Small additions of sludges low in metals often stimulated N_2 fixation, supporting the role of metals in suppression of organisms in the presence of high-metal sludges. These data exemplify long-term biological effects commonly observed in metal-contaminated soils.

Microorganisms may be involved in increasing toxicity of metals to other microorganisms, plants, and animals. Strongly bound cadmium was released into solution by the action of microorganisms (Chanmugathas and Bollag, 1987), suggesting a potential role of microorganisms in metal toxicity. Microbial mobilization of metals has been linked to production of water-soluble ligands (Francis et al., 1980) or changes in the physiochemical condition of the soil. Microorganisms have also been linked to the removal of metals from solution (Kurek et al., 1982; Bollag and Duszota, 1984).

V. Effects of Mining Operations on Soil Biology

The effects of various mining operations on environmental quality have been thoroughly explored. When materials such as ores containing iron sulfides (e.g., pyrite) have been excavated and exposed to the surface environment, oxidation has occurred that lead to the production of sulfate and acidic conditions. The discharge associated with sulfide-bearing mine spoils, particularly those found in coal mining operations, has often been extremely acidic (pH values of 1–2 are common), and has induced profound off-site effects. Drainage from coal mines has often contained high concentrations of potentially toxic metals, such as Zn, Cu, Ni, or Mn. For example, Massey and Barnhisel (1972) reported the release of up to 59 mg kg^{-1} Zn, 85 mg kg^{-1} Cu, and 122 mg kg^{-1} Ni in successive extractions of mine spoils over a 35-week period. The pH of leachates ranged from 1.0 to 4.1, and in some cases declined over the 35-week period, suggesting continued oxidation of sulfide minerals over the duration of the experiment. Not all formations subjected to mining have had the same mineral constituents, and therefore have produced different effects on water chemistry. The composition of geologic material in sites affected by mining influenced the chemistry of sediments transported from associated watersheds (Dick et al., 1986). Materials transported from mined sites and deposited off-site resulted in damage to the productivity of deposition sites (Knabe, 1964). Reclamation of strip-mined sites has not resulted in immediate cessation of off-site effects. The concentration of Mn in stream water receiving runnoff from reclaimed watersheds continued to exceed acceptable limits, and had not come to equilibrium two to three years after reclamation (Dick et al., 1983).

Effects of mining operations on soil microbiology have been noted. Disturbance of soils by mining reduced populations and root colonization by VAM fungi (Reeves et al., 1979). Only 7% of added ammonium was oxidized to nitrate in surface-mined soils (Hons and Hossner, 1980). This was compared to oxidation of 93% of added ammonium in an undisturbed site. The soils under investigation were acidic (pH 4.4-6.6) and had high capacities to retain ammonium in a nonexchangeable form, which was attributed to the presence of lignite. In a subsequent experiment by Reeder (1988), mine spoil materials were found unlikely to adequately supply the N requirements of a reseeded plant community. N mineralization rates in spoil material were very slow (one-fifth the rate of associated topsoil). Nitrate loss due to leaching or volatilization appeared likely in spoil material.

VI. Effects of Land Management Practices on Soil Biology

A. Overview of the Effects of Degradative Processes on the Structure and Function of Microbial Communities

In addition to the introduction of pesticides and other potentially toxic materials, there are numerous mechanical effects of land management that may potentially affect soil biology. Among these effects are erosion, compaction, and changes in patterns of drainage. The removal of significant amounts of surface soil materials by erosion processes results in the loss of organic carbon, inorganic nutrients, and microbial biomass. Subsurface materials may be less conducive to microbial growth, due to inadequate organic carbon, adverse chemical conditions (low pH, etc.), or structural properties that impede oxygen and/or water supply. The detrimental effects of erosion on soil biology are discussed in more detail in Section VI.C.2.

Improper soil drainage results in either water or oxgyen deficiency, both of which affect the structure and function of microbial communities. Anoxic conditions resulting from excess water produce not only changes in community structure (Takai and Kamura, 1966), but also dramatic shifts in the rates or products of soil processes.

B. Effects of Forest Management Practices on Soil Biology

1. Effects on Microbial Communities

Fire is a natural environmental factor in forest ecology, and is also used as a management tool in both forestry and agriculture. Prescribed burning is common practice in the conversion of virgin forests to second-growth forests. Depending on the type of fire, burning may have dramatic effects on physical, chemical, and biological properties of forest soils. Soil pH and the availability of certain inorganic nutrients, such as Ca and Mg, were increased as a result of fire (Viro, 1974). Large amounts of organic carbon (Armson, 1977), as well as sulfur, phosphorus, and boron (Kimmins, 1987), have been lost through hot fires. Some fires have destroyed soil structure, resulting in decreased infiltration and increased surface runoff and erosion (Fuller et al., 1955). Low-temperature fires (260 to 315°C) have often resulted in formation of hydrophobic surface layers (Debano et al., 1967), which reduced soil moisture levels through decreased infiltration. It should be noted that some forest floors have been shown to be naturally hydrophobic due to the presence of certain fungal mycelia (Debano and Rice, 1973), which in some cases have been inhibited by fire, thus decreasing hydrophobicity. Fire has also darkened the soil surface and increased light penetration through the foliage canopy, thus increasing soil temperature (Viro, 1974).

Changes in the soil habitat, as well as direct damage induced by fire, may obviously affect soil organisms. Biotic affects of forest fires have often been concentrated in the surface of the litter layer. Populations of meso- and microfauna have usually been pronounced only in the surface layers of forest litter (Kimmins, 1987). Burning has induced mortality of these organisms in the litter surface, and therefore essentially eliminated their activities, though such effects have usually been temporary, reversing within a few years. Abundance of mites and springtails was significantly decreased by annual burning, but was essentially unaffected by fires at five-year intervals (Metz and Farrier, 1971), which demonstrated the need for a recovery period between prescribed burning events. Effects of fire on soil microorganisms have been difficult to predict (Ahlgreen, 1974), but effects, where observed, have generally been temporary. For example, microbial populations declined immediately following a fire, and then increased after the first rainfall (Ahlgreen and Ahlgreen, 1965).

2. Effects on Nutrient Cycling

Clearcutting of forests results in a termination of litter inputs and an increase in the activity of microorganisms decomposing plant litter. Clearcutting resulted in a 23 % decline in forest floor biomass and a 3-cm reduction of the depth of the forest floor in a hardwood deciduous forest (Dominski, 1971). Similar reports were presented for conifer forests (Cole and Gessel, 1963). Such changes in microbial activity were thought to be brought about by the effects of logging operations and loss of forest cover on soil water content, changes in soil temperature resulting from removal of shading by the overstory, and increase in soil pH that occurred as a result of mineralization processes.

After clearcutting, a normal process of succession is expected as pioneer plants reestablish a foliage canopy and initiate litter inputs to the forest floor. The forest floor builds slowly over an extended period of time until a steady-state condition is reached, unless the process is interrupted by insects, fire, disease, or human activities. It may take 300 years (McFee and Stone, 1965) or longer to reach a steady-state condition of the forest floor.

Under common conditions of soil acidity found in forest soils, autotrophic nitrification (particularly ammonium oxidation) is strongly suppressed. As noted previously, clearcutting of forests often results in an increase in soil pH (Kimmins, 1987). Increases in autotrophic (Likens et al., 1969) and possibly heterotrophic (Duggin, 1984) nitrification rates, resulting from increased soil pH and soil temperature, have been thought to lead to nitrate pollution of nearby tributaries, especially if runoff was increased by the removal of of the overstory, with subsequent reduction of the forest floor by increased activity of decomposer species. Increased production of nitrate resulted in production of nitrous oxide, as a result of either nitrification or denitrification activities (Bowden and Bormann, 1986). Nitrous oxide thus formed was transported off-site to nearby streams where it was degassed from solution to the atmosphere. Because of the large scale of clearcutting operations worldwide, this could have be an important source of nitrous oxide that has been overlooked previously.

Prescribed burning may also affect nutrient cycling processes. Armson (1977) reported increased nitrogen fixation as a result of elevated soil pH induced by fire. Burning has resulted in both long-term increases (Ahlgreen, 1974) and decreases (Meiklejon, 1953 in rates of N mineralization.

C. Effects of Cropping Practices on Soil Biology

Biological effects of cropping practices have been studied for decades. Crop production involves many different activities (plowing, application of fertilizers and/or herbicides, prescribed burning, erosion, wheel traffic, irrigation, etc.) that may directly or indirectly affect soil biology. Attempts have been made to elucidate the impact of these various practices on the biochemical potential of soil organisms. Unfortunately, it has been difficult to separate the effects of individual components of agricultural management programs on soil biology, since many factors appear to be interdependent. Perhaps the best place to start is to observe the results of historical agricultural practices. Some of the most interesting experiments on biological effects of crop production have been those evaluating the long-term effects of monoculture and rotational cropping strategies (10 to 100 years).

In 1843, Lawes and Gilbert established an experiment on Broadbalk Field in which the effects on crop yield of inorganic nutrient sources were compared to farmyard manure. The experiment continued largely unchanged until 1968, when crop rotations were included on some of the experimental units. Among the most noteworthy observations was that the soil contained more nitrogen than was present in the parent rock, and that the nitrogen content of most of the experimental plots remained essentially constant for decades, including those plots not receiving nitrogen fertilizers. For example, plot 5 received P, K, Na, and Mg as inorganic fertilizers, but no amendments containing N. During the period from 1865 to 1967, an average of 24 kg of nitrogen was removed annually through harvesting. During this period, nitrogen content remained between 0.107% and 0.105%, suggesting a nitrogen input eqivalent to crop needs. This input has been attributed largely to biological nitrogen fixation on the order of 40–50 Kg N ha^{-1} annually (Day et al., 1975; Jenkinson, 1971, 1973). The capacity of the soil microbial community to fix atmospheric nitrogen was not measurably changed in over 100 years. The Broadbalk experiment demonstrates the ability of cropping systems to maintain some aspects of soil fertility almost indefinitely, even under continuous monoculture cropping practices. Do these results suggest that crop production has little or no impact on soil biology?

Maintainence of nitrogen fertility in long-term experiments was not unique to Broadbalk Field. Soils planted to rice have also demonstrated long-term recuperative capacity, which has been attributed largely to biological nitrogen fixation (Wada et al.,1978). In lowland rice culture, nitrogen fixation has been associated with cyanobacteria in the sediment and water column (Alimagno and Yoshida (1977), and with heterotrophic organisms in the plant rhizosphere (Sims and Dunigan, 1984). An unfertilized plot planted to rice for 54 years maintained an essentially constant fertility level (1600 kg ha^{-1} constant grain yield), which was attributed to nitrogen fixation (Willis and Green, 1948). It is interesting to note that nitrogen gains through N fixation were absent when nitrogen fertilizers were added.

Sanborn field (University of Missouri, Columbia) was established in 1888 to evaluate the use of farmyard manure and the effects of continuous cropping practices on organic matter content of soil. During most of the first five decades of the experiment, nitrogen content of the plow layer for most plots declined (Smith, 1942). The data suggest that the crop needs exceeded the capacity of the soil microflora to fix and/or release inorganic nitrogen.

1. Effects on Microbial Communities

The above experiments suggest that the capacity of the microbial community in soils to assimilate atmospheric nitrogen and release carbon-bound nitrogen to plants is extremely tolerant of conventional crop productions practices, but falls short or barely meets the nitrogen requirements of crop species. Moreover, the process of cultivating and removing crops does not appear to affect the biochemical potential of soil microflora over the long term, but may affect the inherent fertility of the soil. That the potentiality of soil microorganisms may be unaffected by cropping was supported by Tate and Mills (1983), who examined the structure and function of bacterial communities in a fallow, sugar cane, and grass fields on a muck soil. The fields did not differ significantly in diversity (Shannon-Weaver index), nor in the composition of bacterial communities. The authors suggested that the biochemical potential of the bacterial communities was essentially the same at each site. It should be noted that although types and numbers of bacteria were relatively constant among sites, total microbial biomass was significantly less in fallow soil. This may reflect a decreased mass of fungi, which usually account for a 2.5- to 5.0-fold larger portion of the total microbial biomas than bacteria. Regardless of the application of manure or fertilizers, which resulted in pronounced effects on the carbon and/or nitrogen contents of soil in Broadbalk Field, numbers of bacteria, fungi, and protozoa were essentially identical in all plots (Russell, 1973). The experiments above did not account for potential effects of pesticides, erosion, and other parameters that might potentially affect soil biology. It should be informative at this point to examine the effects of some of these factors on soil microorganisms.

The effects of pesticide usage on soil microbiology has been described above, and will not be repeated here. Suffice it to say that pesticides apparently do measurably affect populations and activities of soil microorganisms, although the effects are usually mild and seem to be of short duration. Other amendments commonly used in crop production include fertilizers and lime, which also may induce short-term effects on soil microorganisms. There is no conclusive evidence to support a long-term degenerative effect of usage of fertilizers or pesticides on the genetic diversity or biochemical potential of soil microorganisms on a global basis. Other aspects of agricultural production that may affect soil biology and therefore bear close examination include erosion, crop rotations (or monocropping), and tillage practices.

2. Effects on Erosion

Erosion of topsoil clearly results in a loss of organic carbon, and therefore a loss of substrates for soil microorganisms. Since soil organic matter is largely associated with the soil surface and is less dense than other soil solids (if not less dense than water), it is usually one of the first major soil components lost by erosion (Lucas et al., 1977). Water erosion is a major source of organic matter decline in modeling the dynamics of carbon in soil (Voroney et al., 1981). This is partly due to the fact that although organic carbon in soil may be protected from degradation by adsorption to clay (Parton et al., 1982), the clay fraction is quite susceptible to water erosion and therefore will carry organic carbon with it. Microbial biomass is usually strongly correlated with organic carbon and/or organic nitrogen content. Thus, highly eroded soils would be expected to posses relatively low levels of biomass and herterotrophic activity, as well as slow rates of associated nutrient transformations. Colonization of roots of bahiagrass (*Paspalum notatum*) by vessicular arbuscular mycorrhizae (VAM) was least in the most erosive positions in the landscape of a pasture in northern Florida (Day et al., 1987). The population density of VAM fungi was apparently significantly reduced by mild erosive forces that left the plant cover intact. Similarly, Powell (1980) reported decreased inoculum potential of VAM fungi in eroded soils lacking plant cover. Simulated erosion of up to 37.5 mm of surface soil did not affect populations of two strains of *Rhizobium* (Habte and El-Swaify, 1988). Two additional strains were apparently intolerant of erosion greater than 15 cm of surface soil removal. Tolerance of erosion was apparently linked to survival in the acidic soil horizons exposed by erosion. Restoring soil pH to approximately 6 alleviated negative effects of erosion on intolerant strains. In this case, the effects of erosion on nitrogen fixation might potentially be remediated in part by the use of acid-tolerant strains or liming.

The extent or strength of soil structure, or the arrangement of primary soil particles into secondary structures, is closely linked to susceptibility of soils to erosion. Soil microorganisms and their products have long been recognized as key players in the aggregation of soils (Lynch and Bragg, 1985). Stability of soil aggregates increases with microbial biomass (Lynch, 1984), and appears to be promoted by the production of extracellular polysaccharides. Polysaccharides, which are common features of soil microorganisms, are produced as a means of attachment to surfaces, or tolerance to dessication or chemical toxicants. Filamentous organisms (including VAM fungi) have also been implicated in aggregate stabilization, primarily through physical binding of microaggregates by hyphae. For example, VAM fungi contributed to aggregation of soil in dune sands (Clough and Sutton, 1978). The significance of mycorrhizae in aggregate stabilization of soil planted to onions was attributed largely to effects of plant roots, the growth of which was stimulated by the presence of mycorrhizae (Thomas et al., 1986).

One might expect that anthropic activities which affect soil microorganisms or their activities might also affect soil structure. The effects of cultivation practices on polysaccharide content of soil have been investigated. A large portion (27% to 43%) of the carbohydrates in soil organic matter was contained in the light fraction (<2 Mg m^{-3}), which declined rapidly after the initiation of cultivation (Dalal and Henry, 1988). Carbohydrate associated with clay was more stable, probably due to protection from biodegradation. Over the 45-year period, total organic carbon declined in approximately the same fashion as polysaccharides, suggesting that carbohydrates were no more labile than the bulk of the soil organic carbon. Decline in aggregate stability as a result of cultivation might therefore be expected to mimic the disappearance of organic carbon, which has been described above. Stability of aggregates in a virgin prairie soil was greater than in soil under cultivation for 14 years (Elliott, 1986). Organic carbon, particularly that associated with macroaggregates declined, suggesting that loss of carbon-binding microaggregates into macroaggregates was a major factor in the degeneration of soil structure observed in this experiment. Conversely, Shiel and Rimmer (1984) reported increased bulk density and deterioration of hydrologic parameters as organic carbon accumulated to produce an organic horizon in a long-term meadow hay plot. The cause of organic matter accumulation was presumed to be acidification resulting from 86 years of application of ammonium sulfate to the plots. Though the data appear to contradict observations above, it should be noted that bridges formed by divalent cations may also contribute to aggregate stabiltiy (Lynch, 1984), which could be adversely affected by acidity. These plots had also become chronically wet, which in combination with acidity, resulted in changes in the microbiology, and most likely the rates of production and decomposition of polysaccharides and other aggregate-stabilizing factors. It is clear that cultivation of soil for crop production may lead to decreased stabiltiy of soil aggregates, and may therefore result in deterioration of soil structure and increased sucsceptibility to erosion.

3. Effects on Nutrient Cycling

Mechanical manipulation of surface soil (e.g., tillage) has long been a part of crop production. The nature and extent of tillage operations has been shown to affect the spatial distribution of microorganisms and their activities in soil. Tillage results in significant changes in physical and chemical properties of the soil environment. It is expected that significant environmental changes could affect the ecology of soil microorganisms. Tillage has been found to result in short-term stimulation of soil microorganisms (Maltby, 1975; Sommers and Biederbeck, 1973), or conversely, significant decline of soil mesofauna (Abbott et al., 1979), as well as dramatic decrease in biomass carbon and nitrogen when virgin lands have been brought into agricultural production (Ayanaba et al., 1976). Although negative effects of tillage on earthworm populations have been observed frequently (Evans and Guild, 1974; Guild, 1951, 1952; Abbott and Parker, 1981), long-term stimulatory responses have also been reported (Gerard and Hay, 1979). Mixing of crop residues (and therefore substrates for microorganisms) has been associated with development of microbial populations to greater depth than if the residues were left on the soil surface (Norstadt and McCalla, 1969). It was therefore not surprizing that populations of aerobic bacteria, facultative anaerobes, and denitrifiers were highly stratified in the surface (0 to 7.5 cm) of soils under long-term no-till practices (Doran, 1980a). In this experiment, conventional tillage operations involving plowing and harrowing resulted in a more uniform distribution of microorganisms throughout the upper 30 cm of the soils under investigation. Data extracted from Doran (1980a) and other sources cited therein clearly demonstrate a decrease in the ratio of populations in soils under reduced tillage to those in conventionally tilled soils (Figure 2). Both organic carbon and nitrogen were also stratified in soils under no-till farming, a likely result of leaving crop residues on the soil surface. Similar observations on the stratification of nutrients in soils under reduced-tillage operations were reported by Dick (1983). As one might predict, placement of crop residue dramatically affected populations of soil microorganisms, particularly nitrifying and denitrifying bacteria (Doran, 1980b). The composition of the bacterial communtiy also appeared to be related to the intensity of tillage operations (Doran, 1980a; Firestone et al., 1987)

Prescribed burning is employed not only in forestry, but also in agricultural production. Fire is most commonly used after clearing forests for cultivation of field crops. The effects of burning in these situations has been essentially the same as in forest management. Organic carbon is lost to the atmosphere, resulting in the release to solution or exchange sites of certain inorganic nutrients previously immobilized in biomass or otherwise associated with organic carbon. Extractable phosphorus, exchangeable cations, and soil pH increased (Nye and Greenland, 1960) as a result of fire. Populations of soil microorganisms, rates of N mineralization, and nit-

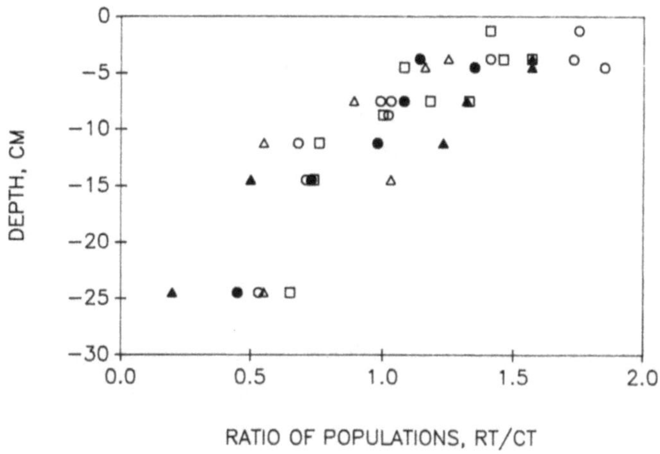

Figure 2. Ratio of populations in soil under reduced tillage (RT) to soil under conventional tillage (CT) as a function of depth. Symbols represent (□) fungi, (○) aerobic bacteria, (●) actinomycetes, (▲) faculative anaerobes, and (△) nitrifiers. Data are adapted from Doran (1980a).

rification have been shown to increase (Corbett, 1934; Dommergues, 1952, 1954), although destruction of nitrifying bacteria and N availability were also observed (Griffiths, 1947). As with other agricultural practices, the long-term effects of prescribed burning have remained essentially unknown. In most cases, burning has been followed by cultivation or some other practice that likely influenced soil microbiology and nutrient cycling as much as the initial fire.

Tillage of previously undisturbed soil has generally resulted in accelerated decomposition of organic carbon (Kononova, 1966), leading to a slow but perceptible decline in the soil organic carbon content (Clement and Williams, 1964; Jenkinson and Johnston, 1977; Russell, 1977). Perspicuous effects of tillage on organic carbon content were demonstrated by Johnston (1973) in an experiment at the Rothamsted Experimental Station (Figure 3). Using natural abundance of ^{13}C, Balesdent et al. (1988) showed the disappearance of carbon derived from native prairie vegetation over a 27-year period after cultivation practices were initiated. This carbon pool then became quite stable, and persisted at near equilibrium levels for the remainder of the approximately 100-year cropping study. The residual prairie carbon was projected to persist for more than 1000 years, whereas the more labile fraction, derived largely from the present crop, had a half-life on the order of 10 to 15 years. This kind of experiment shows promise in elucidating the long-term effects of agricultural practices on carbon cycling by soil microorganisms.

The ability of soil to supply plants with inorganic nitrogen from native

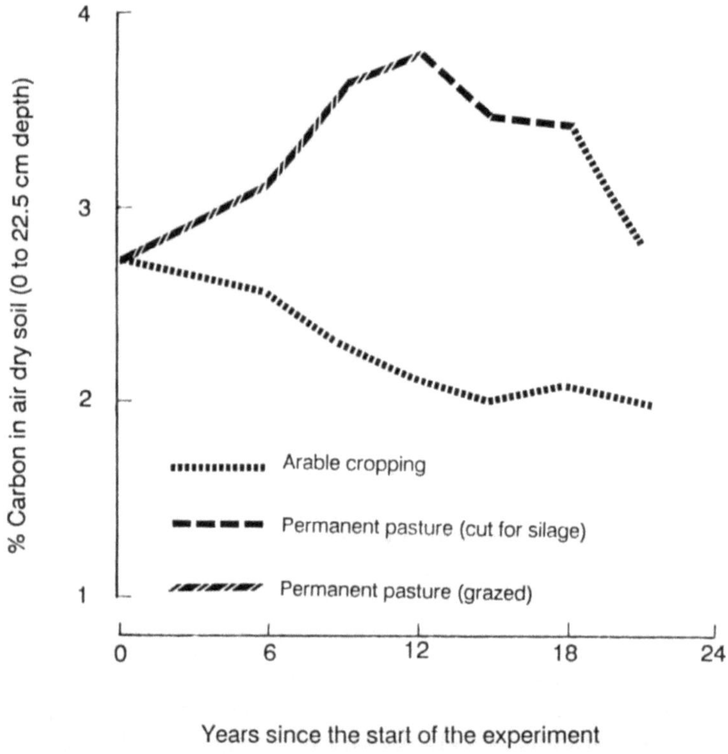

Figure 3. Changes in the organic carbon content of arable and pasture soils in the Highfield ley-arable experiment, Rothamsted, 1949–1972. [From Johnston (1973) with permission.]

organic matter, atmospheric N, or fertilizer amendments is crucial for continued production of crops. Tillage not only reduces organic carbon over the long term, but also affects the organic or total nitrogen content of the soil. This is not particularly surprising in light of the close relationship between carbon and nitrogen content normaly observed in soils. Tillage intesity clearly affects the rates and/or spatial arrangement of numerous aspects of the nitrogen cycle, which merit special attention at this point.

Nitrogen loss associated with tillage is probably due to accelerated decomposition of organic carbon, with concomittant release of nitrogen in excess of the needs of microbial biomass, which also decreases as the available carbon decreases. Inorganic nitrogen released by biomass is eventually lost through plant uptake, leaching, volatilization, etc. What follows will be a comparison of nitrogen cycling under tillage practices that result in mixing of residues within a distinct zone of the soil surface (conventional tillage, CT), and reduced tillage resulting in stratification of residues and nutrients (reduced tillage, RT), as described above in reference to Doran's work.

When inorganic N was limiting, soils under RT supplied less N to plants than those under CT (Blevins et al., 1977; Legg et al., 1979; Meisinger et al., 1985; Moschler and Martens, 1975). Several hypotheses have been evoked to explain this effect. The standing crop of microbial biomass in soils under RT appeared to possess a greater potential for denitrification (Doran, 1980a), which was expressed in denitrification studies (Rice and Smith, 1982) and observations of N_2O flux (Burford et al., 1977). It should be noted that there have been contradictory reports of denitrifier populations under reduced tillage (Staley and Fairchild, 1978), and elevated denitrification in soils under reduced tillage has been attributed to higher water content rather than effects of tillage per se (Rice and Smith, 1982).

Due to the concentration of crop residues at the surface of unplowed agricultural soils, higher net rates of immobilization of native and applied inorganic N relative to mineralization of organic N seemed likely (Mengel et al., 1982; Rice and Smith, 1984; Kitur et al., 1984; Doran, 1980a; Dowdell and Cannell, 1975). This was precedented by observations of the long-term effects of cultivation on carbon and nitrogen content of previously virgin soils, and the higher N and C content of soils under reduced tillage (Blevins et al., 1977; Doran, 1980a; Lal, 1976). Other potential effects of tillage intensity on nitrogen availability have included higher leaching losses under reduced-tillage systems (Thomas et al., 1973) or impairment of function of plant roots under stress from O_2 limitation.

Because of the faster rate of organic matter degradation in plowed soils, the rate constant for N mineralization would be expected to remain higher in these soils relative to soils under reduced tillage. However, for the same reasons, soils under reduced tillage would be expected to accumulate a larger total N pool over the long term, and therefore eventually have a similar capacity to supply available N to plants. This hypothesis has been supported by the observations of Rice et al. (1986) on the changes in nitrogen availability in soils under different tillage practices over a period of 16 years. Figure 4, reprinted from Rice et al. (1986), shows a predictable trend in higher total N content in plots under long-term reduced tillage. Table 4, also from Rice et al. (1986), demonstrates higher N mineralization in soil under reduced tillage for 16 years. The results suggest that problems with N availability commonly associated with reduced-tillage practices may be transitory.

The data available suggest that cultivation of virgin soils results in changes in the structure and functioning of soil microbial communities, though the effects are often subtle. Some of the effects, such as accelerated degradation of organic matter, result in physical damage to the soil ecosystem, as in reduced structural stability and increased erosion. These physical effects may in turn result in additional biological effects, such as decreased microbial activity in eroded soils as a result of organic matter

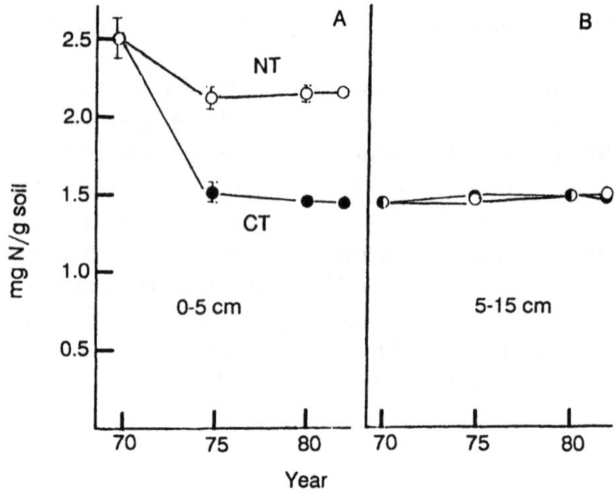

Figure 4. Total soil N as a function of time for no-tillage (○) and conventional tillage (●) system for 0- to 5- (A) and 5- to 15-cm (B) depths on plots receiving 0 kg N ha^{-1} annually. Standard deviations are indicated where they exceed the width of the symbol. Reprinted by permission from Rice, C.W., M.S. Smith, and R.L. Blevins. 1986. Soil nitrogen availability after continuous no-tillage and conventional tillage corn production. *Soil Sci. Soc. Am. J.* 50:1206–1210.

Table 4. Net N mineralization in soil cores from long-term tillage plots and from adjoining undisturbed sod area, which were mixed or left intact and incubated for 2 weeks in the laboratory[a]

	N mineralized by depth[b]		
Cores from	0–7.5cm	7.5–15cm	0–15cm[c]
		—ug N g soil^{-1}—	
Long-term NT	31.8a[d]	13.1a	22.4ab
Long-term CT	13.8b	12.2a	13.0a
Adjoining sod (intact)	—	—	19.3ab
Adjoining sod (mixed)	—	—	23.8b

[a]From Rice et al. (1986).

[b]Values in same column followed by same letter are not significantly different at the 0.05 level by *t*-tests. Values are means of six replicate cores.

[c]Samples were collected in 1982.

[d]0- to 15-cm values for NT and CT were derived by averaging 0- to 7.5- and 7.5- to 15-cm depths.

Source: Rice et al. (1986).

depletion, or exposure of acidic or impermeable subsurface materials. Reduction of organic matter content, stability of soil structure, and susceptibility of soils to erosion appear to be long-term effects of cultivation, particularly when tillage intesity is high. The initial degradative effects of cultivation may stabilize, but will not be expected to reverse in the absence of intervention. Effects of crop production practices on functioning of soil microorganisms appear to be transitory in many cases, and often seem related to organic matter decline, rather than direct effects on the diversity or biochemical potential of soil communities.

It appears likely that structure and function of soil microbial communities will eventually recover after cessation of practices that introduce perturbations. However, the time factor required for recovery is unknown. Although data suggest no long-term toxic effects of recently introduced practices such as pesticide application, little or nothing is known of the sublethal or chronic effects of these activities. Genetic damage, or other chronic effects, may not be expressed for many years. Also unknown are the effects of agricultural practices on fastidious organisms, and on organisms whose function is not known. It is doubtful that the latter two questions will be answered with present research methodology.

VII. Microorganisms as Pollutants

Under certain conditions, microorganisms themselves may become pollutants, particularly in the case of pathogens of animals, plants, or humans. Human and animal pathogens have typically been associated with runnoff from livestock operations (Gary et al., 1983), land application of municipal sewage or sewage sludges (Gerba, 1983), and domestic waste-water treatment in soils with fluctuating or high water tables (Stewart and Reneau, 1988). Most national, regional, or state governments have established standards regarding acceptable concentrations of pathogenic organisms in public waterways and water supplies. Typically, fecal coliform bacteria (FC), the most commonly encountered bacterial contaminants, may not be present in concentrations exceeding about 10^3 liter^{-1} of stream water. Bacterial pollution from livestock operations has been reduced by adopting grazing strategies that produce uniform livestock distribution (Tiedemann et al., 1988), although even ungrazed pastures have produced runoff that exceeded local drinking-water standards (Doran and Linn, 1979). Risk of pathogen contamination from land application of waste has been reduced by applying only treated wastes (Gerba, 1983). The degree and form of waste treatment required has generally been site-specific. Avoiding installation of below-ground septic systems in risky areas has avoided contamination of soils by pathogens associated with domestic wastes, and some useful strategies have been developed to reduce risks when such installation was unavoidable (Stewart and Reneau, 1988).

VIII. Conclusions

It is apparent that, given enough time, most if not all terrestrial microbial communities damaged by activities of humans will eventually return to some healthy condition, although it may not necessarily resemble the original condition, and may involve a very long wait. Clearly, some human activities are much more damaging to soil organisms than others, and the effects of some activities, as in strip mining and metal contamination, impart effects of much longer duration than others, such as pesticide application. Similarly, microbial communities in some econiches, as in cold regions, are particularly sensitive to perturbation, and do not recover quickly. Some measures are available for accelerated recovery of damaged ecosystems, which have not been elaborated herein. It should be noted that remediation usually starts with removing the source of perturabation. Regardless of what cost-effective measures might be introduced for remediation, it seems obvious that if remediation involves discontinuing existing practices, we should begin testing alternatives in the present to assess the long-term effects of proposed "safe" practices.

References

Abbott, I., and C.A. Parker. 1981. Interactions between earthworms and their environment. *Soil Biol. Biochem.* 13:191–197.

Abbott, I., Parker, C.A., and I.D. Sills. 1979. Changes in the abundance of large soil animals and physical properties of soils following cultivation. *Aust. J. Soil Re.* 17:343–353.

Ahlgreen, I.F. 1974. The effect of fire on soil organisms. pp. 47–72. In: T.T. Kozlowski and C.E. Ahlgreen (eds.), *Fire and ecosystems*. Academic Press, NY.

Ahlgreen, I.F., and C.E. Ahlgreen. 1965. Ecology of prescribed burning on soil microorganisms in a Minnesota jack pine forest. *Ecology*. 46:304–310.

Ainsworth, S. 1977. *Steady-state enzyme kinetics*. University Park Press, Baltimore, MD.

Alexander, M. 1969. Microbial degradation and biological effects of pesticides in soil. pp. 209–240. In: *Soil biology: Reviews of research*. UNESCO, Paris.

Alexander, M. 1971. *Microbial ecology*. Wiley, NY.

Alexander, M. 1977. *Introduction to soil microbiology*, 2nd ed. Wiley, NY.

Alimagno, B.V., and T. Yoshida. 1977. In situ acetylene-ethylene assay of biological nitrogen fixation in lowland rice soils. *Plant and Soil* 47:239–244.

Anan'eva, N.D., B.P. Strekozov, and G.K. Tyuryukanova. 1986. Changes in the microbial biomass in soils in response to pesticides. *Agrokhimiya* 5:84–90.

Armson, K.A. 1977. *Forest soils, properties and processes*. Univ. Toronto Press, Toronto.

Atlas, R.M. 1984a. Diversity of microbial communities. pp. 1–47. In: K.C. Marshall (ed.), *Advances in microbial ecology*, Vol.7. Plenum, NY.

Atlas, R.M. 1984b. Use of microbial diversity measurements to assess environmental stress. pp. 540–545. In: M.J. Klug and C.A. Reddy (eds.), *Current perspectives in microbial ecology*. Amer. Soc. Microbiol., Washington, DC.

Atlas, R.M. and R. Bartha. 1987. *Microbial ecology, fundamentals and applications*, 2nd ed. Benjamin/Commings. Menlo Park, CA.

Ayanaba, A., S.B. Tuckwell, and D.S. Jenkinson. 1976. The effects of clearing and cropping on the organic reserves and biomass of tropical forest soils. *Soil Biol. Biochem.* 8:519–525.

Balesdent, J., G.H. Wagner, and A. Mariotti. 1988. Soil organic matter turnover in long-term field experiments as revealed by carbon-13 natural abundance. *Soil Sci. Soc. Amer. J.*52:118–124.

Bartha, R., R.P. Lanzillata, and D. Pramer. 1967. Stability and effects of some pesticides in soil. *Appl. Microbiol.* 15:67–75.

Bell, R., C.S. Evans, and E.R. Roberts. 1988. Decreased incidence of mycorrhizal root tips associated with soil heavy-metal enrichment. *Plant and Soil* 106:143–145.

Bengtsson, G., M. Berden, and S. Rundgren. 1988. Influence of soil animals and metals on decomposition processes: A microcosm experiment. *J. Envir. Qual.* 17:113–119.

Biederbeck, V.O., C.A. Campbell, and A.E. Smith. 1987. Effects of long-term 2,4-D field applications on soil biochemical processes. *J. Envir. Qual.* 16:257–262.

Blevins, R.L., G.W. Thomas, and P.L. Cornelius. 1977. Influence of no-tillage and nitrogen fertilization on certain soil properties after five years of continuous corn. Agron. J. 69:383–386.

Bollag, J.-M., and M. Duszota. 1984. Effect of the physiological state of microbial cells on cadnium sorption. Arch. Environ. Contam. Toxicol. 13:265–270.

Bollag, J.-M., and N.M. Henninger. 1976. Influences of pesticides on denitrification in soil and with an isolated bacterium. *J. Envir. Qual.* 5:15.

Bollag, J.-M., and E.J. Kurek. 1980. Nitrate and nitrous oxide accumulation during denitrification in the presence of pesticide deriatives. *Appl. Envir. Microiol.* 39:845.

Bollen, W.B. 1961. Interactions between pesticides and soil microorganisms. *Ann. Rev. Microbiol.* 15:69–92.

Bowden, W.B., and F.H. Bormann. 1986. Transport and loss of nitrous oxide in soil water after forest clear-cutting. *Science* 233:865–869.

Brand, K.M., and G.K. Sims. 1987. Influence of microbial succession on soil biochemical measurements obtained in lab incubations. *Agron. Abstr.*, p. 179.

Brookes, P.C., and S.P. McGrath. 1984. Effects of metal toxicity on the size of the soil microbial biomass. *J. Soil Sci.* 35:341–346.

Brookes, P.C., and S.P. McGrath. 1987. Adenylate energy charge in metal-contaminated soil. *Soil Biol. Biochem.* 19:219–220.

Brookes, P.C., and S.P. McGrath, and C. Heijnen. 1986. Metals residues in soils previously treated with sewage-sludge and their effects on growth and nitrogen fixation by blue-green algae. *Soil Biol. Biochem.* 18:345–353.

Brookes, P.C., S.P. McGrath, D.A. Klein and E.T. Elliott. 1984. Effects of heavy metals on microbial activity and biomass in field soils contaminated with sewage sludge. pp. 574–583. In: *Environmental contamination*, CEP Ltd., Edinburgh.

Burford, J.R., R.D. Dowdell, and R. Cress. 1977. Denitrification: Effect of cultivation on fluxes of nitrous oxide from the soil surface. pp. 71–72. Ann. Rep. 1977, Agric. Res. Coun. (G.B.) Letcombe Lab.

Cervelli, S., and D.E. Rolston. 1983. Influence of atrazine on denitrification in a soil column. *J. Envir. Qual.* 12:482.

Chamber, M.A., and F.J. Montes. 1982. Effects of some seed disinfectants and methods of rhizobial inoculation of soybeans *Glycine max* (L.). *Plant Soil* 66:353.

Chandra, P. 1964. Herbicidal effects on certain microbial activities in some brown soils of Saskatchewan. *Weed Res.* 4:54–63.

Chandra, P., and W.B. Bollen. 1961. Effects of nabam and mylone on nitrification, soil respiration, and microbial numbers in four Oregon soils. *Soil Sci.* 92:387–393.

Chanmugathas, P., and J.M. Bollag. 1987. Microbial mobilization of cadmium in soil under aerobic and anaerobic conditions. *J. Envir. Qual.* 16:161–167.

Chet, I., and R. Mitchell. 1976. Ecological aspects of microbial chemotactic behavior. *Ann. Rev. Microbiol.* 30:221–239.

Clark, F.E., and E.A. Paul. 1970. The microflora of grasslands. *Adv. Agronomy* 22:405.

Clement, C.R., and T.E. Williams. 1964. Leys and soil organic matter. I. The accumulation of organic carbon in soil under different leys. *J. Agric. Sci.* 63:377–383.

Clough, K.S., and J.C. Sutton. 1978. Direct observation of fungal aggregates in sand dune soil. *Can. J. Microbiol.* 24:333–335.

Cole, D.W., and S.P. Gessel. 1963. Movement of elements through a forest soil as influenced by tree removal and fertilizer additions. pp. 95–104. In: *Forest soil relationships in North America*, C.T. Youngberg (ed.). Oregon State Univ. Press, Corvallis.

Corbett, S.A. 1934. Studies in tropical soil microbiology. II. The bacterial numbers in the soil of the Malay Peninsula. *Soil Sci.* 38:407–416.

Corke, C.T., and F.R. Thompson. 1970. Effects of some phenylamide herbicides and their degradation products on soil nitrification. *Canad. J. Microbiol.* 16:567–571.

Daitloff, A. 1970. The effects of some pesticides on root nodule bacteria and subsequent nodulation. *Aust. J. Exp. Agric. Anim. Husb.* 10:562.

Dalal, R.C., and R.J. Henry. 1988. Cultivation effects on carbohydrate contents of soil and soil fractions. *Soil Sci. Soc. Amer. J.* 52:1361–1365.

Day, J., D. Harris, P.J. Dart, and P. VanBerkum. 1975. The Broadbalk experiment. An investigation into gains from non-symbiotic nitrogen fixation. pp. 71–84. In: W.D.P.Stewart (ed.), *Nitrogen fixation by free-living microorganisms*. Cambridge Univ. Press, Cambridge, U.K.

Day, L.D., D.M. Sylvia, and M.E. Collins. 1987. Interactions among vesicular-arbuscular mycorrhizae, soil, and landscape position. *Soil Sci. Soc. Amer. J.* 51:635–639.

Debano, L.F., and R.M. Rice. 1973. Water-repellent soils:Their implications in forestry. *J. For.* 71:220–223.

Debano, L.F., J.F. Osborn, J.S. Krammes, and J.Lettey. 1967. Soil wettability and wetting agents: Our knowledge of the problem. Rep. PSW-43, U.S. Dept. Agric. For. Serv. Res.

Deshmukh, V.A., and J.G. Shrikhande. 1974. Effect of pre- and postemergence treatment of herbicides on soil microflora and microbial processes. *J. Indian Soc. Soil Sci.* 22:36.

Dick, W.A. 1983. Organic carbon, nitrogen, and phosphorus concentrations and pH in soil profiles as affected by tillage intensity. *Soil Sci. Soc. Amer. J.* 47:102–107.

Dick, W.A., J.V. Bonta, and F. Haghiri. 1986. Chemical quality of suspended sediment from watersheds subjected to surface coal mining. *J. Envir. Qual.* 15:289–293.

Dick, W.A., J.V. Bonta, F. Haghiri, and J.R. Page. 1983. Stream water quality of two small watersheds as affected by surface coal mining. *J. Envir. Qual.* 12:351–358.

Dominski, A. 1971. Accelerated nitrate production and loss in a northern hardwood forest ecosystem underlain by podzol soils following clearcutting and addition of herbicides. Ph.D. thesis, Yale Univ., New Haven, CT.

Dommergues, Y. 1952. Influence du defrinchement de foret suivi d'incendie sur l'activite' biologique du sol'. *Mem. Inst. Sci. Mad. D.* 4:273–296.

Dommergues, Y. 1954, Action de feu sur la microflora des sols de prarie. *Mem. Inst. Sci. Mad. D.* 6:149–158.

Domsch, K.H., and W. Paul. 1974. Simulation and experimental analysis of the influence of herbicides on soil nitrification. *Arch. Microbiol.* 97:283.

Doran, J. W. 1980a. Soil microbial and biochemical changes associated with reduced tillage. *Soil Sci. Soc. Amer. J.* 44:765–771.

Doran, J.W. 1980b. Mocrobial changes associated with residue management with reduced tillage. *Soil Sci. Soc. Amer. J.* 44:518–524.

Doran, J.W., and D.M. Linn. 1979. Bacteriological quality of runoff water from pastureland. *Appl. Envir. Microbiol.* 37:985–991.

Dowdell, R.J., and R.Q. Cannell. 1975. Effect of plowing and direct drilling on soil nitrate content. *J. Soil Sci.* 26:53–61.

Dubey, H.D. 1969. Effects of picloram, diuron, ametryne, and prometryne on nitrification in some tropical soils. *Soil Sci. Soc. Amer. Proc.* 33:893.

Duggin, J.A. 1984. Thesis, Yale Univ., New Haven, CT.

Dunigan, E.P., J.P. Frey, L.D. Allen, and A.McMahon. 1972. Herbicidal effects on the nodulation of *Glycine max* (L.) Merrill. *Agron. J.* 64:806–808.

Elliott, T.E. 1986. Aggregate structure and carbon, nitrogen, and phosphorus in native and cultivated soils. *Soil Sci. Soc. Amer. J.* 50:627–633.

Evans, A.C., and W.J. Guild. 1947. Studies on the relationships between earthworms and soil fertility. I. Biological studies in the field. *Ann. Appl. Biol.* 34:307–330.

Faizah, A.W., W.J. Broughton, and C.K. John. 1980. Rhizobia in tropical legumes. IX. Survival in the seed environment. *Soil Biol. Biochem.* 12:219.

Firestone, M.M., L. Burton, and G.K. Sims. 1987. Effects of tillage practices on diversity of soil bacteria. *Agron. Abstr.*, p. 182.

Fisher, R.A., A.S. Corbet, and C.B. Williams. 1943. The relation between the numbers of species and the number of individuals in a random sample of an animal population. *J. Animal Ecol.* 12:42–58.

Fletcher, W.W. 1960. The effect of herbicides on soil microorganisms. pp. 20–62. In: E.K. Woodford and G.R. Sagar (eds.), *Herbicides and the soil.* Blackwell Scientific, Oxford, England.

Francis, A.J., C.R. Iden, B.J. Nine, and C.K. Chang. 1980. Characterization of organics in leachates from low-level radioactive waste disposal sites. *Nucl. Technol.* 50:158–163.

Fuller, W.H., S. Shannon, and P.S. Burgess. 1955. Effect of burning on certain forest soils of northern Arizona. *For. Sci.* 1:44–50.

Garcia, M., and D.C. Jordan. 1967. Effect of herbicides on nodulation and N-fixation in birdsfoot trefoil. *Bacteriol. Proc.*, p. 8.

Garcia, M., and D.C. Jordan. 1969. Action of 2,4-DB and dalapon on the symbiotic properties of *Lotus corniculatus* (birdsfoot trefoil). *Plant Soil* 30:317.

Gary, H.L., S.R. Steven, and S.L. Ponce. 1983. Cattle grazing impact on surface water quality in a Colorado Front Range stream. *J. Soil Water Conserv.* 38:124–128.

Gerard, B.M., and R.K. Hay. 1979. The effect on earthworms of ploughing, tined cultivation, direct drilling and nitrogen in a barley monocultures system. *J. Agric. Sci.* 93:147–155.

Gerba, C.P. 1983. Pathogens. pp. 147–195. In: A.L. Page T.L. Gleason, J.E. Smith, Jr., I.K. Iskandar, and L.E. Sommers (eds), *1983 workshop on utilization of municipal wastewater and sludge on land*. Univ. California, Riverside.

Gibson, S.A., and J.M. Suflita. 1986. Extrapolation of biodegradation results to groundwater aquifers: Reductive dehalogenation of aromatic compounds. *Appl. Envir. Microbiol.* 52:681–688.

Gottschalk, G. 1986. *Bacterial metabolism*, 2d ed. Springer-Verlag, NY.

Griffiths, A.L. 1947. Effects of artificial soil regeneration by burning. *Indian For.* 73:526–528.

Grossbard, E. 1970. Effect of herbicides on the symbiotic relationship between *Rhizibium trifolii* and white clover. *Occas. Symp. Br. Grassland Soc. White Clover Research* 6:47.

Guild, W.J. 1951. The distribution and population density of earthworms (Lumbicadae) in Scottish pasture fields. *J. Animal Ecol.* 20:88–97.

Guild, W.J. 1952. Variations in earthworm numbers within field populations. *J. Animal Ecol.* 21:169–181.

Habte, M., and S.A. El-Swaify. 1988. Survival of *Rhizobium* in an oxisol subjected to incremental simulated erosion. *Soil Sci. Soc. Amer. J.* 52:1313–1316.

Hauke-Pacewiczowa, T. 1971. The effect of herbicides on the activity of soil microflora. *Pamiet. Pulawski* 46:5.

Heckman, J.R., J.S. Angle, and R.L. Chaney. 1987. Residual effects of sewage sludge on soybean. II. Accumulation of soil and symbiotically fixed nitrogen. *J. Envir. Qual.* 16:118–124.

Helling, C.S., P.C. Kearney, and M. Alexander. 1971. Behavior of pesticides in soils. *Advan. Agron.* 23:147–240.

Hersman, L.E., and D.A. Klein. 1979. Retorted oil shale effects on soil microbiological characteristics. *J. Envir. Qual.* 8:520–524.

Hons, F.M., and L.R. Hossner. 1980. Soil nitrogen relationships in spoil material generated by the surface mining of lignite coal. *Soil Sci.* 129:222–228.

Hossain, A.K.M., and M. Alexander. 1984. Enhancing growth and nitrogen uptake by soybeans using pesticides. *Plant Soil* 81:133.

Ishizawa, S., and T. Matsuguchi. 1966. Effects of pesticides and herbicides upon microorganisms in soil and water under waterlogged conditions. *Bull. Nat. Inst. Agric. Sci., Tokyo, Ser B* 16:1.

Jenkinson, D.S. 1971. The accumulation of organic matter in soil left uncultivated. pp. 113–137. In: Report of Rothamsted Experimental Station for 1970, Part 2.

Jenkinson, D.S. 1973. Organic matter and nitrogen in soils of the Rothamsted Classical Experiments. *J. Sci. Food Agric.* 24:1149–1150.

Jenkinson, D.S., and A.E. Johnston. 1977. Soil organic matter in the Hoosfield continuous barley experiment. pp. 87–112. In: *Report of the Rothamsted Experimental Station for 1976, Part 2.*

Jobson, A., M. McLaughlin, F.D. Cook, and D.W.S. Westlake. 1974. Effect of amendments on the microbial utilization of oil applied to soil. *Appl. Microbiol.* 27:166–171.

Johnston, A.E. 1973. The effects of ley and arable cropping systems on the amount of soil organic matter in the Rothamsted Woburn ley arable experiments. pp. 131–159. In: *Report of the Rothamsted Experimental Station for 1972, Part 2.*

Jones, L.W. 1956. The effects of some pesticides on microbial activities of the soil. Bull. 390, Utah Agr. Exp. Sta., p. 1–17.

Jones, R., and J. Giddens. 1984. Introduction of effective N_2-fixing rhizobial strains into the soybean plants by use of fungicide resistance. *Agron. J.* 76:599.

Joshi, O.P., M.S. Sachdev, K.L. Sahrawat, and B.N. Kohli. 1976. Effect of simazine and atrazine on the mineralization of fertilizer and manure nitrogen. *Plant Soil* 44:367.

Kaspar, H.F., and J.M. Tiedje. 1982. Anaerobic bacteria and processes. pp. 989–1009. In: *Methods of soil analysis, Part 2*, 2d ed. Amer. Soc. Agron., Madison, WI.

Kimmins, J.P. 1987. *Forest ecology*. Macmillan, NY.

Kitur, B.K., M.S. Smith, R.L. Blevins, and W.W. Frye. 1984. Fate of depleted ammonium nitrate applied to no-tillage and conventions tillage corn. *Agron. J.* 76:240–242.

Knabe, W. 1964. A visiting scientist's observations and recommendations concerning strip-mine reclamation in Ohio. *Ohio J. Sci.* 64:132–154.

Kononova, M.M. 1966. *Soil organic matter*. Permagon Press, New York.

Kreutzer, W.A. 1963. Selective toxicity of chemicals to soil microorganisms. *Ann. Rev. Phytopathol.* 1:101–126.

Kreutzer, W.A. 1965. The reinfestation of treated soil. pp. 495–508. In: K.F. Baker and W.C. Snyder (eds.), *Ecology of soil-borne plant pathogens: Prelude to biological control.* Univ. California Press, Berkeley.

Kurek, E., J. Czaban, and J.-M. Bollag. 1982. Sorption of cadmium by microorganisms in competition with other soil constituents. *Appl. Envir. Microbiol.* 43:1011–1015.

Lal, R. 1976. No-tillage effects on soil properties under different crops in western Nigeria. *Soil Sci Soc. Amer. Proc.* 40:762–768.

Lal, R., and S. Lal. 1988. *Pesticides and nitrogen cycle*, Vols 1–3. CRC Press, Boca Raton, Fi.

Legg, J.O., G. Stanford, and O.L. Bannett. 1979. Utilization of labeled-N fertilizer by silage corn under conventional and no-till culture. *Agron. J.* 71:1009–1015.

Likens, G.E., F.H. Bormann, and N.M. Johnson. 1969. *Science* 163:1205.

Linkins, A.E., R.M. Atlas, and P. Gustin. 1978. Effect of surface applied crude oil on soil and vascular plant root respiration, soil cellulase, and hydrocarbon hydroxylase at Barrow, Alaska. *Arctic* 31:355–365.

Logan, T.J. 1989. Chemical degradation of soil. In: *This volume*.

Lucas, R.E., J.B. Holtman, and L.J. Conner. 1977. Soil carbon dynamics and cropping practices. In: W. Lockeretz (ed.), *Agriculture and energy*. Academic Press, NY.

Lynch, J.M. 1984. Interactions between biological processes, cultivation and soil structure. *Plant Soil* 76:307–318.

Lynch, J.M., and E. Bragg. 1985. Microorganisms and soil aggregate stability. *Adv. Soil Sci.* 2:133–171.

Lynch, J.M., and N.J. Poole. 1979. *Microbial ecology, a conceptual approach*. Halsted Press, NY.

Macfadyen, A. 1963. *Animal ecology: Aims and methods*, 2nd ed. Putnam, London.

MacNab, R. 1987. Motility and chemotaxis. pp. 732–759. In: F.C. Neidhart, J.L. Ingraham, K.B. Low, B. Magasanik, M. Schaechter, and H.E. Umbarger (eds.), *Eschericia coli* and *Salomonella typhimurium. Cellular and molecular biology*. ASM, Washington, DC.

Mallik, M.A.B., and K. Tesfai. 1985. Pesticidal effect on soybean-rhizobia symbiosis. *Plant Soil* 85:33.

Maltby, E. 1975. Numbers of soil microorganisms as indicators of changes resulting from moorland reclamation on Exmoor, UK. *J. Biogeograph*. 2:117–136.

Martin, J.P. 1964. Influence of pesticide residues on soil microbiological and chemical properties. *Residue Rev*. 4:96–129.

Martin, J.P. 1966. Influence of pesticides on soil microbes and soil properties. pp. 95–108. In: M.E. Bloodwoth (ed.), *Pesticides and their effects on soils and water*. ASA Spec. Publ. 8. Soil Sci. Soc. Amer., Madison, WI.

Massey, H.F., and R.I. Barnhisel. 1972. Copper, nickel, and zinc released from acid mine spoil materials of Eastern Kentucky. *Soil Sci.* 113:207–212.

McFee, W.W., and E.L. Stone. 1965. Quantity, distribution, and variability of soil organic matter and nutrients in a forest podzol in New York. *Soil Sci. Soc. Amer. Proc*. 29:432–436.

McIntosh, E.F. 1967. An index of diversity and the relation of certain concepts to diversity. *Ecology* 47:392–404.

Meiklejohn. J. 1953. The effect of bush burning on the microflora of some Kenya soils. *Proc. 6th Int. Conf. Microbiol. 1953* 10:317–319.

Meisinger, J.J., V.A. Bandel, G. Stanford, and J.O. Legg. 1985. Nitrogen utilization of corn under minimum tillage and moldboard plow tillage: Four year results using labeled N fertilizer on an Atlantic Coastal Plain soil. *Agron. J*. 77:602–611.

Mekwatanakarn, P. and K. Sivasithamparam. 1987. Effect of certain herbicides on saprophytic survival and biological suppression of the take-all fungus. *New Phytol.* 106:153–159.

Mengel, D.B., D.W. Nelson, and D.M. Huber. 1982. Placement of nitrogen fertilizers for no-till and conventional till corn. *Agron. J*. 74:515–517.

Metz, L.J., and M.H. Farrier. 1971. Prescribed burning and soil mesofauna on the Santee Experimental Forest. pp. 100–106. In: *Prescribed burning symposium*. U.S. Dept. Agric. For. Serv. SE For. Exp. Sta., Ashville, SC.

Metzger, W.C., D.A. Klein, and E.F. Redente. 1986. Bacterial pysiological diversity in the rhizosphere of range plants in response to retorted shale stress. *Appl. Envir. Microbiol.* 52:765–770.

Moschler, N.W., and D.C. Martens. 1975. Nitrogen phosphorus and potassium requirement in no-tillage and conventionally tilled corn. *Soil Sci. Soc. Amer. Proc.* 39:886–891.

Nayak, D.N., and R.V. Rao. 1980. Pesticides and heterotrophic nitrogen fixation in paddy soil. *Soil Biol. Biochem.* 12:1.

Nayak, D.N., and R.V. Rao. 1982. Pesticides and heterotrophic nitrogen fixation in paddy soil. *Soil Biol. Biochem.* 14:207.

Norstadt, F.A., and T.M. McCalla. 1969. Microbial populations in stubble-mulched soil. *Soil Sci.* 107:188–193.

Nye, P.H., and D.J. Greenland. 1960. The soil under shifting cultivation. *Commun. Bur. Soils, Tech. Commun. No. 51.*

Olsen, R.A., and L.R. Bakken. 1987. Viability of soil bacteria: Optimization of plate-counting technique and comparison between total counts and plate counts within different size groups. *Microbial Ecol.* 13:59–74.

Ostwal, K.P., and A.C. Guar. 1971. Effect of seed-dressing on growth, nodulation, and anatomy of soybeans. *Weed Sci.* 19:147.

Pancholy, S.K., and J.Q. Lynd. 1969. Bromacil interactions in plant bioassay, fungi cultures and nitrification. *Weed Sci.* 17:460.

Parkinson, D., R. Gossen, and S. Henderson. 1975. Effects of oil spillage on microorganisms in northern Canadian soils. INA Pub. No. QS-8040-000-EEE-Al. Rep. ALUR 74-75-82 prepared for the Arctic Land Use Res. Program, Northern Nat. Res. and Envir. Branch, Dep. of Indian Affairs Northern Dev, Ottawa, Canada.

Parr, J.F. 1974. Effects of pesticides on microorganisms in soil and water. pp. 315–340. In: W.D.Guenzi (ed.), *Pesticides in soil and water.* Soil Sci. Soc. Amer. Madison, WI.

Parton, W.J., J. Persson, and D.W. Anderson. 1982. Simulation of organic matter changes in Swedish soils. In: *Proc. Third Int. Conf. State of the Art in Ecological Modelling*, Fort Collins, CO.

Pereira, W. E., C.E. Rostad, J.R. Garbarino, and M.F. Hult. 1983. Groundwater contamination by organic bases derived from coal-tar wastes. *Envir. Toxicol. Chem.* 2:283–294.

Powell, C.L. 1980. Mycorrhizal infectivity of eroded soils. *Soil Biol. Biochem.* 12:247–250.

Reeder, J.D. 1988. Transformations of nitrogen-15 labeled fertilizer nitrogen and carbon mineralization in incubated coal mine spoils and disturbed soil. *J. Envir. Qual.* 17:291–298.

Reeves, F.B., D. Wagner, T. Moorman, and J. Kiel. 1979. The role of endomycorrhizea in revegetation practices in the semi-arid West. I. A comparison of incidence of mycorrhizae in severely disturbed vs. natural environments. *Amer. J. Bot.* 66:6–13.

Rice, C.W., and M.S. Smith. 1982. Denitrification in no-till and plowed soils. *Soil Sci. Soc. Amer. J.* 46:1168–1173.

Rice, C.W., and M.S. Smith. 1984. Short-term immobilization of fertilizer nitrogen at the surface of no-till and plowed soils. *Soil Sci. Soc. Amer. J.* 48:295–297.

Rice, C.W., M.S. Smith, and R.L. Blevins. 1986. Soil nitrogen availability after long-term continuous no-tillage and conventional tillage corn production. *Soil Sci. Soc. Amer. J.* 50:1206–1210.

Richards, B.N. 1974. *Introduction to the soil ecosystem.* Longman, Essex, England.

Richards, B.N. 1987. *The microbiology of terrestrial ecosystems*. Longman/Wiley, NY.
Rolston, D.E., and S. Cervelli. 1980. Denitrification as affected by irrigation frequency and applied herbicides pp. 189–199. In: *Agrochemical residue biota interactions in soil and aquatic ecosystems*. Int. Atomic Eneigy Agency, Vienna.
Roper, M.M., and K.C. Marshall. 1978. Effects of a clay mineral on microbial predation and parasitism on *Eschericia*. *Microbial Ecol.* 4:279–290.
Russell, E.W. 1973. The general ecology of the soil population. pp. 214–240. In: *Soil conditions and plant growth*, 10th ed. Longman, London.
Russell, E.W. 1977. The role of organic matter in soil fertility. *Phil. Trans. Roy. Soc. London B* 281:209–219.
Schmidt, E.L. 1982. Nitrification in soil. pp. 253–288. In: *Nitrogen in agricultural soils*. Agronomy Mono. 22, madison, WI.
Segal, W., and R.L. Mancinelli. 1987. Extent of regeneration of the microbial community in reclaimed spent oil shale land. *J. Envir. Qual.* 16:44–48.
Setty, R.A., V.C. Baligar, and S.V. Patil. 1970. Effect of atrazine on the rate of nitrification in black clay loam soil. *Mysore J. Agric. Sci.* 4:11.
Sexstone, A., and R.M. Atlas. 1977. Response of microbial populations in arctic tundra soils to crude oil. *Canad. J. Microbiol.* 23:1327–1333.
Sexstone, A., K. Everett, T. Jenkins, and R.M. Atlas. 1978a. Fate of crude and refined oils in North slope soils. *Arctic* 31:339–347.
Sexstone, A, P. Gustin, and R.M. Atlas. 1978b. Long term interactions of microorganisms and Prudhoe Bay crude oil in tundra soils at Barrow, Alaska. *Arctic* 31:348–354.
Shannon, C.E. 1948. A mathematical theory of communication. *Bell Sys. Technol.* 27:379–423.
Shannon, C.E., and W. Weaver. 1949. *The mathematical theory of communications*. Univ. Illinois Press, Urbana.
Shiel, R.S., and D.L. Rimmer. 1984. Changes in soil structure and biological activity on some meadow hay plots at Cockle Park, Northumberland. *Plant Soil* 76:349–356.
Simpson, E.H. 1949. Measurement of diversity. *Nature* (London) 163:688.
Sims, G.K., and E.P. Dunigan. 1984. Diurnal and seasonal variations in nitrogenase activity (C_2H_2 reduction) of rice roots. *Soil Biol. Biochem.* 16:15–18.
Smith, E.A., and C.I. Mayfield. 1977. Effects of paraquat on selected microbial activities in soil. *Microbial Ecol.* 3:333–343.
Smith, G.E. 1942. Sanborn field: fifty years of field experiments with crop rotations, manure, and fertilizers. *Missouri Agric. Exp. Sta. Bull. 458*.
Smith, O.L. 1982. Soil microbiology: A model of decomposition and nutrient cycling. CRC Press, Boca Raton, Fl.
Sommers, L.E., and V.O. Biederbeck. 1973. Tillage management principles: soil microorganisms. pp. 87–108. In: Soil Conserv. Soc. Amer., *Proc. Natl. Conf. on Conserv. Tillage, Des Moines, lowa, 28–30 Mar. 1973*. Soil Conserv. Soc. Am., Ankeny, IA.
Sparrow, E.B., C.V. Davenport, and R.C. Gordon. 1978. Response of microorganisms to hot crude oil spills on a subarctic taiga soil. *Arctic* 31:324–338.
Sparrow, S.D., and E.B. Sparrow. 1988. Microbial biomass in a subarctic soil ten years after crude oil spills. *J. Envir. Qual.* 17:304–309.
Staley, T.E., and D.M. Fairchild. 1978. Enumeration of denitrifiers in an Appa-

lachian soil. Abstracts of Annual Meeting of Am. Soc. Microbiol., Los Angeles, CA.

Stanier, R.Y., E.A. Adelberg, and J. Ingraham. 1976. *The microbial world*, 4th ed. Prentice-Hall, Englewood Cliffs, NJ.

Stevenson, F.J. 1986. *Cycles of soil. Carbon, nitrogen, phosphorus, sulfur, micronutrients*. Wiley, NY.

Stewart, L.W., and R.B. Reneau, Jr. 1988. Shallowly placed, low pressure distribution system to treat domestic wastewater in soils with fluctuating high water tables. *J. Envir. Qual.* 17:499–504.

Stovold, G.E., and J. Evans. 1980. Fungicide seed dressing, the effects of emergence of soybean and nodulation of pea and soybean. *Aust. J. Exp. Agric. Animal Husb.* 20:497.

Takai, Y., and T. Kamura. 1966. *Folia Microbiol.* 11:304–313.

Tate, R.L., and A.L. Mills. 1983. Cropping and the diversity and function of bacteria in Pahokee muck. *Soil Biol. Biochem.* 15:175–179.

Thomas, G.W., R.L. Blevins, R.E. Phillips, and M.A. MacMahon. 1973. Effect of a killed sod mulch on nitrate movement and corn yield. *Agron. J.* 65:735–739.

Thomas, R.S., S. Dakessian, R.N. Ames, M.S. Brown, and G.J.Bethlenfalvay. 1986. Aggregation of a silty clay loam by mycorrhizal onion roots. *Soil Sci. Soc. Amer. J.* 50:1494–1499.

Thompson, A.R., and C.A. Edwards. 1974. Effects of pesticides on nontarget invertebrates in freshwater and soil. pp. 341–386. In: W.D. Guenzi (ed.), *Pesticides in soil and water*. Soil Sci. Soc. Amer., Madison, WI.

Thorneburg, R.P., and J.A. Tweedy. 1973. A rapid procedure to evaluate the effect of pesticides on nitrification. *Weed Sci.* 21:397.

Tiedemann, A.R., D.A. Higgins, T.M. Quigley, H.R., Sanderson, and C.C. Bohn. 1988. Bacterial water quality responses to four grazing strategies-comparisons with Oregon standards. *J. Envir. Qual.* 17:492–498.

Tiedje, J.M., and F.B. Dazzo. 1982. Soil microbiology. pp. 177–209. In: V.J. Kilmer (ed.), *Handbook of soils and climate in agriculture*. CRC Press, Boca Raton, Fl.

Tu, C.M. 1981. Effects of some pesticides on enzyme activities in an organic soil. *Bull. Envir. Contam. Toxicol.* 27:109.

Tugel, J.B., M.E. Hines, and G.E. Jones. 1986. Microbial iron reduction by enrichment cultures isolated from estuarine sediments. *Appl. Envir. Microbiol.* 52:1167–1172.

van Schreven, D.A., D.J. Lindenbergh, and A. Koridon. 1970. Effect of herbicides on bacterial populations and activity and the persistence of these herbicides in soil. *Plant Soil* 33:513.

Viro, P.J. 1974. Effects of forest fire on soil. pp. 7–45. In: T.T. Kozlowski and C.E. Ahlgree (eds.), *Fire and ecosystems*. Academic Press, NY.

Vlassak, K., K.A.H. Heremans, and A.R. Von Rossen. 1976. Dinoseb as a specific inhibitor of nitrogen fixation in soil. *Soil Biol. Biochem.* 8:91.

Voets, J.P., M. Meerschman, and W. Verstraete. 1974. Soil microbiological and biochemical effects of long-term atrazine applications. *Soil Biol. Biochem.* 6:149.

Voroney, R.P., J.A. VanVeen, and E.A. Paul. 1981. Organic C dynamics in grassland soils. 2. Model validation and simulation of the long-term effects of cultivation and rainfall erosion. *Caned. J. Soil Sci.* 61:211–224.

Wada, H., S.Panichsakpatana, M.Kimura, and Y.Takai. 1978. Nitrogen fixation in paddy soils. I. Factors affecting N_2 fixation. *Soil Sci. Plant Nutr.* 24:357–365.
Willis, W.H., and V.E. Green. 1948. Movement of nitrogen in flooded soil planted to rice. *Soil Sci. Soc. Amer. Proc.* 13:229–237.
Zavarzin, V.I., and T.V. Belyaeva. 1966. The effect of herbicides on the soils content of mineral plant nutrients. *Khim. sel' skh. Khoz.* 4:34.

Need for Action: Research and Development Priorities

R. Lal and B.A. Stewart

I. Introduction

Understanding principles and processes of soil's life support systems is an important pre-requisite to our ability to produce food, feed, and fiber for the world population, and to preserve a healthy status of global environment. Because the problem of soil degradation is very complex, it is important that we approach the issue in a systematic and logical manner. Some important issues to be addressed as researchable priorities are discussed in the following sections.

II. Resource Inventory

An urgent task is to prepare an accurate and reliable inventory of resource base and its productivity status. It is important to realistically assess the soil resources and define their potential and constraints. A more complete characterization of soil resource—physically, chemically, and biologically—is clearly needed. While traditional methods of evaluation of soil resources are available, they are time consuming and labor intensive. Soil scientists must endeavor to use modern technologies including remote sensing. Remote sensing provides a unique opportunity to assess the vegetation, water, and soil resources and their spatial and temporal alterations (Paul et al., 1989; Estes and Cosentino, 1989). These broad scale measurements can be interpreted using a geographic information system (GIS), which is a configuration of computer hardware and software specifically designed for the preparation, presentation, and interpretation of facts pertaining to the surface of the earth (Risser and Iverson, 1989). Remote sensing data of relevance to assess soil resources and their quality are available from a range of sources including Landsat Multi-Spectral Scanner (MSS), Thematic Mapper (TM), SPOT scenes, NOAA-AVHRR, EOS, HIRIS, and MODIS. Before remote sensing techniques can be used to our advantage, however, we must develop methodologies and validate the results by ground truthing. This can be achieved through an interdisciplinary team approach comprising soil scientists, geographers, landscape ecologists, image processors, and terrain analysts. Another important aspect of developing new methodologies is that pertaining to extrapolating results from local scale measurements to regional and global scales. Predicting accurately across scales is a major constraint that must be alleviated through combinations of scale modeling, ground truth surveys, and remote sensing.

III. Separating Emotions from Facts

Despite its widespread severity and global impact, soil degradation remains an emotional rhetoric rather than a precise and quantifiable scientific entity. Soil and plant scientists must work together to delineate critical limits of soil properties beyond which either crop growth is adversely affected or environmental quality is severely and irretrievably jeopardized. The dilemma between emotions and facts is heightened by the fact that production barriers have been broken and high yields are being obtained despite the presumable alarming rates of soil degradation. In fact, global food production has experienced quantum jumps during the last three to five decades. And yet, there are reports of accelerated erosion by wind, water, and desertification; compaction, hard-setting, and laterization; water logging and salinization; and leaching and acidification. It is possible that the scientific community has overdramatized the issue. It is equally plausible, however, that production gains by modern technology have been severely undermined by the widespread degradation of the world's soil resources.

Separating emotions from facts will need long-term research on major soil orders of the world for development of critical limits of soil properties. It is only with the knowledge of these critical limits that soil scientists can adequately interpret the resource inventory to assess their potential and constraints, and to evaluate the rate of soil and environmental degradation by land misuse or resource appreciation through restoration measures. There are many unknowns that require immediate attention. For example, we do not know the rate of new soil formation and cannot precisely calculate the range of soil loss tolerance. Structural attributes of a soil are hard to quantify on the basis of a single-parameter soil test value. Expressing the economic and environmental impact of soil compaction, erosion, salinization, acidification, etc., is not an easy task. An important but weak link in this chain is the lack of knowledge about the contribution of soil degradative processes to the so-called "greenhouse effect." We do not have any reliable estimate of the reservoir of carbon in humified soil organic matter and about the rate of its turnover. Contributions to atmospheric CO_2 by the soil organic matter is anyone's guess.

IV. Restoring Productivity of Degraded Lands

The world community must adopt a strict code of conduct towards the management of the most basic of all resources—the soil. Although coercive measures are rarely successful, planners must explore the use of positive incentive measures for ensuring a proper use of existing lands and in restoring productivity of lands that have been abandoned due to past mismanagement. The long-term policy is to conserve, sustain, and enhance soil's productivity for human welfare and use. For feeding and clothing the global stable population of 10 to 11 billion inhabitants, we have no choice but to restore the productivity of degraded soils.

Legislative measures are difficult to implement and police. Furthermore, legislative measures cannot be adopted successfully in isolation. Once the degradative processes are set in motion they have a snowball effect. These processes are self-perpetuating and are not constrained by ethnic, political, national, or geographic boundaries. Government policies are needed for both corrective and preventive land use planning. UNEP (1982) formulated the "World Soil Policy." This policy, through a United Nations Charter and Promulgation, must be adopted and respected by all nations. The world community of nations can facilitate adoption of these policies through incentives, education, and other humanitarian means. Coercive measures, such as political and economic sanctions against the offenders, may be considered only as a last resort.

Use of restrictive/preventive policies and legislative measures is especially needed for marginal lands and ecosystems that are especially vulnerable to degradation. Ecologically sensitive and marginal ecosystems are also politically sensitive and include tropical rainforest, steeplands, African Sahel, and arid and semi-arid regions prone to desertification. Preventive land use planning for these regions must be made in due consultations with regional/national authorities. Such planning must consider the inherent properties of soil, climate, terrain, and the socioeconomic and cultural factors.

The world community must be able to share the costs of such preventive/restorative planning. Not only will some lands have to be taken out of production, but others will also be subjected to low-yielding conservative technologies. This is essential because these preventive measures are needed in regions plagued by perpetual food crises and impoverished economies. Some of the ecologically sensitive regions that are in dire need of such a policy include countries such as Ethiopia, Nepal, Haiti, the Andean region, countries of the African Sahel, and the Himalyan-Tibetan ecosystem. Farmers in these countries are aware of the long-term consequences of the degradative and productivity-mining practices being used. However, they have adopted these inappropriate measures out of desperation. The international community has to intervene to salvage our common heritage and the global environment.

V. Reaching Out

Soil degradation is a complex phenomena (Figure 1). It is driven by strong interaction among socioeconomic and biophysical factors. It is fueled by increasing population, fragile economy, and dismal farm policies. It is propelled by fragility of soil and harshness of climate. We must recognize that this complex problem has no quick and easy solutions. Although soil scientists have an important role to play in curtailing the problem and in reversing the trend, they cannot tackle this mammoth task by themselves. They need all the help from other professions that they can muster. They must reach out to other disciplines in understanding the problems. The problem is aggravated equally by biophysical and socioeconomic or political factors. However, soil scientists must take an active role in creating awareness and in showing the need for a multidisciplinary approach. Soil scientists must work in cooperation with climatologists, hydrologists, geologists, and ecologists in understanding the basic processes including water and energy balance and the cycling of major elements such as C, N, P, S, etc. Soil scientists must also cooperate with crop scientists, agricultural engineers, and economist to develop productive, profitable, and sustainable agricultural systems. There is also a need to understand the social and political structure that fuels a degradative land use and creates a perpetual crisis. It is a tall order, indeed, but one that the scientific community can no longer afford to neglect.

VI. Conclusions

The world has the capacity to feed itself. It can be done, however, only if soil degradative trends can be reversed and the degraded soils can be restored. We must develop a coordinated and multidisciplinary approach to assess soil resources and evaluate their potential and restraints. We must also understand the processes, causes, and factors of soil degradation and the critical limits of soil properties beyond which productivity decline is

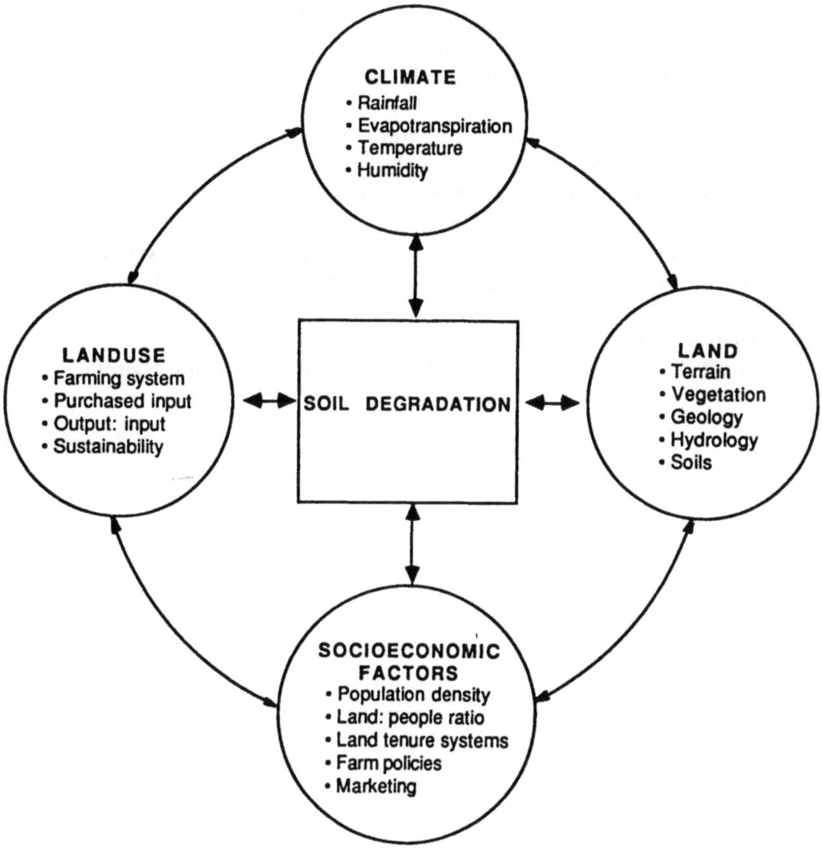

Figure 1. Interdependence of soil degradation on biological and socioeconomic factors.

substantial. We must develop methods to regenerate the productivity of degraded soils. New technological innovations must not only be high yielding but also compatible with sustaining acceptable standards of environmental quality. The world community must develop a resource management policy. This is particularly true when dealing with marginal, fragile, and ecologically sensitive ecosystems. Multidisciplinary teams should devise resource management policies that should be implemented through economic incentives.

References

Estes, J.E., and M.J. Cosentino. 1989. Remote sensing of vegetation. pp. 75–112. In: M.B. Rambler, L. Margulis, and R. Fester (eds.), *Global ecology: Towards a science of the biosphere*. Academic, NY.

Moore, B., M.P. Gildea, C.J. Vorosmarty, D.L. Skole, J.M. Melillo, B.J. Peterson, E.B. Rastetter, and P.A. Steudler. 1989. Biogeochemical cycles. pp. 113–142. M.B. Rambler, L. Margulis, and R. Fester (eds.), *Global ecology: Towards a science of the biosphere.* Academic, NY.

Paul, C.K., M.L. Imhoff, D.G. Moore, and A.M. Sellman. 1989. Remote sensing of environmental change in the developing World. In: D.B. Botkin, M.F. Caswell, J.E. Estes, and A.A. Orio (eds.), *Changing the global environment: Perspectives on human development.* Academic, NY.

Risser, P.G., and L.R. Iverson. 1989. Geographic information systems and the natural resources issue at the sate level. pp. 203–212. D.B. Botkin, M.F. Caswell, J.E. Estes, and A.A. Orio (eds.), *Changing the global environment: Perspectives on human development.* Academic, NY.

UNEP. 1982. World soil policy. United Nations Envir. Program, Nairobi, Kenya.

Index

T

MIX
Papier aus verantwortungsvollen Quellen
Paper from responsible sources
FSC® C105338

If you have any concerns about our products,
you can contact us on
ProductSafety@springernature.com

In case Publisher is established outside the EU,
the EU authorized representative is:
Springer Nature Customer Service Center GmbH
Europaplatz 3, 69115 Heidelberg, Germany

Printed by Libri Plureos GmbH
in Hamburg, Germany